BIOCHEMISTRY RESEARCH TRENDS

CIRCADIAN RHYTHMS AND THEIR IMPORTANCE

BIOCHEMISTRY RESEARCH TRENDS

Additional books and e-books in this series can be found on Nova's website under the Series tab.

BIOCHEMISTRY RESEARCH TRENDS

CIRCADIAN RHYTHMS AND THEIR IMPORTANCE

RAJESHWAR P. SINHA, PHD
EDITOR

nova science publishers
New York

DOI: https://doi.org/10.52305/GXME8274

NOTICE TO THE READER

Library of Congress Cataloging-in-Publication Data

ISBN: 978-1-68507-547-7

Published by Nova Science Publishers, Inc. † New York

This book is dedicated to my parents,
Mr. Maheshwar P. Sinha and Mrs. Asha L. Sinha,
my wife, Dr. Alka Sinha, and my son, Mr. Prakhar
Raj for their continuous love and support.

Contents

PREFACE

Circadian rhythm is the mechanism for measuring time by organisms in units of around 24 h. Almost every organism has the ability to predict daily events such as sunrise and sunset which is essential for their optimal survival strategies. Circadian rhythms persist even in the absence of time cues and ensure that behavioural, physiological and metabolic processes align with changing environmental conditions. Most of the organisms irrespective of whether prokaryotic or eukaryotic have evolved certain mechanisms to sense their external environment in order to prepare a biological response. Often these biological responses result in changes in physiology and biochemistry of organisms for the optimization of their survival. There is no unified mechanism among organisms by which sensory environment can be applied to control their lifespan. In humans, disruptions of the circadian rhythm may have an increased risk of age-related neurodegenerative diseases such as cancer, diabetes, etc.

This book encompasses the most recent knowledge and modern trends in the field of circadian rhythms research. It is also expected to provide a must-read guide for beginners as well as advanced researchers in this field of study. I believe that undergraduate, post-graduate and research scholars of several disciplines such as chemistry, botany, microbiology, biochemistry, biomedical, biotechnology, bioinformatics, life sciences and bionanotechnology would also be immensely benefited from this book.

I have great pleasure in bringing forward this book and express my sincere gratitude to all the contributors. Thanks are also due to Ms. Sonal Mishra and Ms. Neha Kumari for their timely help during the editing of this book. I am also thankful to NOVA Science Publishers Inc. for taking such great efforts to complete the publication of this book so efficiently.

Rajeshwar P. Sinha
October, 2021

In: Circadian Rhythms and Their Importance ISBN: 978-1-68507-547-7
Editor: Rajeshwar P. Sinha

Chapter 1

CIRCADIAN RHYTHMS: AN INTRODUCTION

Rajeshwar P. Sinha ***and Sonal Mishra***
Laboratory of Photobiology and Molecular Microbiology,
Centre of Advanced Study in Botany, Institute of Science,
Banaras Hindu University, Varanasi, Uttar Pradesh, India

ABSTRACT

The term circadian rhythm is defined as the mechanism for measuring time by organisms in units of approximately one day (24 h). Circadian rhythms regulate the temporal schedules of the cell's internal processes naturally repeatedly to facilitate customization to daily environmental fluctuations. Circadian rhythms play a vital role in several biological processes such as body temperature regulation, sleep-wake cycle, cardiovascular health, hormone secretion, glucose homeostasis and ecological adaptation of various organisms etc. It is of utmost importance to understand the mechanism of circadian rhythm so that proper guidelines may be prepared to treat the metabolic imbalances.

* Corresponding Author's E-mail: rpsinhabhu@gmail.com.

Keywords: circadian rhythms, ecological adaptations, innate immunity, metabolic regulator, physiological activities

INTRODUCTION

Adaptation of microorganisms to the environment includes the anticipation of cyclical events such as the daily rhythm of sunlight and night. It has been supposed that the circadian rhythm developed to arrest ultraviolet (UV) sensitive activities, like DNA replication and cell division, during the daytime. Alterations to the circadian output intermediary genes *sasA* and *rpaA* and the glycogen degradation enzyme gene *glgP* annihilate the circadian clocks of UV-C resistance. Metabolism associated with glycogen in the dark influences circadian control in UV sensitivity, while the light hides this effect *via* the photolyase function (Kawasaki and Iwasaki 2020). Most organisms express temporal programs reflecting the 24 h cycle of day and night from microscopic cyanobacteria (Johnson and Golden 1999) to humans. Circadian rhythms have two-fold importance in infectious diseases, acting at the level of both host and pathogen. The molecular level of circadian rhythm machinery is centered on the autoregulatory transcription/translation feedback loop (TTO), wherein transcriptional activators prompt the repressor genes expression and numerous tissue-specific clock-regulated genes (CCGs) (Partch et al., 2014). This mechanism is also focused on the key point of the immune response, by regulating the task of cells and modulators in microorganisms. Its dysregulation may be responsible for immunodeficiency, inflammatory diseases, and elevated menace of contaminations (Curtis et al., 2014). Survey of the circadian rhythmic diversity in established behavioural and physiological processes in various microorganisms such as archaebacteria, cyanobacteria, algae and fungi are discussed (cf. Chapter 2, this volume).

The existing cognitions about circadian rhythms driving by potential molecular components in microorganisms, and also focusing on those that have importance in the regulation of evolutionary life, have been illustrated (cf. Chapter 2 for details, this volume).

Circadian Rhythms and Ecological Adaptations

Endogenous circadian rhythm affects various cellular processes like cell division, expression of genes, cell metabolism, cell development, and behavior of any organism (Johnson et al., 2011). Currently, various mechanisms of circadian regulation have been extensively studied in eukaryotic model organisms including mice, fruit fly and fungi (Bell-Pedersen et al., 2005). However, only few studies have been found on prokaryotic organisms. Effects of various abiotic stresses such as solar radiation, temperature and pressure etc. had an impact on circadian clock. Exposure of solar radiation plays an important role in the balancing of pigment proteins ratio in 24 h alteration in light and dark regimes (Kumar et al., 2018). Circadian rhythm is directly and indirectly linked with thermal fluctuation and pressure for adaptation in harsh environment. In molecular biology, Kai A-B-C proteins are the major players in the regulation of gene expression altered by light and dark regimes. The regulatory oscillators TTFL (transcriptional/translational feedback loop) and PTO (post-translational oscillator) are playing key role in the regulation of circadian clock. TTFL have mostly regulated in eukaryotic circadian while PTO act as cyanobacterial oscillator with cyclic proteins KaiA, KaiB, and KaiC (Hurley et al., 2016; Cohen and Golden 2015). Circadian clock is the key feature for survivability of cyanobacteria under various light/dark, temperature, oxidation and pressure regimes (cf. Chapter 3 for details, this volume).

Circadian Rhythm As Metabolic Regulator in Plant

Plant circadian rhythm is a complex networking within transcription factors that functions in feedback loops. In order to adjust the continuously varying environmental conditions plant develops many strategies for their adaptation and future existence. In relation to diurnal rhythms, plant have developed a time-dependent mechanism which controls the biological events in plants with the help of environmental cues (light and temperature) i.e., the circadian clock. From past to recent several workers have demonstrated the conserved architecture of biological clock with various independent origins (Dunlap 1999; Young and Kay 2001; McClung 2013). Plant biological clock directs reproductive development period so that other events related to this can be facilitated like flowering, nectar synthesis, pollination, leaf development/movement, plant growth, stomatal gaseous exchange, secondary metabolism, enzyme function and photosynthesis etc. (Fenske and Imaizumi 2016). Studies on plant biological clock with the impaired functional circadian clock have resulted in the inference that a systematically organized functioning of biological clock gives a plant a better evolutionary advantage. The gene level studies have given light to the molecular pillars of the circadian clock contains morning genes, evening genes, evening complex, pseudo response regulator family and other clock associated genes. This chapter is framed such, as it apprehends the historical journey to understand the plant circadian rhythms life (cf. Chapter 4 for details, this volume).

Photic Regulation and Health

The circadian rhythm is having a periodicity of approximately 24 h and is governed by a master pacemaker, the suprachiasmatic nucleus (SCN) located in the anterior hypothalamus (Weaver 1998; Mohawk et al., 2012). Light is the most potent environmental cue for the entrainment of endogenous circadian pacemaker that allows it to synchronize most of the behavioural and physiological processes with the 24 h environment cycle (Roenneberg and Foster 1997). Core clock genes are referred to as genes whose proteins are necessary for the generation and regulation of circadian rhythms within individual cells of the organisms (Takahashi 2004). Such autonomously generated SCN rhythms follow the pattern of transcription – translational mechanism of a set of clock genes which includes Period genes (Per1, Per2, Per3), Cryptochrome (Cry1 & 2), Bmal1 and Clock (circadian locomotor output cycles kaput) to evoke behavioural, physiological and molecular response in an intrinsic time-dependent manner (Nagoshi et al., 2004; Gekakis et al., 1998). The resetting of the molecular clock which generates 24 h rhythm requires post-translational modifications such as phosphorylation and ubiquitination of negative limb protein i.e., PER and CRY. The proteasomal degradation of PER and CRY is required to remove the repression to start the next cycle again (Dardente and Cermakian 2007; Ko and Takahashi 2006).

The exposure of light produces a phase-shift effect on melatonin rhythm and acutely inhibits its synthesis (Shanahan and Czeisler 1991). The disruption of circadian rhythm by the prolonged accumulation of irregular LD cycles is the major hallmark of many psychological disorders characterized by sleep disruption. It is known to be critical for the diagnosis of several mood-related disorders including major depressive disorder and bipolar disorder (Hickie et al., 2013; Rosenwasser and Wirz-Justice 1997). In addition to the sleep/wake and feeding/fasting cycle, circadian oscillation of the clock also influences rhythmic change in the downstream molecular and physiological pathway that involves nutrition and metabolism (Ramsey and Bass 2009). Evidence from epidemiological studies suggests a possible correlation between metabolic syndrome and shift work accompanied by light exposure at night (Karlsson et al., 2001). Disruption of circadian rhythm is associated with a risk factor for promoting cardiovascular-related disorder accompanying several metabolic syndromes (Ashby and Louis 2019; Antunes et al., 2010). This chapter explains the role of photic cues in entraining the circadian clock and explains the negative outcomes of the circadian rhythm disruption by untimed light exposure on endocrine rhythm, mood-related behaviour and metabolism life (cf. Chapter 5 for details, this volume).

Circadian Rhythm and Regulation of Spike Activity

The suprachiasmatic nucleus (SCN) contains a variety of neuropeptides (Moore et al., 2002; Irwin and Allen 2010; Ono et al., 2021), that might contribute to the network functions and therefore influence the parameters of circadian rhythms and their entrainment. It is a well-established fact that in addition to light, nonphotic cues, such as meal timing, exercise, temperature, social contacts, can entrain the circadian clock (Mistlberger and Skene 2005; Challet, 2007). Neuropeptide Y (NPY) can play a modulatory role in the mammalian circadian system (cf. Chapter 6 for details, this volume).

Circadian Rhythm and the Expression of Disorders

Circadian rhythms disturbances were hypothesized to play a role in the pathophysiology of affective disorders, also underpinning alterations in the hypothalamic-pituitary-adrenal (HPA) axis. Circadian rhythms can be defined as endogenous (i.e., unprompted) daily variations in physical, mental and behavioral activity, which can nevertheless synchronize with geophysical time, thus showing entrainability (Zauderer and Gnazer 2015; Garbazza and Benedetti 2018). This is possible through the detection of external time givers, the so-called zeitgebers, which are mainly represented by photoperiod but also by other factors, such as meal times and exercise. In mammals, the circadian system has a hierarchical organization, headed by a central clock, the suprachiasmatic nucleus (SCN), composed of nearly 20,000 neurons and situated in the anterior hypothalamus (Mohawk et al.,

2012; Steardo et al., 2019). Important features of the SCN as a whole are robustness and plasticity.

Main types of circadian rhythms disrupted in psychiatric disorders are sleep-wake cycle, hormonal circadian rhythms, alimentary rhythms, activity rhythms, social rhythms etc. When taking into account the pathophysiology of hormonal rhythms in psychiatric disorders, the most important molecules to keep in mind are melatonin and glucocorticoids, also considering their role in synchronizing central and peripheral clocks; anyway, the others must not be underestimated, especially when it comes to psychiatric disorders comorbidities and outcomes (Lopresti et al., 2013; Potter et al., 2016). Circadian rhythm disturbances have been widely demonstrated in people affected by bipolar disorder (BD) (Takaesu 2018). In bipolar subjects, irregular circadian rhythms contribute to the pathophysiology of episodes of mania and depression (Gold and Kinrys 2019). Circadian rhythms in BD subjects can be restored by both pharmacologic and non-pharmacologic interventions (Benedetti and Dallaspezia 2009). Controlled exposures to environmental stimuli that act on biological rhythms are labelled as chronotherapies. According to a recent cohort study, 50% of BD subjects exhibit abnormal sleep, with 50% of them presenting further circadian rhythm disturbances (cf. Chapter 7 for details, this volume).

Circadian Rhythm and Innate Immunity

Cyclical environmental changes generated by Earth's rotation and revolution have imposed adaptive pressure on living organisms through evolution. This has led to the development of an internal system of temporal organization that allows physiology and behaviour to be synchronized with the environment to optimize energy and favour vital functions. The circadian system of a complex organism is composed of various oscillators (cells or cellular processes) that repeat themselves with a period close to 24 h, synchronization pathways (of phase and period with respect to the light-dark cycle), transmission or communication pathways towards rhythm effector systems, and oscillator-coupling mechanisms (Aguilar et al., 2008). The mammalian eye, light is detected by retinal specialized photoreceptors (Berson et al., 2002). The immune response is a series of processes carried out in a delicately concerted manner, basically including from the antigen recognition to its elimination. The innate immune response encompasses the epithelial barriers that prevent the entry of microorganisms or exogenous particles into the body, as well as the effector mechanisms of soluble molecules and cells that specifically recognize these antigens if they manage to enter the sub-epithelial tissues or the bloodstream. Succinctly, the innate immune response prevents and controls all type of infections, eliminates damaged, transformed or dead cells and stimulates the adaptive immune response. Adaptive immunity can be cellular or humoral depending on the type of effectors that are produced (Abbas et al., 2018). Macrophages play a key role in modulating the immune response as they have a wide repertoire of receptors that allow them to identify various damage signals.

Circadian regulation of innate immunity like biochemical modulators rhythmically released that affect immune cells. Modulation of spleen macrophage physiology and immune response by melatonin and enkephalins are also discussed in this chapter (den Haan and Kraal 2012; Nikbakht et al., 2013). Both melatonin and PENK-derived opioid peptides, specifically MEnk, are important biochemical messengers with important immunomodulatory functions. The aim of this chapter is to analyse the evidence that suggests the interaction between these systems by two broadly released modulators, melatonin and endogenous opioid peptides, in particular, the MEnk-containing bioactive peptides (cf. Chapter 8 for details, this volume).

Circadian Rhythm and Physiological Activities

Circadian rhythms are controlled by circadian clocks, which drive day-night oscillations with a free running period of ~24 hrs. The molecular machinery of the circadian clock has been extensively characterized and reviewed elsewhere (Takahashi 2017). At its core is a negative feedback loop involving four basic helix-loop-helix or per-arnt-sim domain transcription factors: CLOCK, BMAL1, PER, and CRY3. Circadian clocks have been identified in the majority of mammalian cells. The suprachiasmatic nucleus (SCN) in the hypothalamus contains the central circadian clock and is synchronized with the environment by external cues such as light. Continuous 24-h ECG recordings from healthy volunteers have shown a circadian rhythm in the ECG.

Interestingly, direct recordings of stellate ganglion nerve activity and vagal nerve activity in an ambulatory dog model of heart failure show a circadian rhythm in sympathetic but not vagal tone to the heart (Ogawa et al., 2007). This chapter talks about the diabetes mellitus and circadian rhythms of heart rate, arterial stiffness and blood pressure are discussed in this chapter. Diabetic cardiac autonomic neuropathy and circadian rhythm's disorders are also discussed in this chapter. Here, changes in vegetative tone can increase the arterial stiffness due to trophic effects, which contribute to changes in the structure of the vessels (Christen et al., 2010; Stoner et al., 2012). A great amount of data demonstrates that hyperglycemia and IR activate number of mechanisms triggering the structural and functional changes in the arterial wall, which are likely to contribute to accelerated vascular aging and increased CV risk in T2DM (Prenner et al., 2015). Several pathophysiological changes in cardiomyocytes underlie the risk factors for IR and HI, resulting in a potentially vicious cycle, which prompts the following question: does heart failure cause DM? (Negishi 2018) (cf. Chapter 9 for details, this volume).

CIRCADIAN RHYTHM AND *DROSOPHILA* LIFESPAN

The *Drosophila* is a perfect model system to study connection between neurosensory systems and circadian biology due to the considerable genetic control, short generation time and simple nervous system. *Drosophila* avails of photosensitive neurons layer and retina, by which both stationary objects and motion can be sensed. The electrochemical information from the retina is utilized to generate an image of the environment. Moreover, retinal information comes up with the embus of the circadian rhythm (Helfrich-Forster 2020). The minimal nexus light perception procedure used by *Drosophila* is performed by the non-retinal photopigments cryptochrome and Rh7 functioning as light sensors directly in circadian clock neurons and more cells over the body (Stanewsky et al., 1998, Ni et al., 2017). Counteracting circadian decreases aging via a suprachiasmatic nucleus (SCN) transplant procured from fetal tissue able to reinstate rhythms in rats and enhance life in the golden hamster (Hurd and Ralph 1998, Li and Satinoff 1998). The function of circadian rhythm was delivered by a brain-specific *Bmal1*, adequate to reinstate peripheral transcriptional clocks in the liver and enhance survival (McDearmon et al., 2006, Hughes et al., 2012). Thus, circadian rhythms may be thought about as a sensory tone that permits organisms to recognize time and to predict meal times, predator activity periods and seasonal changes (Johnson 2021).

CONCLUSION

The circadian rhythm plays a pivotal role in several biological processes and long-term circadian disruptions due to modern lifestyle may be associated with many unavoidable conditions such as obesity, diabetes, anxiety, depression, cancer progression and premature mortality, etc. Our knowledge related to the genetic and molecular mechanisms of circadian rhythm is expanding exponentially. However, the functional part of these mechanisms is still unknown to a large extent. Therefore, there is a need to understand the interactions between the basics of circadian rhythm and its manifestation in diverse organisms as well as to prepare a roadmap to mitigate the disorders.

Acknowledgments

S. Mishra is thankful to the University Grants Commission (UGC), New Delhi, India, for the financial assistance in the form of a fellowship as Junior Research Fellow (Joint CSIR-UGC JRF-2019/NTA Ref. No.: 191620046790).

References

Abbas A. K., Lichtman A. H., Pillai S. (2018) In: *Cellular and Molecular Immunology*, Ninth Edition: Elsevier, Inc, Philadelphia, PA, USA.

Aguilar R., Carmona V., Fanjul M. L. (2008) Circadian rhythms. In: *Functional biology of animals. Comparative neurophysiology*: Fanjul M. L., Hiriart M., editors. Siglo XXI, Mexico, DF, pp. 327-352.

Antunes L. C., Levandovski R., Dantas G., Caumo W., Hidalgo M. P. (2010) Obesity and shift work: chronobiological aspects. *Nutr Res Rev* 23(1): 155-168.

Ashby T., Louis M. (2019) Circadian misalignment and cardiovascular risk. *Cardiovasc Innov Appl* 3(4): 435-440.

Bell-Pedersen D., Cassone V. M., Earnest D. J., Golden S. S., Hardin P. E., Thomas T. L., Zoran M. J. (2005) Circadian rhythms from multiple oscillators: lessons from diverse organisms. *Nat Rev Genet* 6(7): 544-556.

Benedetti F., Dallaspezia S. (2009) Melatonin, circadian rhythms, and the clock genes in bipolar disorder. *Curr Psychiatry Rep* 11(6): 488-493.

Berson D. M., Dunn F. A., Takao M. (2002) Phototransduction by retinal ganglion cells that set the circadian clock. *Science* 295: 1070-1073.

Challet E. (2007) Mini review: entrainment of the suprachiasmatic clockwork in diurnal and nocturnal mammals. *Endocrinology* 148: 5648-5655.

Christen A. I., Armentano R. L., Miranda A., Graf S., Santana D. B., Zócalo Y., Baglivo H. P., Sánchez R. A. (2010) Arterial wall structure and dynamics in type 2 diabetes mellitus: methodological aspects and pathophysiological findings. *Curr Diabetes Rev* 6(6): 367-377.

Cohen S. E., Golden S. S. (2015) Circadian rhythms in cyanobacteria. *Microbiol Mol Biol Rev* 79(4): 373-385.

Curtis A. M., Bellet M. M., Sassone-Corsi P., O'Neill L. A. (2014) Circadian clock proteins and immunity. *Immunity* 40: 178-186.

Dardente H., Cermakian N. (2007) Molecular circadian rhythms in central and peripheral clocks in mammals. *Chronobiol Int* 24(2): 195-213.

den Haan J. M., Kraal G. (2012) Innate immune functions of macrophage subpopulations in the spleen. *J Innate Immun* 4: 437-445.

Dunlap J. C. (1999) Molecular Bases for Circadian Clocks. *Cell* 96(2): 271-290.

Fenske M. P., Imaizumi T. (2016) Circadian rhythms in floral scent emission. *Front Plant Sci* 7: 462.

Garbazza C., Benedetti F. (2018) Genetic Factors Affecting Seasonality, Mood, and the Circadian Clock. *Front Endocrinol* 9: 481.

Gekakis N., Staknis D., Nguyen H. B, Davis F. C., Wilsbacher L. D., King D. P., Takahashi J. S., Weitz C. J. (1998) Role of the CLOCK protein in the mammalian circadian mechanism. *Science* 280(5369): 1564-1569.

Gold A. K., Kinrys G. (2019) Treating Circadian Rhythm Disruption in Bipolar Disorder. *Curr Psychiatry Rep* 21(3): 1-8.

Helfrich-Förster C. (2020) Light input pathways to the circadian clock of insects with an emphasis on the fruit fly Drosophila melanogaster. *J Comp Physiol A Neuroethol Sens Neural Behav Physiol* 206:259-272.

Hickie I. B., Naismith S. L., Robillard R., Scott E. M., Hermens D. F. (2013) Manipulating the sleep-wake cycle and circadian rhythms to improve clinical management of major depression. *BMC Med* 11: 79.

Hughes M. E., Hong H. K., Chong J. L., Indacochea A. A., Lee S. S., Han M., Takahashi J. S., Hogenesch J. B. (2012) Brain-specific rescue of Clock reveals system-driven transcriptional rhythms in peripheral tissue. *PLoS genetics* 8(7): e1002835.

Hurd M. W., Ralph M. R. (1998) The significance of circadian organization for longevity in the golden hamster. *J Biol Rhythms* 13(5): 430-436.

Hurley J. M., Loros J. J., Dunlap J. C. (2016) Circadian oscillators: around the transcription-translation feedback loop and on to output. *Trends Biochem Sci* 41(10): 834-846.

Irwin R. P., Allen, C. N. (2010) Neuropeptide mediated calcium signaling in the suprachiasmatic nucleus network. *Eu. J Neurosci* 32:1497-1506.

Johnson C. H., Golden S. S. (1999) Circadian programs in cyanobacteria: adaptiveness and mechanism. *Annu Rev Microbiol* 53: 389-409.

Johnson C. H., Stewart P. L., Egli M. (2011) The cyanobacterial circadian system: from biophysics to bioevolution. *Annu Rev Biophys* 40: 143-167.

Johnson J. (2021) *Circadian Rhythms, Light, and Social Perception: Sensory Regulation of Lifespan in Drosophila* (Doctoral dissertation).

Karlsson B., Knutsson A., Lindahl B. (2001) Is there an association between shift work and having a metabolic syndrome? Results from a population based study of 27,485 people. *Occup Environ Med* 58(11): 747-752.

Kawasaki K., Iwasaki H. (2020) Involvement of glycogen metabolism in circadian control of UV resistance in cyanobacteria. *PLoS genetics* 16(11): e1009230.

Ko C. H., Takahashi J. S. (2006) Molecular components of the mammalian circadian clock. *Hum Mol Genet* 15: R271-277.

Kumar D., Kannaujiya V. K., Pathak J., Sundaram S., Sinha R. P. (2018) Composition and functional property of photosynthetic pigments under circadian rhythm in the cyanobacterium *Spirulina platensis*. *Protoplasma* 255(3): 885-898.

Li H., Satinoff E. (1998) Fetal tissue containing the suprachiasmatic nucleus restores multiple circadian rhythms in old rats. *Am J Physiol Regul Integr Comp Physiol* 275(6): R1735-1744.

Lopresti A. L., Hood S. D., Drummond P. D. (2013) A review of lifestyle factors that contribute to important pathways associated with major depression: diet, sleep and exercise. *J Affect Disord* 148(1): 12-27.

McClung C. A. (2013) How might circadian rhythms control mood? Let me count the ways. *Biol Psychiatry* 74(4): 242-249.

McDearmon E. L., Patel K. N., Ko C. H., Walisser J. A., Schook A. C., Chong J. L., Wilsbacher L. D., Song E. J., Hong H. K., Bradfield C. A., Takahashi J. S. (2006) Dissecting the functions of the mammalian clock protein BMAL1 by tissue-specific rescue in mice. *Science* 314(5803): 1304-3108.

Mistlberger R. E., Skene D. J. (2005) Nonphotic entrainment in humans? *J Biol Rhythms* 20: 339-352.

Mohawk J. A., Green C. B., Takahashi J. S. (2012) Central and peripheral circadian clocks in mammals. *Annu Rev Neurosci* 35: 445-462.

Mohawk J. A., Green C. B., Takahashi J. S. (2012) Central and peripheral circadian clocks in mammals. *Annu Rev Neurosci* 35: 445-462.

Moore R. Y., Speh J. C., Leak R. K. (2002) Suprachiasmatic nucleus organization. *Cell Tissue Res* 309: 89-98.

Nagoshi E., Saini C., Bauer C., Laroche T., Naef F., Schibler U. (2004) Circadian gene expression in individual fibroblasts: cell-autonomous and self-sustained oscillators pass time to daughter cells. *Cell* 119(5): 693-705.

Negishi K. (2018) Echocardiographic feature of diabetic cardiomyopathy: where are we now? *Cardiovasc Diagn Ther* 8(1): 47-56.

Ni J. D., Baik L. S., Holmes T. C., Montell C. (2017) A rhodopsin in the brain functions in circadian photoentrainment in *Drosophila*. *Nature* 545(7654): 340-344.

Nikbakht N., Shen S., Manser T. (2013) Cutting edge: Macrophages are required for localization of antigen-activated B cells to the follicular perimeter and the subsequent germinal center response. *J Immunol* 190: 4923-4927.

Ogawa M., Zhou S., Tan A. Y., Song J., Gholmieh G., Fishbein M. C., Luo H., Siegel R. J., Karagueuzian H. S., Chen L. S., Lin S. F., Chen P. S. (2007) Left stellate ganglion and vagal nerve activity and cardiac arrhythmias in ambulatory dogs with pacing-induced congestive heart failure. *J Am Coll Cardiol* 50(4): 335-343.

Ono D., Honma K. I., Honma S. (2021) Roles of neuropeptides, VIP and AVP, in the mammalian central circadian clock. *Front Neurosci* 15: 650154.

Partch C. L., Green C. B., Takahashi JS (2014) Molecular architecture of the mammalian circadian clock. *Trends Cell Biol* 24: 90-99.

Potter G. D., Skene D. J., Arendt J., Cade J. E., Grant P. J., Hardie L. J. (2016) Circadian Rhythm and Sleep Disruption: Causes, Metabolic Consequences, and Countermeasures. *Endocr Rev* 37(6): 584-608.

Prenner S. B., Chirinos J. A. (2015) Arterial stiffness in diabetes mellitus. *Atherosclerosis* 238(2): 370-379.

Ramsey K. M., Bass J. (2009) Obeying the clock yields benefits for metabolism. *Proc Natl Acad Sci USA* 106(11): 4069-4070.

Roenneberg T., Foster R. G. (1997) Twilight times: light and the circadian system. *Photochem Photobiol* 66(5): 549-561.

Rosenwasser A. M., Wirz-Justice A. (1997) Circadian rhythms and depression: Clinical and experimental models. In: Redfern PH, Lemmer B, editors. *Physiology and Pharmacology of Biological Rhythms*: Berlin, Springer, 457-486. *Sens Neural Behav Physiol* 206(2): 259-272.

Shanahan T. L., Czeisler C. A. (1991) Light exposure induces equivalent phase shifts of the endogenous circadian rhythms of circulating plasma melatonin and core body temperature in men. *J Clin Endocrinol Metab* 73(2): 227-235.

Stanewsky R., Kaneko M., Emery P., Beretta B., Wager-Smith K., Kay S. A., Rosbash M., Hall J. C. (1998) The cryb mutation identifies cryptochrome as a circadian photoreceptor in *Drosophila*. *Cell* 95(5): 681-692.

Steardo Jr. L., De Filippis R., Carbone E. A., Segura-Garcia C., Verkhratsky A., De Fazio P. (2019) Sleep Disturbance in Bipolar Disorder: Neuroglia and Circadian Rhythms. *Front Psychiatry* 10(501).

Stoner L., Young J. M., Fryer S. (2012) Assessments of arterial stiffness and endothelial function using pulse wave analysis. *Int J Vasc Med* 2012: 903107.

Takaesu Y. (2018) Circadian rhythm in bipolar disorder: A review of the literature. *Psychiatry Clin Neurosci* 72(9) 673-682.

Takahashi J. S. (2004) Finding new clock components: past and future. *J Biol Rhythms* 19(5): 339-347.

Takahashi J. S. (2017) Transcriptional architecture of the mammalian circadian clock. *Nat Rev Genet* 18(3): 164-179.

Weaver D. R. (1998) The suprachiasmatic nucleus: a 25-year retrospective. *J Biol Rhythms* 13(2): 100-112.

Young M. W., Kay S. A. (2001) Time zones: comparative genetics of circadian clocks. *Nat Rev Genet* 2(9): 702-715.

Zauderer C., Ganzer C. A. (2015) Seasonal affective disorder: an overview. *Ment Health Pract* 18(9): 21-24.

In: Circadian Rhythms and Their Importance ISBN: 978-1-68507-547-7
Editor: Rajeshwar P. Sinha

Chapter 2

MOLECULAR MECHANISMS AND IMPORTANCE OF CIRCADIAN RHYTHMS IN MICROORGANISMS

Sonal Mishra, Neha Kumari, Jyoti Jaiswal, Prashant R. Singh, Amit Gupta, Niharika Sahu and Rajeshwar P. Sinha*

Laboratory of Photobiology and Molecular Microbiology,
Centre of Advanced Study in Botany, Institute of Science,
Banaras Hindu University, Varanasi, Uttar Pradesh, India

ABSTRACT

Circadian rhythms regulate the temporal schedules of the cell's physiology to facilitate customization to environmental alterations per day which is almost ubiquitous. The molecular mechanisms of circadian rhythms in microorganisms are subject to environmental stimuli such as light intensity, temperature, and humidity generated by Earth's rotation. In the absence of environmental stimuli, how microorganisms maintain

* Corresponding Author's E-mail: rpsinhabhu@gmail.com.

24 h circadian rhythms had been a mystery for a long period. The most comprehensive understanding of how synchronization of microorganisms circadian oscillator with the environment is generated and how information is transmitted from an oscillator to generate bio-activity rhythms is a breakthrough for the improvement of recent and future work. These consequences have changed the view about circadian rhythms. Rhythm differs from a linear model in that it is considered as a collective organization of multidisciplinary elements that are integrated to accelerate and reset the oscillator within the context of the cell. This chapter discusses how this mechanism of circadian rhythms varies in many microorganisms and how it becomes important for microbes. Over the past few years, we have seen numerous advances in the understanding of the molecular mechanisms of circadian rhythms in microorganisms, and this chapter attempts to outline these mechanisms.

Keywords: circadian rhythm, circadian clock, microorganisms, environmental stimuli

INTRODUCTION

Almost all life on Earth has evolved in a cyclic environment, exposed to daily rhythms of light intensity, temperature, humidity, and other environmental stimuli. The organisms adapt and look for the daily environmental changes due to the selective pressure imposed by these rhythms. The adaptation ability of organisms to deal with these daily changes is called a circadian rhythm. The hypothalamic suprachiasmatic nucleus acquires an environmental cue from a circadian rhythm that coordinates the performance of the oscillatory clock existing in peripheral tissues (Costantini et al. 2020).

The term circadian rhythm is given in the sense of the mechanism for measuring time by organisms in units of approximately one day. Circadian rhythms produce daily rhythmic actions in the organism to take advantage of the daily environmental cycle. Approximately ever-present circadian rhythms are found in almost all microorganisms, including both prokaryotes and eukaryotes. The most remarkable thing is that these circadian rhythms generate rhythmic actions even in a constant condition situated organisms, in the absence of any environmental changes. This is evidence that the organism's circadian rhythmicity is intrinsic, self-reliant, and not entirely driven by the direct response to the environmental changes. Therefore, circadian rhythms must include at least one or more self-reliant oscillators that can generate rhythmic outputs to drive the rhythmic actions that we observe even in the absence of external time signals.

These properties of circadian rhythms have given rise to a model for the circadian clock that consists of three parts. First: an input route to receive time signals from the environment, second: a self-contained oscillator that interprets time signals, and third: an output route by which the oscillator affects the behavior of the organism (Figure 1). This respected model has served circadian biologists well for many years. With most models, it is certainly an oversimplification and the above three-part model has become progressively inadequate to learn more about the mechanisms underlying circadian rhythmicity.

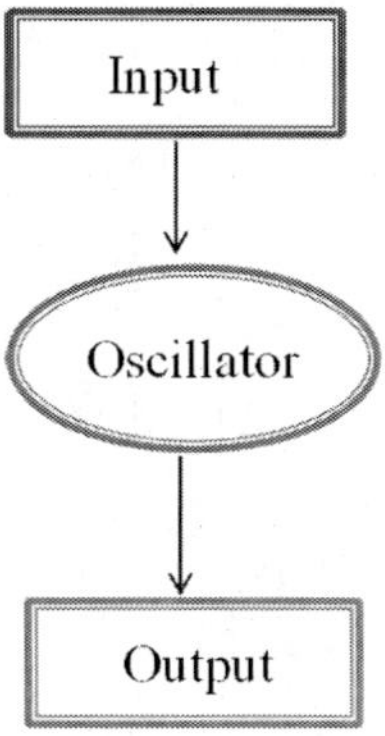

Figure 1. The simple three-part model in potential complexity of the circadian system is as: input > oscillator > output.

The molecular level of circadian rhythm machinery is centered on the autoregulatory transcription/translation feedback loop (TTO), wherein transcriptional activators prompt the repressor genes expression and numerous tissue-specific clock-regulated genes (CCGs) (Partch et al. 2014). This mechanism is also focused on the key point of the immune response, by regulating the task of cells and modulators in microorganisms. Its dysregulation may be responsible for immunodeficiency, inflammatory diseases, and elevated menace of contaminations (Curtis et al. 2014). Circadian rhythms in pathogenic microorganisms such as bacteria and fungi have a role in controlling the antibacterial and antifungal host defenses, and speculation on the effect of circadian drug application in infectious diseases. Furthermore, circadian rhythms have two-fold importance in infectious diseases, acting at the level of both host and pathogen. In this chapter, we have discussed the existing cognitions about circadian rhythms driven by potential molecular components in microorganisms, and have also focused on those that have importance in the regulation of evolutionary life.

MOLECULAR BASES OF CIRCADIAN RHYTHMS IN MICROORGANISMS

Adaptation of organisms to the environment includes the anticipation of cyclical events such as the daily rhythm of sunlight and night. Most organisms express temporal programs reflecting the 24 h cycle of day and night from microscopic cyanobacteria (Johnson and Golden 1999) to humans. This daily temporal organization is not simply driven by the external rotation of day and night but is regulated by an endogenous biochemical oscillator, a circadian clock that controls multiple behavioral, physiological, and biochemical processes (Dunlap 1999). Next to, all eukaryotes where clocks are marked shared a general molecular map. The key clock has an inside self-determining oscillator centered on a negative TTO (Bell-Pedersen et al. 2005; Montenegro-Montero et al. 2015; Dunlap 1999) accompanied by further loops in plants (Romanowski and Yanovsky 2015). However, some researchers have focused on the cyanobacterium *Synechococcus elongatus* (Mackey et al. 2011). Archaeal peroxiredoxin oxidation by circadian rhythms proposes three domains of life and, as such, plays an evolutionary role in the origin of circadian rhythms (Edgar et al. 2012).

Table 1. Circadian rhythmic diversity in established behavioral and physiological processes in microorganisms

Microorganism	Circadian rhythm	Time	References
Archaebacteria			
Halobacterium salinarum	Anticipatory behavior	Diurnal	Whitehead et al. 2009
Haloferax volcanii	Timing of cell division	Light/dark cycles	Maniscalco et al. 2014
Halobacterium salinarum	Peroxiredoxin oxidation protein	Light/dark cycles	Edgar et al. 2012
Cyanobacteria			
Synechococcus strain WH7803	Cell division cycle	Nocturnal	Sweeney and Borgese 1989
Cyanothece sp. ATCC 51142	Formation of granules for carbohydrate storage, alternation of photosynthesis, and nitrogen fixation	Diurnal	Schneegurt et al. 1994
Trichodesmium thiebautii	Photosynthetic genes	Diurnal	Roenneberg and Carpenter 1993
Synechococcus sp. PCC 7942	An oxygen-sensitive enzyme in purine biosynthetic pathways	Nocturnal	Liu et al. 1995
Synechococcus sp. PCC 7942	Photosynthetic genes	Diurnal	Liu et al. 1996
Oscillatoria sp. strain 23	Nitrogenase activity	Diurnal	Stal and Krumbein, 1985
Algae			
Euglena gracilis	Phototaxis behavior	Diurnal	Pohl 1948
Chlamydomonas reinhardtii	Phototaxis behavior	Diurnal	Johnson et al. 1992
Chlamydomonas reinhardtii	Cell division	Diurnal	Bruce 1970; Goto and Johnson 1995
Chlamydomonas reinhardtii	Chemotaxis	Nocturnal	Byrne et al. 1992
Chlamydomonas reinhardtii	Methylammonium uptake	Nocturnal	Byrne et al., 1992
Chlamydomonas reinhardtii	Survival from ultraviolet radiation	Diurnal	Nikaido and Johnson, 2000
Chlorella spp.	Photosynthesis	Diurnal	Hesse 1972

Table 1. (Continued)

Microorganism	Circadian rhythm	Time	References
Fungi			
Pilobolus spp.	Zygospore discharge	Diurnal	Uebelmesser 1954
Pilobolus spp.	Sporangiosphore development	Diurnal	Bruce et al. 1960
Pellicularia filamentosa	Basidiospore discharge	Nocturnal	Carpenter 1949
Sphaerobolus stellatus	Basidiospore discharge	Nocturnal	Engel and Friederichen 1964
Apiosordaria verruculosa	Ascospore discharge	Diurnal	Ingold and Marshall 1963
Daldinia concentrica	Ascospore discharge	Nocturnal	Ingold and Cox 1955
Sordaria fimicola	Ascospore discharge	Diurnal	Austin 1968
Neetria spp.	Ascospore discharge	Nocturnal	Ingold 1971
Neurospora crassa	Conidiospore development	Diurnal	Pittendrigh et al. 1959
Neurospora crassa	Ascospore discharge	Not determined	Lakin-Thomas et al. 1990

Multiple actions can be controlled by the central oscillator through its synchronization when it senses climatic cues (Bell-Pedersen et al. 2005; Montenegro-Montero et al. 2015; Dunlap 1999). The fascination of biology derives largely from the diversity of its phyla and species and their specific modes of action and behavior. However, in each phylum, some model organisms are studied extensively to understand the basic and typical features of this group. An overview of the diversity of circadian rhythms in behavioral and physiological processes found in microalgae is given in Table 1.

Over the last decades, major discoveries in circadian rhythmicity have been made using microbes. Two eukaryotes that are frequently studied these days for circadian investigations were first reported by Pittendrigh et al. (1959) and Bruce (1970) in the fungus *Neurospora crassa* and the Chlorophycean member *Chlamydomonas reinhardtii*, respectively. These microbes have been indisputably chosen for extended research because of their potential approaches in genetics, either molecular or classical. But, at the end of all, today we have reports on circadian rhythmicity in both prokaryotic and eukaryotic microorganisms which is discussed below.

CIRCADIAN RHYTHM IN PROKARYOTES

Cyanobacteria

Cyanobacteria are among the most primitive prokaryotic organisms, which exhibit circadian rhythm. One such example is exhibited in diazotrophic cyanobacteria. It can perform oxygenic photosynthesis as well as nitrogen fixation. In oxygenic photosynthesis, oxygen is released as a by-product, which acts as an inhibitor for nitrogen fixation by acting on the nitrogenase enzyme, creating a conflicting situation. To avoid this, multicellular filamentous cyanobacteria may have a spatial and temporal separation, and unicellular bacteria exhibit temporal variation to exhibit both photosynthesis and nitrogen fixation processes. Photoprotective compounds such as mycosporine-like amino acids synthesis (Mishra et al. 2019a; 2019b) and photosynthesis takes place during the daylight, while nitrogen fixation peaks during night/dark. Interestingly, this rhythmicity is maintained even when they are shifted to constant light, albeit with temperature rhythmicity (Bhadra et al. 2017).

The genetically tractable circadian rhythm has been extensively studied in cyanobacterium *S. elongatus* PCC 7942, which appeared as a key model organism for the research and study of molecular mechanism descriptions and modulation of circadian rhythms. Introductory work was carried out in cyanobacteria to establish the reporter gene luciferase in the rhythmic monitoring of gene expression, which facilitated genetic testing and also led to the identification of the first clock mutants (Ishiura et al. 1998; Kondo et al. 1993). *S. elongatus* PCC 7942 is with three chief proteins *kaiA, kaiB, and kaiC*, responsible for the circadian rhythm-based transcription, and translation, and its inactivation ensues in arrhythmicity (Ishiura et al. 1998; Nakahira et al. 2004).

Mechanism of Circadian Rhythm in Cyanobacteria

The molecular mechanism of circadian rhythm in the cyanobacterial oscillator is a post-translational oscillator (PTO), in contrast to that of TTO of eukaryotic circadian rhythm and this regulatory mechanism consists of KaiA, KaiB, and KaiC proteins. In these proteins, KaiC possesses the properties of autokinase, autophosphatase, and ATPase. Phosphorylation of KaiC protein at Ser431 and Thr432 residue and ATPase activity plays a key role in circadian timekeeping mechanism (Figure 2) (Nishiwaki et al. 2004; Xu et al. 2004). KaiC is a bi-lobed homohexamer ring structure, each monomer of which is made up of two domains such as CI and CII which interreact to form a ring and a C-terminal tail called an A-loop (Pattanayek et al. 2004). Phosphorylation by these rings is carried out by binding to and hydrolyzing ATP, which is confined to the CII ring (Nishiwaki et al. 2000). The autokinase activity of KaiC is aided by KaiA binding to its C-terminal segment during the day (Kim et al. 2008), whereas its autophosphatase activity is stimulated by the KaiB binding to its A-loop by dissociating KaiA.

Phosphorylation and dephosphorylation of KaiC take place in an ordered and sequential manner. During the phosphorylation step of KaiC, threonine-432 (Thr432) phosphorylates first and then serine-431 (Ser431) residue, and thus it becomes a twice phosphorylated species (Nishiwaki et al. 2004; Nishiwaki et al. 2007, Rust et al. 2007). During the subjective night process of dephosphorylated first occur with Thr432 followed by Ser431, ensuing in a nonphosphorylated species and completing a 24-h cycle (Figure 2). Structural changes which are responsible for the clockwise direction of the phosphorylation cycle in KaiC protein, point up recently with delicate nuclear magnetic resonance technique. The interaction and conformation of the CI and CII rings are interrelated and altered by the purpose of the phosphorylation state of KaiC. They drive transitions from a KaiA-associated phosphorylation state to a KaiB-associated dephosphorylation state (Chang et al. 2011; Chang et al. 2012).

KaiA (dimer) interacts with A-loop of nonphosphorylated KaiC, the exposed state of A-loop, promotes phosphorylation of KaiC during the day. The phosphorylation at Ser431 results from compact interaction between CI and CII rings, displaying B-loops (located on the CI ring) which serve as a binding site for KaiB. KiaB binds to exposed B-loop promoting dephosphorylation phase of the cycle (Chang et al. 2011; Chang et al. 2012) KaiC autophosphatase activity occurs unusually due to deficit of motifs related through the characteristic protein phosphatases. The phosphate group gets transferred from S431 and T432 to ADP, to form ATP. This ATP molecule then hydrolyzes to form ADP and Pi. In the KaiC phosphorylation cycle, ordered dephosphorylation events are promoted by this proposed distinctive mechanism.

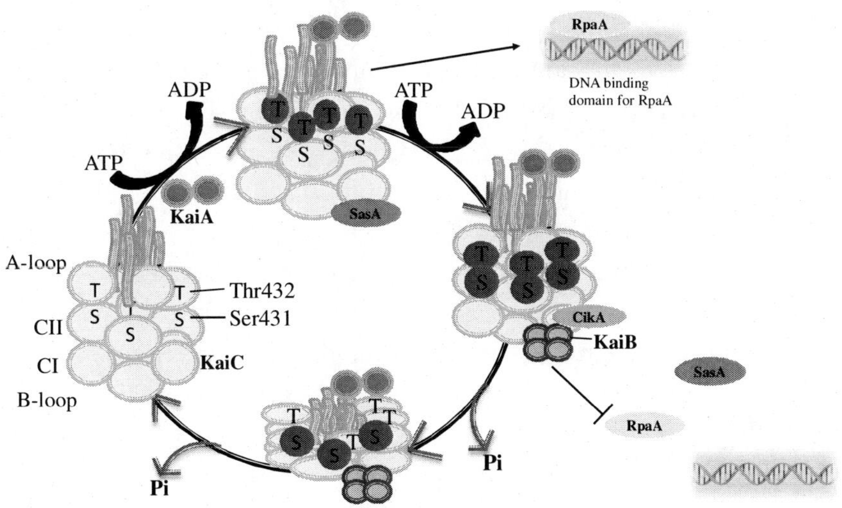

Figure 2. Molecular mechanism of the circadian rhythm in the cyanobacterium *Synechococcus elongatus*. The KaiA dimer binds to the KaiC hexamers, which contain CI and CII rings with an A-loop segment, and assists its phosphorylation. Following phosphorylation of KaiC, SasA competes with KaiB and binds to phosphorylated KaiC and its binding persuades activation of RpaA. RpaA is an output protein, which regulates global gene expression by binding to the DNA binding domain. The KaiB tetramer is formed from the accumulation of its monomer and binds to KaiC in place of SasA, inactivating KaiA, resulting in KaiC dephosphorylation and inactivation. Furthermore, the interaction of CikA with KaiB, resulting in RpaA dephosphorylation and inactivation. The red-filled circles 'T' and 'S' represent the phosphorylated form of the Thr432 and Ser431 residues, respectively.

Input Mechanism in Cyanobacteria

Cyanobacterial circadian rhythms can exhibit rhythmicity by either their independent oscillators or dependent environmental stimuli. Instead of sensing through photoreceptors, the bacterial cell senses climate change indirectly through a redox reaction. Components like CikA and LdpA helps in redox sensitivity (Ivleva et al. 2005; Ivleva et al. 2006). Key systematic constituents such as the dimer body *KaiA* gene and hexamer moiety *KaiC* gene, spot the photosynthetic metabolites, which can be used as an input to modulate the clock with the outer environment (Rust et al. 2011; Wood et al. 2010).

CikA was one of the first to be identified redox sensing input components (Schmitz et al. 2000), and it became mutant after achieving a profound oscillation of 5 h dark. This oscillation was attributable to a mutation in the CikA, altering its phenotype by increasing the effect of low-amplitude pulses and a failure to resume oscillations in it (Schmitz et al. 2000; Miyagishima et al. 2005). Indeed, the CikA mutant does not reset the oscillatory phase, it is only responsible for reinforcing the effect of CikA as a member of an input pathway responsible for conveying messages from the environment to the oscillator. However now we know that CikA is additionally responsible for modifying output activities (Gutu and O'Shea 2013).

CikA, KaiA, and cellular quinone pools interact among themselves to achieve environmental sensation (Kim et al. 2012). Photosynthetic performance in cyanobacteria affects the abundance and redox status of cellular metabolites, including the plastoquinone pool. Upon onset of darkness, this pool becomes oxidized. Hence the presence of oxidized quinone (Q_0) can act as a proxy of darkness for the system. With the addition of Q_0, even on the subjective day, the phase shift can be achieved (Kim et al. 2012). Pseudoreceiver (PsR) of CikA and KaiA binds only with the oxidized form (Ivleva 2006; Wood et al. 2010). The two outermost membrane proteins such as CikA and KaiA are situated at the pole of the cells, which may aid in the association with Q_0 in the thylakoid membrane (Cohen et al. 2014). Furthermore, the strength of CikA protein is dependent on the rhythmic control and external changes, where both CikA and light intensity are inversely proportional to each other and CikA is at peak during the dark (Ivleva et al. 2006). The CikA and KaiA aggregation occurs *in vitro* and rapid disappearance *in vivo* due to the addition of oxidized quinone suggesting that degradation dependent aggregation of CikA and KaiA may be a practical mechanism, leading to the clock corresponds concurs with the cycle of day-night (Ivleva et al. 2006; Wood et al. 2010).

Apart from CikA/KaiA, another component LdpA is a Fe-S center-binding redox-active protein, which can also monitor changes in cellular redox status and act as a clock input component (Katayama et al. 2003; Ivleva et al. 2005). Mutants of *ldpA* show a 1 h duration and a lack of sensitivity to the light gradient that generally controls the circadian period (Ktayama et al. 2003). LdpA is able to sense the change in electron transport pathway that is dependent on the light intensity, it also alters CikA abundance in a light-dependent manner and is consequently responsible for fine-tuning the circadian period, this suggests cooperation among CikA and LdpA, which is determined by changing environmental alerts to the oscillator.

Output Mechanism in Cyanobacteria

Photosynthetic metabolites also act as key input signals for circadian rhythm. Across the day, KaiC can directly sense the change in the ratio of ATP/ADP (Rust et al. 2011). Although the cell's response to darkness was alike throughout the circadian rhythm, the oscillator considers the dissimilar level for alerts at different rhythms in the cycle. Notably, the oscillator was subjectively responsive to switches in ATP/ADP in the mid-day, when the KaiC phosphorylation was increasing, corresponding to the day point. Taking both the data into account, redox status with quinone pool and ATP/ADP ratio with KiaC, exact photoperiod of the circadian clock can be encoded (Kim et al. 2012).

Additional links between cellular metabolism and the circadian rhythm have recently come to the light. Under continuous illumination rhythms of glycogen, abundance can be regulated by the circadian clock (Diamond et al. 2015, Pattanayak et al. 2014), where during day-time glycogen accumulation occurs as a result of photosynthesis. After receiving a dark phase, mutants showed the copious abilities to shift the circadian phase and disclose that the glycogen level in the cell is contrarily equivalent to the phase shift potential (Pattanayak et al. 2014). These upshots evince that the clock controls the metabolic rhythm that can give feedback to the oscillator and enable the cell to predict darkness.

An output signaling from the KaiABC-based oscillator relay on the twain component controlling complex comprising of a sensor histidine kinase (HK) SasA and alike feedback director RpaA (Iwasaki and Kondo 2000). The worldwide rhythmic expression of the gene is driven by the SasA/RpaA output pathway in conjunction with cell division (Dong et al. 2010). Nevertheless, rhythms of chromosome concentration are with self-contained SasA (Smith et al. 2006), which suggests the existence of other output pathways. RpaA (response regulator) is regulated by two cognate HK, SasA and CikA. The sensor domain of SasA consists of an N-terminal (B-loop) KaiB alike and a received HK domain. Earlier there is a contradiction in the sequence similarities between the B-loop of KaiB and SasA, since the divergence that takes place generated the typical distinction betwixt secondary and tertiary structures (Vakonakis et al. 2004). Which results in autophosphorylation of His-161 in the SasA kinase domain (Chang et al. 2015; Valencia et al. 2012) and subsequent transfer of that phosphate to Asp-53 on RpaA, rendering the transcription factor competent for DNA binding (Markson 2013).

At the biochemical level, SasA competes with KaiB for the B-loops on the CI domain of phosphorylated KaiC (Tseng et al. 2014). The binding of SasA is faster than that of KaiB, which may be due to the refolding requirement of KaiB (fs-KaiB) for binding to KaiC (Chang et al. 2015; Valencia et al. 2012). Period-reliant initiation and transmission of rhythm occur in cyanobacteria by the interaction of SasA with KaiC. KaiA also plays a role in the competition between KaiB and SasA for the CI domain of KaiC. CikA is a multidomain protein that plays role in dephosphorylating RpaA, and ties the clock to subcellular localization and entrainment functions (Cohen et al. 2014; Mackey et al. 2008; Schmitz et al. 2000; Ivleva et al. 2006). Refolding requirement of KaiB can link to the pseudo-receiver (PsR) domain of CikA (Chang et al. 2015). The phosphatase activity of CikA is induced by its interaction with KaiBC complex concerning RpaA (Chang et al. 2015). Through the interactions of its PsR domain, CikA is incorporated into the night complex with fold-switched KaiB at an interface that overlaps with the KaiA binding site. Robust rhythms of RpaA phosphorylation can be created through the antagonistic effect of SasA and CikA. This suggests that RpaA is a cue for incorporating and triggering circadian inputs and rhythmic outputs, respectively.

Mechanism of Circadian Rhythm in Archaebacteria

Prefatory evidence suggests that Archaea, the second major prokaryotic group, also exhibit rhythmic activities. In the saline or halophilic archaeon *Halobacterium salinarum* NRC-1, the expression of 12% of genes oscillates periodically from ultradian (13-h) to circadian (22-h) under constant conditions (Whitehead et al. 2009). The halophilic archaebacteria *Haloferax volcanii* possess four cryptic genes alike *kaiC*, such as *cirA, cirB, cirC,* and *cirD,* extremely related to the cyanobacterial *kaiC* family genes. These four genes of *H. volcanii* contribute the 28 - 33% identity and 48 - 55% similarity by *S. elongatus kaiC* and procure anticipated walker motifs A and B, which are responsible for the phase-dependent expression of gene and schedule of cell division (Maniscalco et al. 2014).

However, there is no clear evidence of the presence of homologs of *kaiB* and *kaiA* in archaea. Diurnal rhythmic expression of *Cir* genes was abolished in the Cir mutants due to the lacking of Cir genes except *cirA* gene (Maniscalco et al. 2014). Despite all this, detailed research will be needed to ascertain whether these fluctuations constitute a precise circadian rhythm. This consequently leads to the panoramic notion of circadian rhythms thinking that it may also be a feature of some archaea.

Circadian Rhythm In Eukaryotes

Circadian Rhythmic Mechanism in Fungi

Fungi have a dominant circadian rhythm with increasing complexity and circadian timekeeping has appeared to manage per diem rhythms in the development and release of spore. For the study of the circadian rhythm, *N. crassa* remains to be an excellent model system. The annual cycle in airborne spore, diurnal or nocturnal spore release, and per diem circadian in development of asexual spore (Pittendrigh et al. 1959: Sargent et al. 1966) are some examples showing a process of the circadian rhythm. Some rhythms are exclusively affected by periodic climatic changes (Ingold 1971), while some are endogenous and do not depend on external signals. Some factors such as light, humidity, wind velocity, temperature, etc. are periodic and affect circadian rhythm in fungi. Circadian rhythms may arise in spore development and variability due to these factors. Light, temperature, wind current is high during daytime while humidity peaks at night. Thin-walled spores, which are more likely to be injured by light or dryness, and spores that require turgor pressure to release, liberated during the night. Whereas thick-walled spores which are resistant to ultraviolet radiation and hydrophobic are released during the day (diurnal) (Bell-Pedersen et al. 1996).

In the dung fungus *Pilobolus* spp. spore release is regulated by circadian rhythm and direction is affected during the light (Uebelmesser 1954). This suggests that light conditions come up with a way to make sure that spore release take place on nearby moist grasses (Ingold 1971), therefore, diurnal rhythm was chosen for the spore release during development to limit spore diffusion. Although, in the case of basidiospore of *Pellicularia filamentosa* spore are discharged at night under the control of the circadian clock when they are not subjected to desiccation and injury from harmful ultraviolet radiation (Carpenter 1949).

A very significant role in the sporulation process is played by light in most fungi such as in *Asperigillus nidulans* (Mooney and Yager 1990). This proposes that day light gives an external cue that induces growth beyond the ground level at the relevant times for release. These fungi having circadian rhythm may secret signal that initiates growth in the lack of light and captivating the growth system in feedback to the light.

In *Neurospora crassa*, the rhythmic response was first recorded in 1959, which comes with an excellent system of studying the circadian clock. It also exhibits circadian rhythms in asexual conidia formation (Pittendrigh et al. 1959; Sargent et al. 1966; Larkin-Thomas et al. 1990). In a mid-subjective night in the course of vegetative growth on the agar surface, the clock starts macrocondition, begins to produce aerial hyphae that grow and give rise to conidiospore. During the day, this growth cycle stops, and the growth carries on with alike vegetative hyphae. A real rhythmic quality is persisting in the conidiation of *N. crassa*, where primarily, the is sustained at 24°C in continuous darkness with a periodicity of 22 h (Pittendrigh et al. 1959) and thereafter, the rhythm period is temperature compensated in the physiological temperature cover 18-30°C. At last, the circadian rhythm in conidiation can be directed by the light/dark cycle (Sargent et al. 1966; Nakashima and Feldman 1980) and quick temperature pulses (Gooch et al. 1994; Rensing et al. 1995). All the above factors in conidiation rhythm show characteristics of the circadian rhythm.

Mechanism of Circadian Rhythm in Neurospora

Circadian clock regulation in *Neurospora* occurs by TTO. Three CCGs, frequency (*frq*), and two white-collar (*wc*) genes, *wc-1* and *wc-2* have been reported as the core components of a transcription/translation oscillator. In the TTO model (Dunlap 1999; Dunlap and Loros 2004), the Frequency (FRQ) protein act negative element that inhibits its transcription, and the WC proteins are identified as the positive elements (Figure 3a). The chief constituent of *Neurospora* circadian rhythm is FRQ, which interacts with RNA helicase and forms FRQ-FRH complex (FFC), and this is the negative regulator of the circadian negative feedback loop (Aronson et al. 1994; Cheng et al. 2005). The White-Collar Complex (WCC) is established in animals by the interplay between the pairs of PER-

ARNT-SIM (PAS) domain holding transcription factors WC-1 and WC-2, as CLOCK and BMAL1.

The positive element WCC rhythmically binds to Clock Box (c-box) in the *frq* promoter and activates transcription of *the frq* gene during the early subjective day in constant darkness time. This increases the FRQ level protein. FRQ forms a homodimer and interacts with FRH to form the FFC complex. Proper structure of FRQ and steady-state level can be maintained through this interaction (Cheng et al. 2001; Cheng et al. 2005; Guo et al. 2010). In the absence of FRH, FRQ is unstable. Synthesis of FRQ is followed by the developing post-translational moderations such as casein kinases (CKI and CKII) dependent phosphorylations. These phosphorylations also occur by some other kinases during the day and evening, finally leading to the ubiquitin/proteasome degeneration (He and Liu 2005; Pregueiro et al. 2006).

E3 ubiquitin ligase FWD-1 is responsible for the ubiquitination of FRQ. FRQ steadiness controls by the COP9 signalosome by controlling the action of SCF-FWD-1 complex that is specific for the degeneration of FRQ. Interacting with WCC, FFC promotes phosphorylation of WCs. There is WC independent regulation of *frq* gene transcription has also been found where RCO-1 repressor complex represses *frq* transcription. In addition, FFC loads the RNA exosome complex to the *frq* mRNA to manage its half-life (Guo et al. 2009). CKs promote degradation of FRQ, antagonistically FRQ phosphorylates and sustained by the protein kinase A (PKA). Collective protein phosphatases, like PP1, PP2A, and PP4 countered the roles of these kinases (Yang et al. 2004; Huang et al. 2007). A series of regulators fine-tunes phosphorylation of FRQ to determine the stability and length of the clock. Indicators of circadian rhythm in *Neurospora* are the excess of *frq* mRNA and FRQ protein and the phosphorylation account of FRQ.

Rhythms are now being classified as '*frq*-less oscillators' (FLO) in the absence of a functional *frq* gene (Iwasaki and Dunlap 2000). The presumption is that only the 'true circadian rhythm' is driven by the FRQ/WC-based TTO. Apart from TTO, all the other rhythms are driven by separate mechanisms. These oscillatory mechanisms must interact in some way, and various schemes have been proposed to be linked together (Dunlap and Loros 2004; Lakin-Thomas and Brody 2004; Bell-Pedersen et al. 2005). There are now several proposed FLOs that are dubiously complicating these coupling schemes (Figure 3b, c).

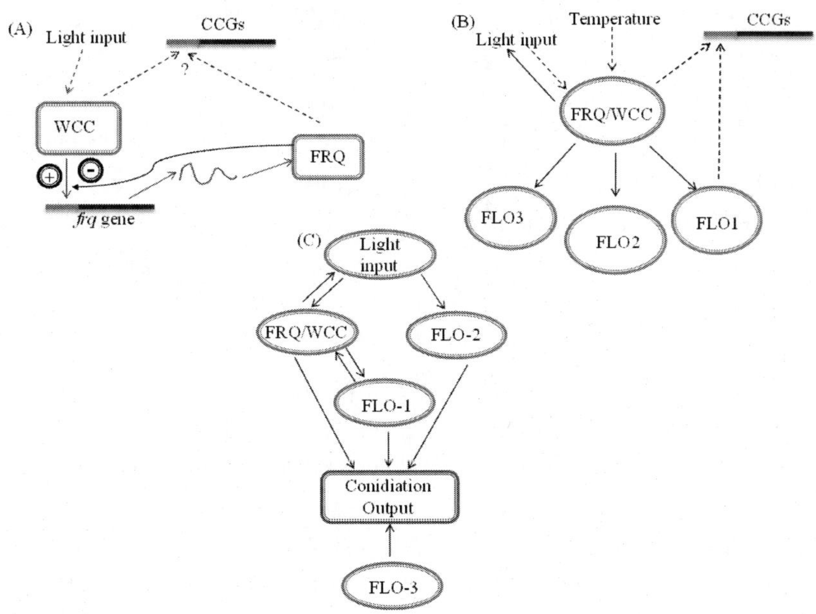

Figure 3. Molecular mechanism of the circadian complexity in the *Neurospora* sp. The transcription/translation feedback loop-based circadian rhythm in *Neurospora* sp. The negative element, the *frq* gene codes for the FRQ protein, is activated by the positive element WCC that initiates transcription of the *frq* gene. WCC is a complex of two white-collar proteins such as WC-1 and WC-2, with WC-1 being the major light receptor and monitoring light input to the circadian clock in *Neurospora* sp. Circadian clock output occurs *via* transcriptional regulation of clock-controlled genes (CCGs) (Source: Lakin-Thomas, 2006) (A); the FRQ/WCC-dependent transcription/translation feedback loop (Figure 3a) is considered to be the central oscillator and consists of many FLOs (FLO1, FLO2, FLO3) for the circadian clock in *Neurospora* sp. FRQ and WCC respond to light input pathways to make the input rhythmic, and supply a rhythmic output to the CCGs to guide the observed conidiation rhythm, but how do the FLOs interact with the transcription/translation feedback loop, It is still not known. In frq and wc null mutants, the transcription/translation feedback loop is absent, allowing FLOs to process arrhythmic output (Source: Adapted and modified from Lakin-Thomas and Brody, 2004; Lakin-Thomas, 2006) (B, C).

Input and Output Mechanism in Neurospora

The blue light photoreceptor WC-1 is either directly or indirectly answerable for all light reactions in *Neurospora* (Ballario et al. 1996; Froehlich et al. 2002; He et al. 2002). The direct binding of light-induced WCC to the hundreds of genes initiates transcription (Chen et al. 2009; Smith et al. 2010). Flavin adenine dinucleotide, with an association of which WC-1 expresses light activity, is linked covalently to the sustained light observing domain such as LOV domain constitute for light, oxygen and voltage. Light induction by the WCC results in increases in *frq* levels and when the amount of *frq* is short and subjectively arises before dawn, quick light exposure elevates *frq* mRNA to top levels and thus proceeds the clock from mid to late morning corresponding to the highest *frq* expression. It comes with a molecular explanation in terms of light penetration and phase remodeling (Crosthwaite et al. 1997).

The VVD protein can adapt to clock inputs. It is a small (21kDa) flavin-binding blue-light photoreceptor with a PAS domain and an N-terminal Cap (Heintzen et al. 2001; Zoltowski et al. 2007), which conduct communicating light adaptation in *Neurospora* (Heintzen et al. 2001; Schwerdtfeger and Linden 2001, 2003; Shrode et al. 2001). Photoadaptation is a phenomenon where the transcription level of the light-induced gene gets elevated when wild type *Neurospora* get exposed to light and the elevation revert to the pre-induction level within 2-4 h, following light exposure in wild type. *Neurospora* can notice alteration in the strength of light in place of directly detecting 'on versus off' (Schwerdtfeger and Linden 2001, 2003). The molecular basis of light adaptation comes after the successive procedure is that VVD is swiftly and distinctly induced by light and which also causes removal of the N-terminal cap from the LOV domain, in reaction to which, a fast-exchanging dimer is produced by VVD in light (Zoltowski et al. 2007). Photo-induced VVD swiftly proceeds into the nucleus, link with the WC-1, and reduces its capability to activate transcription. Through all these processes, VVD leads to photoadaptation (Chen et al. 2010; Hunt et al. 2010; Malzahn et al. 2010). Along with these, VVD plays a key role in modulating WCC mediated circadian clock properties, such as regulating light response and entraining the clock running. Although VVD decays in dark conditions (night) and increases in light, and ensures the running of a clock (Malzahn et al. 2010).

Another environmental factor is temperature which can entrain the clock, however, temperature primarily influences FRQ (Liu et al. 1997; Pregueiro et al. 2005). The *frq* amplitude and *frq* expression are similar at mean level of changing climatic temperatures, while the level of FRQ is higher at higher level of temperature. The entire FRQ communicate with a specific time and temperature of the day. However, if a culture is transferred from one temperature to another, the FRQ level will not alter that quickly (Liu et al. 1998).

In addition to the light, temperature also regulate FRQ at the post-translational level. A short and long interval of FRQ (S-FRQ and L-FRQ) of the protein vary by hundred amino acids is formed by the *frq* pre-mRNA, by undergoing a nexus temperature sensitive replacement splicing in the 50 UTR. Both FRQ can rhythm across a temperature range from 20-30 °C (Liu et al. 1997; Colot et al. 2005; Diernfellner et al. 2007). At low temperatures, both S-FRQ and L-FRQ occur at comparable levels of concentration, whereas at higher temperatures, L-FRQ appears to be at higher concentrations with a higher FRQ. These FRQ forms can be required to support circadian rhythmicity. Circadian output is the biology's true source of what makes the rhythm itself significant and recalling above that core clock operation results in cyclic activation and inactivation of the WCC transcription factor.

Mechanism of Circadian Rhythm in Algal Members

Similar to the many other microorganisms, algae have a well rhythmic potential due to the synchronous internal day-night clock and this circadian rhythm regulates various biological processes such as growth, reproduction, and cell structure in algae.

Circadian Mechanism in Chlorophyta

Chlamydomonas reinhardtii is a unicellular green alga that belongs to the Chlorophyta and originated more than 1×10^9 years ago (Yoon et al. 2004). *C. reinhardtii* is usually referred to as a 'green yeast,' as it possesses several things that are beneficial for the molecular based genetic studies, like rapid growth, haploid genome formation, genetic manipulation proficiency, and high-frequency changes in the genome of the nucleus, mitochondria, and chloroplast (Harris 2001).

ROC genes such as *ROC15, ROC40, ROC55, ROC66, ROC75, and ROC114* are in charge of the continuation of the circadian rhythmicity of chloroplast bioluminescence. Mutation in these genes generates critical alteration of circadian rhythm and causes circadian alteration of growth rate in *C. reinhardtii* (Matsuo et al. 2008). As reported by Matsuo et al. (2008), mutants of ROC55 and ROC15 showed shorter intervals of circadian rhythms such as 18 and 21 h, respectively, whereas mutants of ROC75 and ROC114 displayed arrhythmic. Additionally, dependent mutants such as ROC40 and ROC66 had only a longer circadian interval, which was 30 h under continuous conditions of light and dark, respectively.

The reported transcription factors were encoded by the *ROC15, ROC40, ROC66*, and *ROC75* genes, whereas F-box proteins by the *ROC114* gene (Figure 4). Moreover, casein (Ser/Thr protein) kinase is also a key regulator of phosphorylation, involved in the circadian clock of *C. reinhardtii* (Figure 4) (Schmidt et al. 2006). These determined that the circadian clock of *C. reinhardtii* and eukaryotic models such as *Neurospora crassa, Drosophila melanogaster, and Arabidopsis thaliana,* implicated with transcription, phosphorylation, and ubiquitination processes (Mizoguchi et al. 2006; Dunlap 1999; Gallego and Virshup 2007). The amount of mRNA of ROC15, ROC40, ROC66, and ROC75 genes were observed to fluctuate with different aspects in a circadian manner (Matsuo et al. 2008) and speculated that transcriptional feedback loops were formed by their genetic interaction with each other.

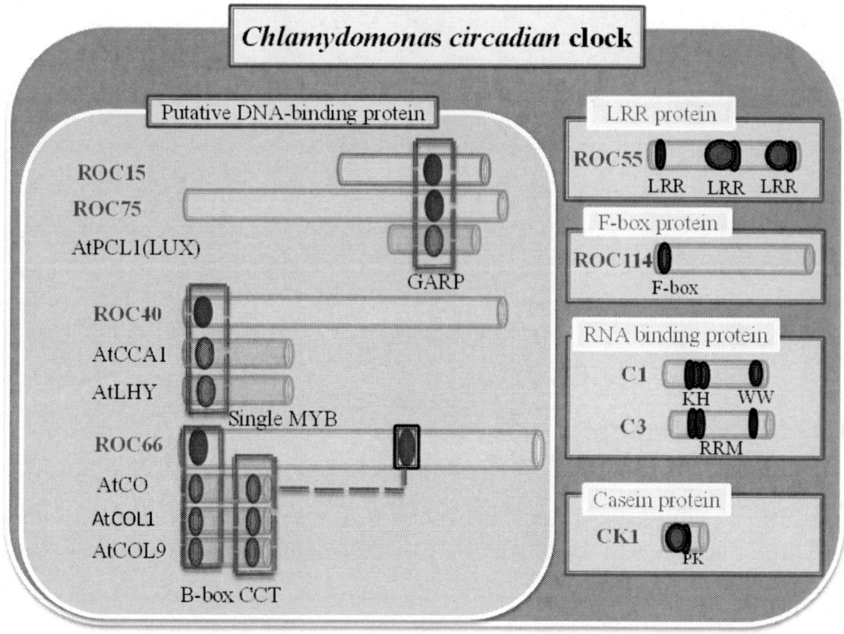

Figure 4. Diagrammatic elucidation of circadian clock proteins in *Chlamydomonas reinhardtii*. The forecasted operating domains are shown with two differently colored boxes, where DNA-binding proteins are lined up along the homologous *Arabidopsis thaliana* proteins. Domains with important sequence resemblance are indicated through dashed gray lines (Source: Adapted and modified from Matsuo and Ishiura, 2011).

Additionally, an important regulatory mechanism such as post-transcriptional mechanisms is regulated in the circadian clock of *C. reinhardtii*, which was due to their association with CHLAMY1 (Iliev et al. 2006), which binds to UG-repeat sequences in the 30-UTR of ROC40 mRNAs (Waltenberger et al. 2001). These UG-repeat sequences are involved in the concordance binding sequence of CHLAMY1 (Waltenberger et al. 2001; Matsuo et al. 2008). Therefore, regulation of ROC40 translation may be occurred by CHLAMY1, a multimeric protein complex composed of two subunits (C1 and C3) (Zhao et al. 2004), and it may be one of the important parts in the mechanism of circadian rhythm in *C. reinhardtii*. The C1 subunit is comprised of a lysine homology (KH) domain and a WW protein-protein interaction domain, whereas the CUG-BP-ETR-3-like factor (CELF) family protein contains the C3 subunit-containing RNA recognition motifs (RRMs) (Figure 4). Knockdown of the C3 subunit by RNA interference elevated the circadian rhythm in phototactic activity by approximately 6 h, while alike shifts in the C1 subunit have no response (Iliev et al. 2006). This type of circadian rhythmicity was also observed in nitrate reductase activity but was thought to be independent of phototaxis (Iliev et al. 2006). Additionally, knockdown of C1 ensued in the altered extent of C3 expression, making it likely that the regulation of these circadian protein expressions is connected (Iliev et al. 2006).

Circadian Mechanism in Pyrrophyta

Dinoflagellates are single-celled eukaryote, Pyrrophyta members and many of its light-emitting species show a predominant circadian rhythm of bioluminescence. In photosynthetic dinoflagellates, the photosynthetic activity is maximum during the day, whereas highest level of bioluminescence is occurred at night phase. Previous research on dinoflagellates regarding the clock was significantly facilitated by photo-radiation, as it provided an underlying reporter. Regulating bioluminescence seems biologically reasonable as it would be of negative merit in the day in multiple species, which living or moving close to the surface. Nevertheless, numerous further processes are regulated likewise, few with invisible practical value. In *Lingulodinium polyedrum*, the per diem synthesis and demolition of proteins are used to regulate the activity of circadian timekeeping and it is a process to conserve nitrogen (Hastings 2013).

Many Quasi-Independent Oscillators in Dinoflagellates

In the case of dinoflagellates, circadian timekeeping is regulated at the translational level, which is distinct from most other organisms (Hastings 2007). Two distinct rhythms exhibited by bioluminescence are unforced light up and a persistent luminescence. These rhythms have non-identical times at which they have their peaks (acrophases). Photosynthetic capacity and time of day are some of the other rhythms at which dividing cells have quite further acrophases. Circadian rhythms have also been reported at the synthesis grades of multiple enzymes (Markovic et al. 1996) and put forward at different rhythm has its quasi-independent oscillator (Nassoury et al. 2005). Depending on monitoring, the acrophase relationships of these rhythms can vary under different growing conditions, such as the amount of light in a light-dark (LD) 24-h cycle (for example, LD6:18 or LD18:6), or the extent of the expression duration. Furthermore, what is compelling is that under continuous conditions, distinct clock may have dissimilar independent durations, that cannot be assisted effortlessly in a single rhythmic model (Roenneberg and Morse 1993). The variable rhythms seem to be linked, though almost relatively, with unrevealed coupling mechanism.

Synthesis and Destruction of Protein in Dinoflagellates

Both luciferase (LCF) and its binding protein (LBP) are depleted at the close of each subjective night and a new one is synthesized before the latter's activity peaks, reflecting the *in vivo* rhythm in the *L. polyedrum* (Dunlap and Hastings 1981). These mechanisms are utilised to control actions on the face of it, and it might expect an additional conventional method, like phosphorylation and dephosphorylation and action of control is generally linked by advanced transcription. The amplitude of the abundance rhythm of the glyceraldehyde dehydrogenase (GAPDH) is not as appreciable as with LBP or LCF, which means that not all GAPDH is destroyed in one cycle (Hastings 2013).

Thus, synthesis can be strongly circadian because it takes place at most through the tiny timeframe, but if the lifetime of the protein lasts for several days, the synthesis will only account for a small fraction of the total cellular amount of that protein, therefore its abundance will not exhibit high amplitude rhythms. In *L. polyedrum*, several other proteins reveal plenty of rhythmic alterations and regulated at the translational level (Milos et al. 1990). However, the equivalent protein was further incorporated by *in vitro* translation from poly(A)+ RNA, taken out of cells during the day and night period, representing the entire transcription of competent at all.

Circadian Rhythmic Mechanism in Protozoa

Research in the field of physiological behavior has led to the selection of protozoa specifically *Paramecium* (Kung and Saimi 1982) and it has been known for more than 20 years to have a circadian rhythm of mating activity (Ehret 1953; Barnett 1966; Karakashian 1968). *P. multimicronucleatum* reveals circadian rhythms of dynamics or transverse frequencies described by Hasegawa and Tanakadate 1987 but does not convey rhythmic photo-accumulation.

The photo-accumulation is directed close to a circadian oscillator in *P. bursaria* as similar in *Euglena* (Bruce and Pittendrigh 1956) and the analysis of this rhythm is simply self-operating with the help of a microcomputer's data collection system. Circadian photo-accumulation exhibits all three main properties of circadian rhythm that is temperature compensation, persistence in constant conditions, and entrainment by light and dark pulses. *P. bursaria* manifests mating functions' circadian rhythms in single cells populations (Miwa et al. 1987) and it dispense these attributes with additional multicellular organisms, although it is unicellular. Therefore, complex intracellular interactions are not involved in circadian rhythm production in these cells. The photo-accumulation rhythm of *P. bursaria* is a property on the dynamics rhythm of *P. multimicronucleatum*. So, the photo-accumulation rhythm in this way is clear enough that the fresh data can be applied to the survey at all. According to the report of Ehret 1953 and Miwa et al. 1987, mating action together with photo-accumulation automatable rhythm can be assayed in populations of cells.

Importance of Circadian Rhythm in Microorganisms

Life on the earth has evolved around 3.5 billion years ago which are affected by daily changes in light, temperature, humidity, etc. that are the result of the Earth's rotation around its axis in about 24 h. Approximately all organisms evolved a system to cope up these daily changes, due to earth rotations which controlled organismal physiology with the help of particular timekeeper called the "circadian clock" (Bhadra et al. 2017). Circadian rhythms are endogenous processes found in all living organisms from prokaryotes (cyanobacteria) to eukaryotes (protozoa, fungi, algae, plants, birds, and mammals). Cyanobacteria are the simplest organism used for rigorously testing and robust study of circadian clock. *S. elongatus* PCC 7942 is a most studied model organism. It is used for detailed molecular study and the genes involved in circadian clock regulation (Cohen and Golden 2015). The core clock encodes three genes *kaiA*, *kaiB*, and *kaiC*. These genes regulate photosynthesis, amino-acid uptake and cell division in *S. elongatus,* and also enhanced overall fitness of organism (Kondo et al. 1994). Chromosome compaction, global gene expression, and it also plays a crucial role in recovering from the harmful effects of ultraviolet radiation and photosynthetically active radiation (Liu et al. 1995; Mori et al. 1996; Smith and Williams 2006; Dong et al. 2010; Maniscalco et al. 2014; Diamond et al. 2015; Kannaujiya and Sinha 2017).

Peroxiredoxin is a homolog of the kai gene and other proteins in bacteria and may work as clock components, redox state of this class of protein in archaea and cyanobacteria oscillate autonomously. Occurrence of peroxiredoxin proteins in most of bacterial species, might have conserved redox and act as the timekeeper in all these organisms (Edgar et al. 2012). Besides those different types of photoreceptors have been described in numerous non-photosynthetic species. Different plant symbionts consist of red light-sensing phytochrome-like photoreceptors and in bacteria showing symbiotic association with roots of plant, like *Bacillus subtilis* consist blue light-sensing photoreceptors (Losi et al. 2002). Both red and blue-light photoreceptors are fond in *E. coli* which regulates day-night cycles in *E. coli* (Davis et al. 1999; Gomelsky and Klug 2002).

Host and Microbes Cross-Talk Between Their Circadian Rhythms

Microbes and hosts show a cross-talk between their circadian rhythms, which is either pathogenic or symbiotic in nature, and exact reason is still unknown. Examples, of symbiotic relationship between plant roots and fungi is constituted by the arbuscular mycorrhizal (AM), such a hypothesis has been recently reviewed by Lee (Lee et al. 2019). Different mechanisms proved a relationship between the circadian rhythms of the AM fungi and the plant, including carbon, nitrogen, and phosphorus exchanges that are regulated by symbiotic mycorrhiza which is under circadian control in plants (Lee et al. 2019). In another example, *B. cinerea* is a necrotrophic plant pathogen shows its interaction with *Arabidopsis thaliana* (Larrondo and Canessa 2019).

Previously it is known that plant have different defense mechanism which response to daily changes but Hevia, in his study found that it was fungal clock that play significant role which decides the response of daily changes as a result of the interaction between plant (Hevia et al. 2015).

Microbes Regulates Circadian Physiology

In recent years it was found that commensal bacteria play a key role regulating the host circadian physiology. Mukherji in his work found that gut microbes which recognized the intestinal epithelial clock and produces glucocorticoid shows interaction with Toll-like receptors (TLRs), which regulated the circadian oscillators (Mukherji et al. 2013), relative abundance and composition of gut microbial communities exhibit diurnal oscillations, which induce rhythmic production and hence circadian activity is influenced by microbial metabolites released by microbes. (Thaiss et al. 2014). The rhythmicity of microbiome dependents on feeding of host and a functional host circadian clock. An inverted phase of the cycling microbiome was seen by the reversal of feeding habits. In addition, timed feeding recovered the loss of microbial oscillations in mice which lack circadian clock (Thaiss et al. 2014).

Circadian Clock Regulates the Immune System

Different studies reported that the immune system is regulated by circadian clock. There many immune cells vary over a day and the functional activities of various immune cells show daily rhythmicity (Scheiermann et al. 2018). Internal clock maximizes the immunity of host when exposed to pathogen and excess immune responses were prevented by saved energy when infection risk is expected to below.

One study shows that cross-talk between the microbes and host immune response is often mediated and enhanced by the circadian clock. In the body's microbial ecosystem, the most studied ecosystem is the gut ecosystem. For example, in *Salmonella typhimurium* infection inflammatory response is regulated by the circadian clock, and pro-inflammatory genes were induced by functional clock (Bellet et al. 2013). Studies show that some pathogens get benefitted by the host clock. In case of sleeping sickness caused by the infection of *Trypanosoma brucei*, leads to reduction of the host circadian clock both at the molecular and behavior level (Rijo-Ferreira et al. 2018). Circadian network is affected by microbial products or components. For example, endotoxins produced by bacterial modified the lung circadian transcriptome and metabolome.

Diurnal Oscillation of Gut Microbiota Regulates the Central Nervous System Function

Recently, it was found that the diurnal oscillation of gut microbiota regulates the central nervous system (CNS). The intestinal microbes not only maintain the health of the GI tract, but also regulate the metabolic homeostasis and immune system (Sandhu et al. 2017). The central nervous system (CNS) has been attacked by signals produced by the Microbiota and which regulate different functions, like stress, anxiety, cognitive processes, and emotional behavior microglia maturation, neurogenesis, and integrity of blood-brain barrier (BBB) (Sharon et al. 2016; Dinan and Cryan 2017). The circulatory system also shows another important way for sign exchange between the CNS and gut bacteria. The metabolites produced by intestinal commensals which work as factory, micronutrients, and neurotransmitters, could certainly travel to the brain and passes through the BBB, and its function is influenced. Some strains of bacteria can synthesize or promotes to the metabolism of neuromodulators (Strandwitz 2018). Surprisingly, the intestinal microflora is participated in all of the above-mentioned processes. Hormonal changes consort sleep disturbances and cause depression, obesity, and diabetes (Depner et al. 2014; Riemann et al. 2019). Turnbaugh found that the intestinal microflora of depressed and obese patients shows significant difference among healthy individuals (Turnbaugh et al. 2009; Valles-Colomer et al. 2019). However, the possible reason which interplay between circadian rhythms, mood alterations, sleep, and diurnal fluctuation in intestinal microbes have not been resolved yet.

Conclusion

Critical resources of circadian rhythms are similar in all microorganisms (cyanobacteria, archaebacteria, fungi, algae, protozoa, etc.) like duration of approximately 24 h, sensitivity to light, and temperature. Circadian rhythms do not only come with limited environmental conditions; they remain important even in the absence of external stimuli. Interestingly, microorganisms exhibit many convolutions in their circadian system, even with their simple morphology and development. They perform their daily circadian rhythms differently from one another, even though they have some resemblances in sequences between clock genes such as the development and expansion of TTOs in these organisms. Cyanobacteria and *Neurospora* may have more than one input pathway, depending on the presence or absence of an environmental stimulus. At the rhythmic level, *Synechococcus elongatus*, *Neurospora* sp., and *Chlamydomonas reinhardtii*, come up with the proof of multiple oscillators inside the coupled cell. Output pathways may also have complexities with the regulation of gene expression at different levels, such as the transcription and post-transcription levels. *Chlamydomonas reinhardtii* and dinoflagellate species show a predominant circadian rhythm of bioluminescence. Mating reactivity and photo-accumulation automatable rhythm can be assayed in the population of *Paramecium* sp. cells. The circadian clock may regulate the immune system and is, to some extent, responsible for metabolic homeostasis. Furthermore, the gut microbiota influences the CNS by diurnal oscillations. Studies on the mechanisms of circadian rhythms in microorganisms can give us an understanding of how diverse those mechanisms can be and how it plays an important role in the evolution of life.

Acknowledgments

S. Mishra and J. Jaiswal are thankful to the University Grants Commission (UGC), New Delhi, India, for the financial assistance in the form of fellowship as Junior Research Fellow. P. R. Singh is grateful to CSIR, New Delhi, India, for providing financial support in the form of Senior Research Fellow. N. Kumari, N. Sahu, and A. Gupta are grateful to CSIR, New Delhi, India, for providing financial support in the form of a Junior Research Fellowship.

References

Aronson BD, Johnson KA, Loros JJ, Dunlap JC (1994) Negative feedback defining a circadian clock: autoregulation of the clock gene frequency. *Science* 263(5153): 1578-1584.

Austin B (1968) An endogenous rhythm of spore discharge in *Sordaria fimicola*. *Ann Bot* 32: 261-278.

Ballario P, Vittorioso P, Magrelli A, Talora C, Cabibbo A, Macino G (1996) White collar-1, a central regulator of blue light responses in *Neurospora* is a zinc finger protein. *EMBO J* 15(7): 1650-1657.

Barnett A (1966) A circadian rhythm of mating type reversals in *Paramecium multimicronucleatum* syngen 2 and its genetic control. *J Cell Physiol* 67(2): 239-270.

Bellet MM, Deriu E, Liu JZ, Grimaldi B, Blaschitz C, Zeller M, Edwards RA, Sahar S, Dandekar S, Baldi P, George MD (2013) Circadian clock regulates the host response to *Salmonella*. *Proc Natl Acad Sci USA* 110: 9897-9902.

Bell-Pedersen D, Cassone VM, Earnest DJ, Golden SS, Hardin PE, Thomas TL, Zoran MJ (2005) Circadian rhythms from multiple oscillators: lessons from diverse organisms. *Nat Rev Genet* 6(7): 544-556.

Bell-Pedersen D, Garceau N, Loros JJ (1996) Circadian rhythms in fungi. *J Genet* 75(3): 387-401.

Bhadra U, Thakkar N, Das P, Bhadra MP (2017) Evolution of circadian rhythms: from bacteria to human. *Sleep Med* 35: 49-61.

Bruce VG (1970) The biological clock in *Chlamydomonas reinhardtii*. *J Protozool* 17: 328- 334.

Bruce VG, Pittendrigh CS (1956) Temperature independence in a unicellular "clock." *Proc Natl Acad Sci USA* 42(9): 676.

Bruce VG, Weight F, Pittendrigh CS (1960) Resetting the sporulation rhythm in *Pilobolus* with short light flashes of high intensity. *Science* 131(3402): 728-730.

Byrne TE, Wells MR, Johnson CH (1992) Circadian rhythms of chemotaxis to ammonium and methylammonium uptake in *Chlamydomonas*. *Plant Physiol* 98: 879-886.

Carpenter JB (1949) Production and discharge of basidiospores by *Pellicularia filamentosa* (pat) rogers on *Hevea* rubber. *Phytopathology* 39(12): 980-985.

Chang YG, Cohen SE, Phong C, Myers WK, Kim YI, Tseng R, Lin J, Zhang L, Boyd JS, Lee Y, Kang S, Lee D, Li S, Britt RD, Rust MJ (2015) Circadian rhythms. A protein fold switch joins the circadian oscillator to clock output in cyanobacteria. *Science* 349: 324-328.

Chang YG, Kuo NW, Tseng R, LiWang A (2011) Flexibility of the C-terminal, or CII, ring of KaiC governs the rhythm of the circadian clock of cyanobacteria. *Proc Natl Acad Sci USA* 108: 14431-14436.

Chang YG, Tseng R, Kuo NW, LiWang A (2012) Rhythmic ring-ring stacking drives the circadian oscillator clockwise. *Proc Natl Acad Sci USA* 109: 16847-16851.

Chen CH, DeMay BS, Gladfelter AS, Dunlap JC, Loros JJ (2010) Physical interaction between VIVID and white collar complex regulates photoadaptation in *Neurospora. Proc Natl Acad Sci USA* 107(38): 16715-16720.

Chen CH, Ringelberg CS, Gross RH, Dunlap JC, Loros JJ (2009) Genome-wide analysis of light-inducible responses reveals hierarchical light signaling in *Neurospora. EMBO J* 28(8): 1029-1042.

Cheng P, He Q, Wang L, Liu Y (2005) Regulation of the *Neurospora* circadian clock by an RNA helicase. *Genes Dev* 19: 234-241.

Cheng P, Yang Y, Heintzen C, Liu Y (2001) Coiled-coil domain-mediated FRQ–FRQ interaction is essential for its circadian clock function in *Neurospora. EMBO J* 20(1-2): 101-108.

Cohen SE, Erb ML, Selimkhanov J, Dong G, Hasty J, Pogliano J, Golden SS (2014) Dynamic localization of the cyanobacterial circadian clock proteins. *Curr Biol* 24: 1836-1844.

Cohen SE, Golden SS (2015) Circadian rhythms in cyanobacteria. *Microbiol Mol Biol Rev* 79(4): 373-385.

Colot HV, Loros JJ, Dunlap JC (2005) Temperature-modulated alternative splicing and promoter use in the circadian clock gene frequency. *Mol Biol Cell* 16(12): 5563-5571.

Costantini C, Renga G, Sellitto F, Borghi M, Stincardini C, Pariano M, Zelante T, Chiarotti F, Bartoli A, Mosci P, Romani L (2020) Microbes in the era of circadian medicine. *Front Cell Infect Microbiol* 10: 30.

Crosthwaite SK, Dunlap JC, Loros JJ (1997) *Neurospora* wc-1 and wc-2: transcription, photoresponses, and the origins of circadian rhythmicity. *Science* 276(5313): 763-769.

Curtis A M, Bellet MM, Sassone-Corsi P, O'Neill LA (2014) Circadian clock proteins and immunity. *Immunity* 40: 178-186.

Davis SJ, Vener AV, Vierstra RD (1999) Bacteriophytochromes: phytochrome-like photoreceptors from nonphotosynthetic eubacteria. *Science* 286: 2517-2520.

Depner CM, Stothard ER, Wright KP (2014) Metabolic consequences of sleep and circadian disorders. *Curr Diabetes Rep* 14: 507.

Diamond S, Jun D, Rubin BE, Golden SS (2015) The circadian oscillator in *Synechococcus elongatus* controls metabolite partitioning during diurnal growth. *Proc Natl Acad Sci USA* 112: 1916-1925.

Diernfellner A, Colot HV, Dintsis O, Loros JJ, Dunlap JC, Brunner M (2007) Long and short isoforms of *Neurospora* clock protein FRQ support temperature-compensated circadian rhythms. *FEBS Lett* 581(30): 5759-5764.

Dinan TG, Cryan JF (2017) The microbiome-gut-brain axis in health and disease. *Gastroenterol Clin* 46: 77-89.

Dong G, Yang Q, Wang Q, Kim YI, Wood TL, Osteryoung KW, van Oudenaarden A, Golden SS (2010) Elevated ATPase activity of KaiC applies a circadian checkpoint on cell division in *Synechococcus elongatus*. *Cell* 140: 529-539.

Dunlap J, Hastings JW (1981) The biological clock in *Gonyaulax* controls luciferase activity by regulating turnover. *J Biol Chem* 256: 10509-10518.

Dunlap JC (1999) Molecular bases for circadian clocks. *Cell* 96(2): 271-290.

Dunlap JC, Loros JJ (2004) The *Neurospora* circadian system. *J Biol Rhythms* 19: 414-424.

Edgar RS, Green EW, Zhao Y, Van Ooijen G, Olmedo M, Qin X, Xu Y, Pan M, Valekunja UK, Feeney KA, Maywood ES (2012) Peroxiredoxins are conserved markers of circadian rhythms. *Nature* 485: 459-464.

Ehret CF (1953) An analysis of the role of electromagnetic radiations in the mating reaction of *Paramecium bursaria*. *Physiol Zool* 26(3): 274-300.

Engel H, Friederichsen I (1964) Der abschuss der sporangiolen von *Sphaerobolus stellatus* (Thode) Pers in *kontinuierlicher dunkelheit*. *Planta* 61: 361-70. [The shot of the sporangioles of *Sphaerobolus stellatus* (Thode) Pers in continuous darkness. *Plant*]

Froehlich AC, Liu Y, Loros JJ, Dunlap JC (2002) White Collar-1 a circadian blue light photoreceptor, binding to the frequency promoter. *Science* 297(5582): 815-819.

Gallego M, Virshup DM (2007) Post-translational modifications regulate the ticking of the circadian clock. *Nat Rev Mol Cell Biol* 8: 139-148.

Gomelsky M, Klug G (2002) BLUF: a novel FAD-binding domain involved in sensory transduction in microorganisms. *Trends Biochem Sci* 27: 497-500.

Gooch VD, Wehseler RA, Gross CG (1994) Temperature effects on the resetting of the phase of the *Neurospora* circadian rhythm. *J Biol Rhythms* 9(1): 83-94.

Goto K, Johnson CH (1995) Is the cell division cycle gated by a circadian clock? The case of *Chlamydomonas reinhardtii*. *Cell Biol* 129: 1061-1069.

Guo J, Cheng P, Liu Y (2010) Functional significance of FRH in regulating the phosphorylation and stability of *Neurospora* circadian clock protein FRQ. *J Biol Chem* 285(15): 11508-11515.

Guo J, Cheng, P, Yuan H, Liu Y (2009) The exosome regulates circadian gene expression in a posttranscriptional negative feedback loop. *Cell* 138: 1236-1246.

Gutu A, O'Shea EK (2013) Two antagonistic clock-regulated histidine kinases time the activation of circadian gene expression. *Mol Cell* 50: 288-294.

Harris EH (2001) *Chlamydomonas* as a model organism *Annu Rev Plant Biol* 52(1): 363-406.

Hasegawa K, Tanakadate A (1987) The Paramecium circadian behavioral rhythm: light phase response curves and entrainment. *J Biol Rhythms* 2(4): 269-278.

Hastings J (2013) Circadian Rhythms in Dinoflagellates: What Is the Purpose of Synthesis and Destruction of Proteins? *Microorganisms* 1(1): 26-32.

Hastings JW (2007) The *Gonyaulax* clock at 50: Translational control of circadian expression. *Cold Spring Harb Symp Quant Biol* 72: 141-144.

He Q, Cheng P, Yang Y, Wang L, Gardner KH, Liu Y (2002) White collar-1, a DNA binding transcription factor and a light sensor. *Science* 297(5582): 840-843.

He Q, Liu Y (2005) Degradation of the *Neurospora* circadian clock protein frequency through the ubiquitin–proteasome pathway. *Biochem Soc Trans* 33(5): 953-956.

Heintzen C, Loros JJ, Dunlap JC (2001) The PAS protein VIVID defines a clock-associated feedback loop that represses light input, modulates gating, and regulates clock resetting. *Cell* 104(3): 453-464.

Hesse M (1972) Endogenous rhythm of the productivity in *Chlorella* and the influence of light. *Z Pjlanzenphysiol* 67: 58-77.

Hevia MA, Canessa P, Muller-Esparza H, Larrondo LF (2015) A circadian oscillator in the fungus *Botrytis cinerea* regulates virulence when infecting Arabidopsis thaliana. *Proc Natl Acad Sci USA* 112: 8744-8749.

Huang G, Chen S, Li S, Cha J, Long C, Li L, He Q, Liu Y (2007) Protein kinase A and casein kinases mediate sequential phosphorylation events in the circadian negative feedback loop. *Genes Dev* 21(24): 3283-3295.

Hunt SM, Thompson S, Elvin M, Heintzen C (2010) VIVID interacts with the WHITE COLLAR complex and FREQUENCY-interacting RNA helicase to alter light and clock responses in *Neurospora*. *Proc Natl Acad Sci USA* 107(38): 16709-16714.

Iliev D, Voytsekh O, Schmidt EM, Fiedler M, Nykytenko A, Mittag M (2006) A heteromeric RNA-binding protein is involved in maintaining acrophase and period of the circadian clock. *Plant Physiol* 142: 797-806.

Ingold CT (1971) Periodicity. In: *Fungal spores* (*London: Oxford University Press*) 215-238.

Ingold CT, Cox B J (1955) Periodicity of spore discharge in *Daldinia*. *Ann Bot* 29: 201-209.

Ingold CT, Marshall B (1963) Further observations on light and spore discharge in certain pyrenomycetes. *Ann Bot* 27: 481-491.

Ishiura M, Kutsuna S, Aoki S, Iwasaki H, Andersson CR, Tanabe A, Golden SS, Johnson CH, Kondo T (1998) Expression of a gene cluster kaiABC as a circadian feedback process in cyanobacteria. *Science* 281(5382): 1519-1523.

Ivleva NB, Bramlett MR, Lindahl PA, Golden SS (2005) LdpA: a component of the circadian clock senses redox state of the cell. *EMBO J* 24: 1202-1210.

Ivleva NB, Gao T, LiWang AC, Golden SS (2006) Quinone sensing by the circadian input kinase of the cyanobacterial circadian clock. *Proc Natl Acad Sci USA* 103: 17468-17473.

Iwasaki H, Dunlap JC (2000) Microbial circadian oscillatory systems in *Neurospora* and *Synechococcus*: models for cellular clocks. *Curr Opin Microbiol* 3: 189-196.

Iwasaki H, Kondo T (2000) The Current State and Problems of Circadian Clock Studies in Cyanobacteria. *Plant Cell Physiol* 41(9): 1013-1020.

Johnson CH, Golden SS (1999) Circadian programs in cyanobacteria: adaptiveness and mechanism. *Annu Rev Microbiol* 53: 389-409.

Johnson CH, Kondo T, Goto K (1992) Circadian rhythms in *Chalmydomonas*. In: Hiroshige T, Homna K, Editors. *Circadian Clocks from Cell to Human*: Hokkaido Press, Sapporo, pp. 139-155.

Kannaujiya VK, Sinha RP (2017) Impacts of diurnal variation of ultraviolet-B and photosynthetically active radiation on phycobiliproteins of the hot-spring cyanobacterium *Nostoc* sp. strain HKAR-2. *Protoplasma* 254: 423-433.

Karakashian MW (1968) The rhythm of mating in *Paramecium aurelia*, syngen 3. *J Cell Physiol* 71(3): 197-209.

Katayama M, Kondo T, Xiong J, Golden SS (2003) ldpA encodes an iron-sulfur protein involved in light-dependent modulation of the circadian period in the cyanobacterium *Synechococcus elongatus* PCC 7942. *J Bacteriol* 185: 1415-1422.

Kim YI, Dong G, Carruthers CW, Golden SS, LiWang A (2008) The day/night switch in KaiC, a central oscillator component of the circadian clock of cyanobacteria. *Proc Natl Acad Sci USA* 105(35): 12825-12830.

Kim YI, Vinyard DJ, Ananyev GM, Dismukes GC, Golden SS (2012) Oxidized quinones signal onset of darkness directly to the cyanobacterial circadian oscillator. *Proc Natl Acad Sci USA* 109: 17765-17769.

Kondo T, Strayer CA, Kulkarni RD, Taylor W, Ishiura M, Golden SS, Johnson CH (1993) Circadian rhythms in prokaryotes: luciferase as a reporter of circadian gene expression in cyanobacteria. *Proc Nat Acad Sci USA* 90(12): 5672-5676.

Kondo T, Tsinoremas NF, Golden SS, Johnson CH, Kutsuna S, Ishiura M (1994) Circadian clock mutants of cyanobacteria. *Science* 266: 1233-1236.

Kung C, Saimi Y (1982) The physiological basis of taxes in *Paramecium. Annu Rev Physiol* 44(1): 519-534.

Lakin-Thomas PL (2006) New models for circadian systems in microorganisms. *FEMS Microbiol Lett* 259(1): 1-6.

Lakin-Thomas PL, Brody S (2004) Circadian rhythms in microorganisms: new complexities. *Annu Rev Microbiol* 58: 489-519.

Lakin-Thomas PL, Cote GG, Brady S (1990) Circadian rhythms in *Neurospora crassa* biochemistry and genetics. *Crit Rer Microbial* 17: 365-416.

Larrondo LF, Canessa P (2019) The clock keeps on ticking: emerging roles for circadian regulation in the control of fungal physiology and pathogenesis. *Curr Top Microbiol Immunol* 422: 121-156.

Lee SJ, Morse D, Hijri M (2019) Holobiont chronobiology: mycorrhiza may be a key to linking aboveground and underground rhythms. *Mycorrhiza* 29: 403-412.

Liu Y, Garceau NY, Loros JJ, Dunlap JC (1997) Thermally regulated translational control of FRQ mediates aspects of temperature responses in the *Neurospora* circadian clock. *Cell* 89(3): 477-486.

Liu Y, Tsinoremas NF, Johnson CH, Lebedeva NV, Golden SS, Ishiura M, Kondo T (1995) Circadian orchestration of gene expression in cyanobacteria. *Genes Dev* 9: 1469-1478.

Liu Y, Tsinoremas, NF, Golden SS, Kondo T, Johnson CH (1996) *Mol Microbiol* 20: 1071-1081.

Losi A, Polverini E, Quest B, Gartner W (2002) First evidence for phototropin-related blue-light receptors in prokaryotes. *Biophys J* 82: 2627-2634.

Mackey SR, Choi JS, Kitayama Y, Iwasaki H, Dong G, Golden SS (2008) Proteins found in a CikA interaction assay link the circadian clock, metabolism, and cell division in *Synechococcus elongatus. J Bacteriol* 190: 3738-3746.

Mackey SR, Golden SS, Ditty JL (2011) The itty-bitty time machine: genetics of the cyanobacterial circadian clock. *Adv Genet* 74: 13-53.

Malzahn E, Ciprianidis S, Káldi K, Schafmeier T, Brunner M (2010) Photoadaptation in *Neurospora* by competitive interaction of activating and inhibitory LOV domains. *Cell* 142(5): 762-772.

Maniscalco M, Nannen J, Sodi V, Silver G, Lowrey P, Bidle K (2014) Light-dependent expression of four cryptic archaeal circadian gene homologs. *Front Microbiol* 5: 79.

Markson JS, Piechura JR, Puszynska AM, O'Shea EK (2013) Circadian control of global gene expression by the cyanobacterial master regulator RpaA. *Cell* 155: 1396-1408.

Matsuo T, Ishiura M (2011) *Chlamydomonas reinhardtii* as a new model system for studying the molecular basis of the circadian clock. *FEBS Letters* 585(10): 1495-1502.

Matsuo T, Okamoto K, Onai K, Niwa Y, Shimogawara K, Ishiura M (2008) A systematic forward genetic analysis identified components of the *Chlamydomonas* circadian system. *Genes Dev* 22(7): 918-930.

MatsuoT, Okamoto K, Onai K, Niwa Y, Shimogawara K, Ishiura M (2008) A systematic forward genetic analysis identified components of the *Chlamydomonas* circadian system. *Genes Dev* 22: 918-930.

Milos P, Morse, D, Hastings JW (1990) Circadian control over synthesis of many *Gonyaulax* proteins is at the translational level. *Naturwissenschaften* 77: 87-89.

Mishra S, Pandey A, Ahmed H, Kumar D, Singh V, and Sinha R P (2019a) Mycosporine-like amino acids (MAAs) profile of cyanobacteria from different historical kunds of Varanasi, India. *Int J Appl Sci Bio* **7**: 317-326.

Mishra S, Pandey A, Sinha RP (2019b) Characterization of photoprotective and pharmaceutically important compounds from cyanobacteria and algal assemblages from historical kunds of Varanasi India. *J Chem Pharm Res* 11(9): 24-36.

Miwa I, Nagatoshi H, Horie T (1987) Circadian rhythmicity within single cells of *Paramecium bursaria. J Biol Rhythms* 2(1): 57-64.

Miyagishima SY, Wolk CP, Osteryoung KW (2005) Identification of cyanobacterial cell division genes by comparative and mutational analyses. *Mol Microbiol* 56: 126-143.

Mizoguchi T, Putterill J, Ohkoshi Y (2006) Kinase and phosphatase: the cog and spring of the circadian clock. *Int Rev Cytol* 250: 47-72.

Montenegro-Montero A, Canessa P, Larrondo LF (2015) Around the fungal clock: recent advances in the molecular study of circadian clocks in *Neurospora* and other fungi. *Adv Genet* 92: 107-184.

Mooney JL, Yager LN (1990) Light is required for conidiation in *Aspergillus nidulans*. *Genes & Dev* 4(9): 1473-1482.

Mori T, Binder B, Johnson CH (1996) Circadian gating of cell division in cyanobacteria growing with average doubling times of less than 24 hours. *Proc Natl Acad Sci USA* 93: 10183-10188.

Mukherji A, Kobiita A, Ye T, Chambon P (2013) Homeostasis in intestinal epithelium is orchestrated by the circadian clock and microbiota cues transduced by TLRs. *Cell* 153: 812-827.

Nakahira Y, Katayama M, Miyashita H, Kutsuna S, Iwasaki H, Oyama T, Kondo T (2004) Global gene repression by KaiC as a master process of prokaryotic circadian system. *Proc Natl Acad Sci USA* 101(3): 881-885.

Nakashima H, Feldman JF (1980) Temperature-sensitivity of light-induced phase shifting of the circadian clock of *Neurospora. Photochem Photobiol* 32(2): 247-251.

Nassoury N, Hastings JW, Morse D (2005) Biochemistry and Circadian Regulation of Output from the Gonyaulax Clock: Are There Many Clocks or Simply Many Hands? In: Kippert F, Editor. *The Circadian Clock in Eukaryotic Microbes: Landes Bioscience*: Austin, TX, USA, pp. 1-18.

Nikaido SS, Johnson CH (2000) Daily and circadian variation in survival from ultraviolet radiation in *Chlamydomonas reinhardtii. Photochem Photobiol* 71(6): 758-765.

Nishiwaki T, Iwasaki H, Ishiura M, Kondo T (2000) Nucleotide binding and autophosphorylation of the clock protein KaiC as a circadian timing process of cyanobacteria. *Proc Natl Acad Sci USA* 97: 495-499.

Nishiwaki T, Satomi Y, Kitayama Y, Terauchi K, Kiyohara R, Takao T, Kondo T (2007) A sequential program of dual phosphorylation of KaiC as a basis for circadian rhythm in cyanobacteria. *EMBO J* 26(17): 4029-4037.

Nishiwaki T, Satomi Y, Nakajima M, Lee C, Kiyohara R, Kageyama H, Kitayama Y, Temamoto M, Yamaguchi A, Hijikata A, Go M, Iwasaki HTakao T, Kondo T (2004) Role of KaiC phosphorylation in the circadian clock system of *Synechococcus elongatus* PCC 7942. *Proc NatlAcad Sci USA* 101: 13927-13932.

Partch CL, Green CB, Takahashi JS (2014) Molecular architecture of the mammalian circadian clock. *Trends Cell Biol* 24: 90-99.

Pattanayak GK, Phong C, Rust MJ (2014) Rhythms in energy storage control the ability of the cyanobacterial circadian clock to reset. *Curr Biol* 24: 1934-1938.

Pattanayek R, Wang J, Mori T, Xu Y, Johnson CH, Egli M (2004) Visualizing a circadian clock protein: crystal structure of KaiC and functional insights. *Mol Cell* 15: 375-388.

Pittendrigh CS, Bruce BG, Rosensweig NS, Rubin ML (1959) Growth patterns in *Neurospora. Nature* 184: 169-170.

Pohl R (1948) Tagesrhythmus im phototaktischen Verhalten der *Euglena gracilis. Z Naturforsch B* 3(9-10): 367-74. [Diurnal rhythm in the phototactic behavior of the *Euglena gracilis.*]

Pregueiro AM, Liu Q, Baker CL, Dunlap JC, Loros JJ (2006) The *Neurospora* checkpoint kinase 2: a regulatory link between the circadian and cell cycles. *Science* 313(5787): 644-649.

Pregueiro AM, Price-Lloyd N, Bell-Pedersen D, Heintzen C, Loros JJ, Dunlap JC (2005) Assignment of an essential role for the *Neurospora* frequency gene in circadian entrainment to temperature cycles. *Proc Natl Acad Sci USA* 102(6): 2210-2215.

Rensing L, Kallies K, Gebauer G, Mohsenzadeh S (1995) The effects of temperature change on the circadian clock of *Neurospora. Circadian Clocks and their Adjustments* 1: 26-50.

Riemann D, Krone LB, Wulff K, Nissen C (2019) Sleep, insomnia, and depression. *Neuropsychopharmacology* 45: 74-89.

Rijo-Ferreira F, Carvalho T, Afonso C, Sanches-Vaz M, Costa RM, Figueiredo LM, Takahashi JS (2018) Sleeping sickness is a circadian disorder. *Nat Commun* 9: 62.

Roenneberg T, Carpenter EJ (1993) Daily rhythm of O_2 evolution in the cyanobacterium *Trichodesmium thiebautii* under natural and constant conditions. *Mar Biol* 117(4): 693-697.

Romanowski A, Yanovsky MJ (2015) Circadian rhythms and post-transcriptional regulation in higher plants. *Front plant sci* 6: 437.

Rust MJ, Golden SS, O'Shea EK (2011) Light-driven changes in energy metabolism directly entrain the cyanobacterial circadian oscillator. *Science* 331: 220-223.

Rust MJ, Markson JS, Lane WS, Fisher DS, O'Shea EK (2007) Ordered phosphorylation governs oscillation of a three-protein circadian clock. *Science* 318: 809-812.

Sandhu KV, Sherwin E, Schellekens H, Stanton C, Dinan TG, Cryan JF (2017) Feeding the microbiota-gut-brain axis: diet, microbiome, and neuropsychiatry. *Transl Res* 179: 223-244.

Sargent ML, Briggs WR, Woodward DO (1966) Circadian nature of a rhythm expressed by an invertaseless strain of *Neurospora crassa*. *Plant Physiol* 41(8): 1343-1349.

Scheiermann C, Gibbs J, Ince L, Loudon A (2018) Clocking in to immunity. *Nat Rev Immunol* 18: 423-437.

Schmidt M, Geßner G, Luff M, Heiland I, Wagner V, Kaminski M, Geimer S, Eitzinger N, Reißenweber T, Voytsekh O, Fiedler M (2006) Proteomic analysis of the eyespot of *Chlamydomonas reinhardtii* provides novel insights into its components and tactic movements. *Plant Cell* 18(8): 1908-1930.

Schmidt M, Geßner G, Luff M, Heiland I, Wagner V, Kaminski M, Geimer S, Eitzinger N, Reißenweber T, Voytsekh O, Fiedler M (2006) Proteomic analysis of the eyespot of *Chlamydomonas reinhardtii* provides novel insights into its components and tactic movements. *Plant Cell* 18(8): 1908-1930.

Schmitz O, Katayama M, Williams SB, Kondo T, Golden SS (2000) CikA, a bacteriophytochrome that resets the cyanobacterial circadian clock. *Science* 289: 765-768.

Schneegurt MA, Sherman DM, Nayar S, Sherman LA (1994) *J Bacteriol* 176: 1586-1597.

Schweiger HG, Schweiger M (1977) Circadian rhythms in unicellular organisms: An endeavor to explain the molecular mechanism. *Int Rev Cytol* 51: 315.

Schwerdtfeger C, Linden H (2001) Blue light adaptation and desensitization of light signal transduction in *Neurospora crassa*. *Mol Microbiol* 39(4): 1080-1087.

Schwerdtfeger C, Linden H (2003) VIVID is a flavoprotein and serves as a fungal blue light photoreceptor for photoadaptation. *EMBO J* 22(18): 4846-4855.

Sharon G, Sampson TR, Geschwind DH, Mazmanian SK (2016) The central nervous system and the gut microbiome. *Cell* 167: 915-932.

Shrode LB, Lewis ZA, White LD, Bell-Pedersen D, Ebbole DJ (2001) vvd is required for light adaptation of conidiation-specific genes of

Neurospora crassa but not circadian conidiation. *Fungal Genet Biol* 32(3): 169-181.

Smith KM, Sancar G, Dekhang R, Sullivan CM, Li S, Tag AG, Sancar C, Bredeweg EL, Priest HD, McCormick RF, Thomas TL (2010) Transcription factors in light and circadian clock signaling networks revealed by genome wide mapping of direct targets for *Neurospora* white collar complex. *Eukaryotic cell* 9(10): 1549-1556.

Smith RM, Williams SB (2006) Circadian rhythms in gene transcription imparted by chromosome compaction in the cyanobacterium *Synechococcus elongatus*. *Proc Natl Acad Sci USA* 103: 8564-8569.

Stal LJ, Krumbein WE (1985) Nitrogenase activity in the nonheterocystous cyanobacterium *Oscillatoria* sp. grown under alternating light dark cycles. *Arch Microbiol* 143: 67-71.

Strandwitz P (2018) Neurotransmitter modulation by the gut microbiota. *Brain Res* 1693: 128-133.

Sweeney BM, Borgese MB (1989) A circadian rhythm in cell division in a prokaryote, the cyanobacterium *Synechococcus* Wh7803 1 NOTE. *J Phycol* 25(1): 183-186.

Thaiss CA, Zeevi D, Levy M, Zilberman-Schapira G, Suez J, Tengeler AC, Abramson L, Katz MN, Korem T, Zmora N, Kuperman Y (2014) Transkingdom control of microbiota diurnal oscillations promotes metabolic homeostasis. *Cell* 159: 514-529.

Tseng R, Chang YG, Bravo I, Latham R, Chaudhary A, Kuo NW, Liwang A (2014) Cooperative KaiA-KaiB-KaiC interactions affect KaiB/SasA competition in the circadian clock of cyanobacteria. *J Mol Biol* 426: 389-402.

Turnbaugh PJ, Hamady M, Yatsunenko T, Cantarel BL, Duncan A, Ley RE, Sogin ML, Jones WJ, Roe BA, Affourtit JP, Egholm M (2009) A core gut microbiome in obese and lean twins. *Nature* 457: 480-484.

Uebelmesser ER (1954) Uber den endonomen Tagesrhythmus der Sporangientragerbildung bon *Pilobolus*. *Arch Mikrobiol* 20: 1-33. [On the endonomic circadian rhythm of the formation of sporangia carriers on pilobolus]

Vakonakis I, Klewer DA, Williams SB, Golden SS, LiWang AC (2004) Structure of the N-terminal domain of the circadian clock-associated histidine kinase SasA. *J Mol Biol* 342(1): 9-17.

Valencia SJ, Bitou K, Ishii K, Murakami R, Morishita M, Onai K, Furukawa Y, Imada K, Namba K, Ishiura M (2012) Phase-dependent generation and transmission of time information by the KaiABC circadian clock oscillator through SasA-KaiC interaction in cyanobacteria. *Genes Cells* 17: 398-419.

Valles-Colomer M, Falony G, Darzi Y, Tigchelaar EF, Wang J, Tito RY, Schiweck C, Kurilshikov A, Joossens M, Wijmenga C, Claes S (2019) The neuroactive potential of the human gut microbiota in quality of life and depression. *Nat Microbiol* 4: 623-632.

Waltenberger H, Schneid C, Grosch JO, Bareiss A Mittag M (2001) Identification of target mRNAs for the clock-controlled RNA-binding protein Chlamy 1 from *Chlamydomonas reinhardtii. Mol Gen Genom* 265: 180-188.

Whitehead K, Pan M, Masumura KI, Bonneau R, Baliga NS (2009) Diurnally entrained anticipatory behavior in archaea. *PloS One* 4(5): e5485.

Wood TL, Bridwell-Rabb J, Kim YI, Gao T, Chang YG, LiWang A, Barondeau DP, Golden SS (2010) The KaiA protein of the cyanobacterial circadian oscillator is modulated by a redox-active cofactor. *Proc Natl Acad Sci USA* 107: 5804 -5809.

Xu Y, Mori T, Pattanayek R, Pattanayek S, Egli M, Johnson CH (2004) Identification of key phosphorylation sites in the circadian clock protein KaiC by crystallographic and mutagenetic analyses. *Proc Natl Acad Sci USA* 101: 13933-13938.

Yang Y, He Q, Cheng P, Wrage P, Yarden O, Liu Y (2004) Distinct roles for PP1 and PP2A in the *Neurospora* circadian clock. *Genes Dev* 18(3): 255-260.

Yoon HS, Hackett JD, Ciniglia C, Pinto G, Bhattacharya D (2004) A molecular timeline for the origin of photosynthetic eukaryotes. *Mol Biol Evol* 21: 809-818.

Zhao B, Schneid C, Iliev D, Schmidt EM, Wagner V, Wollnik F, Mittag M (2004) The circadian RNA-binding protein CHLAMY 1 represents a novel type heteromer of RNA recognition motif and lysine homology domain containing subunits. *Eukaryot Cell* 3: 815-825.

Zoltowski BD, Schwerdtfeger C, Widom J, Loros JJ, Bilwes AM, Dunlap JC, Crane BR (2007) Conformational switching in the fungal light sensor Vivid. *Science* 316(5827): 1054-1057.

In: Circadian Rhythms and Their Importance ISBN: 978-1-68507-547-7
Editor: Rajeshwar P. Sinha

Chapter 3

CYANOBACTERIAL ENDOGENOUS CLOCK: MOLECULAR MECHANISMS AND ECOLOGICAL ADAPTATIONS

Vinod K. Kannaujiya[1,*], Saumi Pandey[1], Nasreen Amin[1] and Rajeshwar P. Sinha[2]

[1]Department of Botany, MMV, Banaras Hindu University, Varanasi, Uttar Pradesh, India

[2]Laboratory of photobiology and Molecular Microbiology, Centre of Advanced Study in Botany, Institute of Science, Banaras Hindu University, Varanasi, Uttar Pradesh, India

ABSTRACT

Cyanobacteria are geologically the most primitive group of Gram-negative prokaryotes during the development of life on Earth's surface. They played indispensable role in the creation of hospitable oxygenic environment on our green planet. However, excessive level of free oxygen inevitably altered the growth and metabolic function of

* Corresponding Author's E-mail: vinod.kannaujiya@gmail.com.

cyanobacteria. Photosynthetically active radiation and ultraviolet radiation of solar spectrum also had effects on growth and survivability of cyanobacteria. To avoid impact of environmental free-oxygen and solar radiation, cyanobacteria have developed strategies of light-dark oscillation or circadian rhythm for sustainable growth and healthy metabolic regulation. Circadian oscillation also makes homeostatic coordination between their internal physiology and variable environment cycles. Circadian oscillation not only regulates cellular adaptability but also controls many vital functions notably photosynthesis, photo-recovery of light harvesting complex, amino acid uptake, nitrogen fixation, carbohydrate synthesis and respiration. In the past few years, many studies have been conducted to understand the cyanobacterial chronobiology. Recently, modern techniques of genomics and proteomics have identified key regulatory proteins such as Kai A-B-C produced from native gene cluster which contribute key role in maintaining endogenous clock. However, detailed study about the circadian oscillation in cyanobacteria is still scarce. Here, we have documented a detailed knowledge of circadian regulated stress adaptability and their molecular mechanisms in cyanobacteria. This chapter also outlines the current understanding of circadian regulation and their importance in life science.

Keywords: cyanobacteria, endogenous clock, circadian rhythm, abiotic stress

INTRODUCTION

Circadian rhythm is a genetically regulated 24 h periodic oscillation in living organisms (Huang 2018). This is an internal autonomous clock that foresees ambient changes in the environment of Earth's diurnal cycle. Endogenous circadian rhythm affects various cellular processes such as cell division, expression of genes, cell metabolism, cell development, and behavior of any organism (Johnson et al. 2011). Currently, various mechanisms of circadian regulation have been extensively studied in eukaryotic model organisms including mice, fruit fly and fungi (Bell-Pedersen et al. 2005). However, only few studies have been conducted on prokaryotic organisms.

Cyanobacteria are phylogenetically primitive organisms that might have played an indispensable role in the establishment of aerobic environment on the early Earth's surface. To synchronize the effects of environmental stress and adaptivity in various conditions, cyanobacteria use their own endogenous circadian oscillation for internal homeostasis and nitrogen fixation (Grobbelaar et al. 1986). Most of the periodic rhythm is regulated by temperature alteration and its phases can be modulated by light (Kondo and Ishiura 2000). The molecular mechanism of endogenous clock in the model cyanobacterium *Synechococcus elongatus* PCC 7942 has already been clearly demonstrated (Nishiwaki et al. 2004; Kim et al. 2020).

Mostly, there are three key features that play an important role in circadian rhythm: (1) persistence of oscillations in day/night alteration (2) compensation of circadian rhythm by temperature, and (3) regulation of internal homeostasis (Iwasaki and Kondo 2000; Johnson et al. 2011). However, functional aspects of the circadian systems are divided into three pathways such as input pathway, a central oscillator and output pathway. Circadian clock in cyanobacteria also regulates the gene expression, carbohydrate synthesis, amino acids uptake, metabolites regulation, and cell division (Liu et al. 1995; Golden et al. 1997; Dong et al. 2010b; Diamond et al. 2015; Taton et al. 2020).

Interestingly, rhythmic oscillation in cyanobacterial growth alternate 24 h light and dark cycle. Exposure of solar radiation (photosynthetically active radiation, ultraviolet-A and ultraviolet-B radiations) also plays an important role in balancing pigment proteins ratio in 24 h alteration in light and dark regimes (Kumar et al. 2018). Thus, circadian mechanism clearly maintains the internal fitness under exposure of various environmental conditions (Figure 1). However, maintenance of fitness is not constant for all organisms (Johnson et al. 2008). In cyanobacteria, most of the changes in the day/night cycle are not sensed directly by the photoreceptors as in eukaryotic organisms rather than they are sensed indirectly by redox status (quinone pool) of the cells that changes with the photosynthetic status of the organisms.

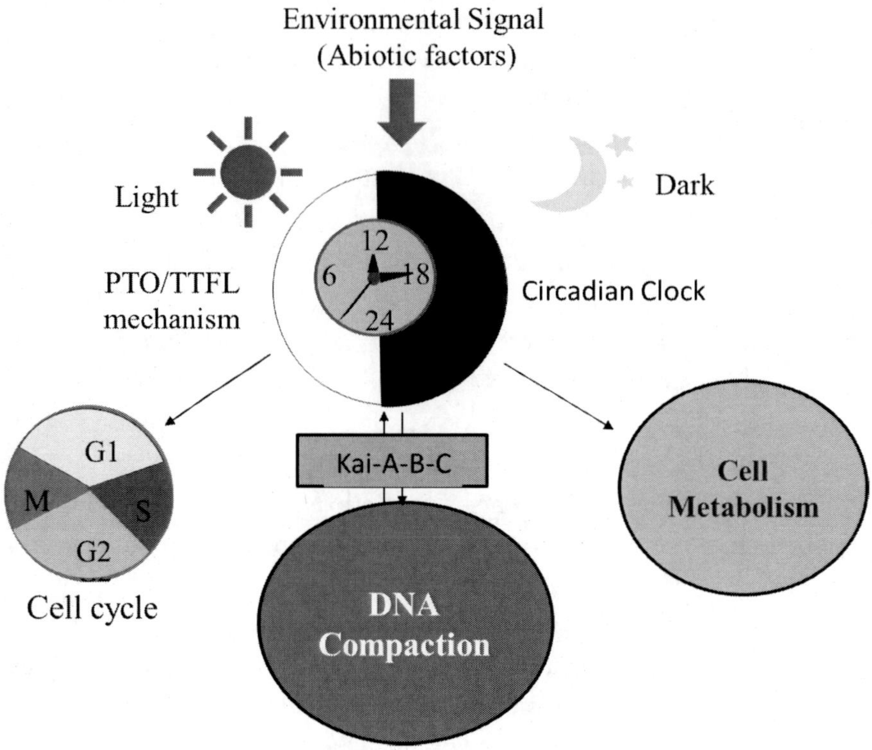

Figure 1. Schematic model for environmental triggered circadian oscillation with possible interaction between cell cycle, cellar metabolism and DNA compaction. For more details see reference (Wijnen and Young 2006).

In order to synchronize the clock with external stimuli, the core oscillator components are known to directly detect metabolites of photosynthesis (Wood et al. 2010; Rust et al. 2011). It is also evident that circadian clocks regulate amino acid uptake (Chen et al. 1991), gene expression pattern (Liu et al. 1995), chromosomal compaction (Smith and Williams 2006), and cell division (Mori et al. 1996) other than photosynthesis and nitrogen fixation. In addition, day/night circadian oscillation plays a key role in the development of cyanobacteria as bio-factory for the production of various biotechnological and industrial products (Wijffels et al. 2013; Diamond et al. 2015). In molecular biology, Kai A-B-C proteins are the major players in the regulation of gene expression altered by light and dark regimes. Kai C-related proteins are found in more than 40 cyanobacterial strains which indicates that circadian clock system is ubiquitous in all organisms (Lorne et al. 2000). However, time tuning mechanism of KaiC proteins remains unclear in various environmental conditions (Ito-Miwa et al. 2020). Therefore, it is of utmost importance to study various circadian clock genes for a clear understanding of the internal regulation in cyanobacteria. This chapter outlines the details about circadian regulation in cyanobacteria.

Effects of Various Abiotic Stress on Circadian Clock

Solar Radiation

Solar radiation mainly consists of photosynthetically active radiation (PAR; 400-700 nm), ultraviolet-A, B, and C radiations (UV-A; 315-400 nm, UV-B; 280-315 nm, and UV-C; 100-280 nm). These are obligate physical parameters that affect the health of plants and microbes directly in 24 h cycle. The rhythmic fluctuations of light and dark show remarkable effects on damage and recovery in photosynthesis, cell division, metabolic gene expression, enzymatic activities and fitness (Kono and Terashima 2014; Maniscalco et al. 2014; Kannaujiya and Sinha 2017). However, unregulated circadian clock in cyanobacteria cannot anticipate the fluctuations in an environment (Woelfle and Johnson 2009). It has been well established that it exhibits deleterious impacts on the aquatic and terrestrial food chain of flora-fauna in Earth ecosystem.

In addition, UV radiation also affects the nitrogenase enzyme activity, primary production, motility, DNA and protein (Kannaujiya and Sinha 2015). However, continuous release of ozone depleting pollutants that further increases the intensity of harmful radiations on the Earth's surface can cause more deleterious effects on living organisms. Impacts of UV radiation on clock regulation has been well established on various cyanobacteria (Kannaujiya and Sinha 2015; Kumar et al. 2018; Kolowrat et al. 2010a; Kawasaki and Iwasaki 2020; Liao and Rust 2021). Inhibition in DNA replication was observed in late light phase and promote cytokinesis in dark phase under 24 h oscillation in *Prochlorococcus* (Holtzendorff et al. 2001; Jacquet et al. 2001). This might be due to an internal photoprotective mechanism that protects DNA from damage under high UV radiation (Kolowrat et al. 2010b; Pathak et al. 2016; Rajneesh et al. 2017; Pathak et al. 2019). In addition, large variations are observed in adenine nucleotide pool on endogenous DNA superhelix (Rust et al. 2011). The length of light and dark periods and combinations of photosynthetically active radiation and UV radiations have tremendous impacts on photosynthetic efficiency of photosystem (Kumar et al. 2018). Apart from photosynthetic efficiency, glycogen metabolism is also affected in various cyanobacteria under circadian rhythm (Diamond et al. 2015; Puszynska and O'Shea 2017; Welkie et al. 2019). These studies have revealed that accumulation and degradation of glycogen is necessary for the adaptive fitness in day/night cycle (Puszynska and O'Shea 2017; Welkie et al. 2019).

Recently, it was found that accumulation and degradation of glycogen was dependent on the UV irradiation in *Synechococcus elongatus* PCC 7942 (Kawasaki and Iwasaki 2020). This suggests that lower glycogen degradation occurs under UV irradiation (day time) for higher adaptive resistance, whereas, more glycogen degradation occurs during night for energy production and maintenance of cellular metabolism under the regulation of CikA, SasA and KaiABC oscillators (Kawasaki and Iwasaki 2020).

Temperature

Temperature fluctuation is a basic phenomenon in natural climate which directly affect the physiology of organisms in an ecosystem (Häder et al. 2011). Solar radiation is a key factor for fluctuation of temperature in the ecosystem. Currently, anthropogenic and non-anthropogenic gases (greenhouse gases) have caused reduction in ozone thickness and increase in temperature on the Earth's surface (Hansen et al. 2006; Kannaujiya et al. 2014). Continuous depletion of ozone thickness leads to alarming increase in UV radiation along with increase in temperature which exhibit deleterious impacts on living organisms including cyanobacteria. These factors can cause species extinction, ecosystem perturbance and alteration in physiology of organisms. However, hot spring cyanobacteria can tolerate high temperature (Paerl and Huisman 2008).

Circadian rhythm is directly and indirectly linked with thermal fluctuation for adaptation in harsh environments. The decrease in temperature up to 15 °C extends < 2 h circadian periods in cyanobacteria which directly relate the impact of temperature in biological system (Kondo et al. 1993; Johnson and Egli 2014). Any deviation from optimum temperature for growth had affected both growth rate as well as metabolism of the cyanobacterium *Synechococcus elongatus* (Xu et al. 2013). The lowering of optimum temperature leads to inhibition or stop of circadian clock due to oscillation death/oscillation arrest in certain stages (Murayama et al. 2017). The ratio of phosphorylated KaiC to total KaiC functions as a pacemaker in cyanobacteria that controls global gene regulation (Murayama et al. 2017).

Oxidative Stress

Reactive oxygen species (ROS) are produced due to change in ambient conditions such as high light intensity and alteration in temperature. Daily fluctuation in light and temperature leads to production of various ROS during the process of cellular metabolism. It consists of hydrogen peroxide (H_2O_2), hydroxyl radical (OH^-), superoxide anion (O_2^-), and singlet oxygen ($^1O_2$) as well as organic radicals (Kannaujiya et al. 2014). For tolerance to oxidative stress, redox level of antioxidative enzyme is maintained by the circadian proteins which play important role in maintaining the limits of ROS in cellular system (Lai et al. 2012; Diamond et al. 2017).

In addition, it has also been reported that the level of ROS keeps low during energy metabolism in circadian regulation (Diamond et al. 2017).

The level of oxidative stress tolerance and circadian rhythms in the cyanobacterium *Synechococcus elongatus* PCC 7942 has been well analyzed by the application of methyl viologen under light irradiation (Tanaka et al. 2020). To counteract the damage caused by free radicals, the cells have an inbuilt mechanism such as antioxidative enzymatic (superoxide dismutase, catalase, and peroxidase, etc.) and non-enzymatic (tocopherols, and ascorbate, etc.) processes involved in controlling the oxidative level (Stangherlin and Reddy 2013) and help in genetic regulation of circadian rhythm. The levels of peroxidase, superoxide dismutase, and ascorbate peroxidase have been found to be increased during the light phase and decreased in the dark phase under various combinations of PAR and UV radiations (Kumar et al. 2018).

Pressure

Pressure is a necessary physical criterion that is required for survivability of cyanobacteria in deep water column. It plays a crucial role in maintaining the activity, stability, and structure of proteins. A native 3-D structure of a protein is conformationally stable by many interactions which may change due to deleterious environmental factors (Akasaka 2006). The conformation of proteins may alter during destabilization of folded condition to unfolded and partially folded proteins (Williamson and Kitahara 2019). The inequality between these two states lies in structural changes, bulk water density, water molecules attached and cavity loss in proteins (Roche et al. 2012). Pressure is a common phenomenon for destabilization and accelerating phosphorylation of proteins in cellular environment. The role of pressure in the regulation of phosphorylating cycle between KaiC and ATPase activity has been well established (Kitahara et al. 2019). Pressure enhances KaiC phosphorylating cycle up to 22 h at 1 bar and decline in 14 h at 200 bar due to ATP action on CI and CII domain (Kitahara et al. 2019). Hence, the pressure play crucial role in the circadian oscillation by KaiC phosphorylation.

Operation and Cyclic Interplay of Circadian Proteins

Circadian clock are well documented in single-celled cyanobacterium *Synechococcus elongatous* PCC 7942 (Hörnlein et al. 2020; Kondo and Ishiura 2000). In circadian, most of the input signal comes from external stimuli by changes in light, temperature and other abiotic factors that are transferred to central oscillators for redox mediated harmonization of the daily cycle. The cyanobacterium *Synechococcus elongates* PCC 7942 has key components of input pathway such as CikA, LdpA, and Pex proteins (Katayama et al. 2003; Ivleva et al. 2006). The oscillatory central proteins are functionalizing output pathway that triggered downstream gene processing for release of metabolic biomolecules and maintain homeostasis in physiology of organism. Certain proteins like SasA (*Synechococcus* adaptive sensor A: sensory histidine kinase) (Masahiro et al. 1993), RpaA (regulator of phycobilisome association A) (Ashby and Mullineaux 1999) and LabA (low-amplitude and bright) (Taniguchi et al. 2007) are the key players of *Synechococcus elongatus* PCC 7942 (Takai et al. 2006). A close relationship of kinase enzyme (SasA) between oscillator and output pathway has been shown (Smith and Williams 2006). Most of the effector signals from external environment are perceived by redox-sensitive proteins Ldp A (Light dependent period A) and CikA (Circadian input kinase A) that helps in maintaining the harmony of circadian oscillator in organisms (Ivleva et al. 2005; 2006; Cohen and Golden 2015) through sensing the changes in light intensity *via* redox state of the plastoquinone pool (Ivelva et al. 2006) and phosphatases activity (Gutu and O'Shea 2013). The oscillatory regulatory proteins Kai-A-B-C (Ishiura et al. 1998; Nakajima et al. 2005) are the key components for timekeeping cycle (Swan et al. 2018) (Figure 2). Genetically, Kai proteins are transcribed from clustered gene *kaiABC* and categorized into three form such as *kaiA, kaiB*, and *kaiC* (Iwasaki and Kondo 2000; Kurosawa et al. 2006). More interestingly, *kaiB* and *kaiC* genes are found in most of the prokaryotic and eukaryotic organisms whereas *kaiA* is only found in cyanobacteria. However, *kaiA* gene is absent in some cyanobacterial species such as *Prochlorococcus marinus* MED4 and *Prochlorococcus* PCC 9511 that might have been lost during the process of evolution (Kawamoto et al. 2020).

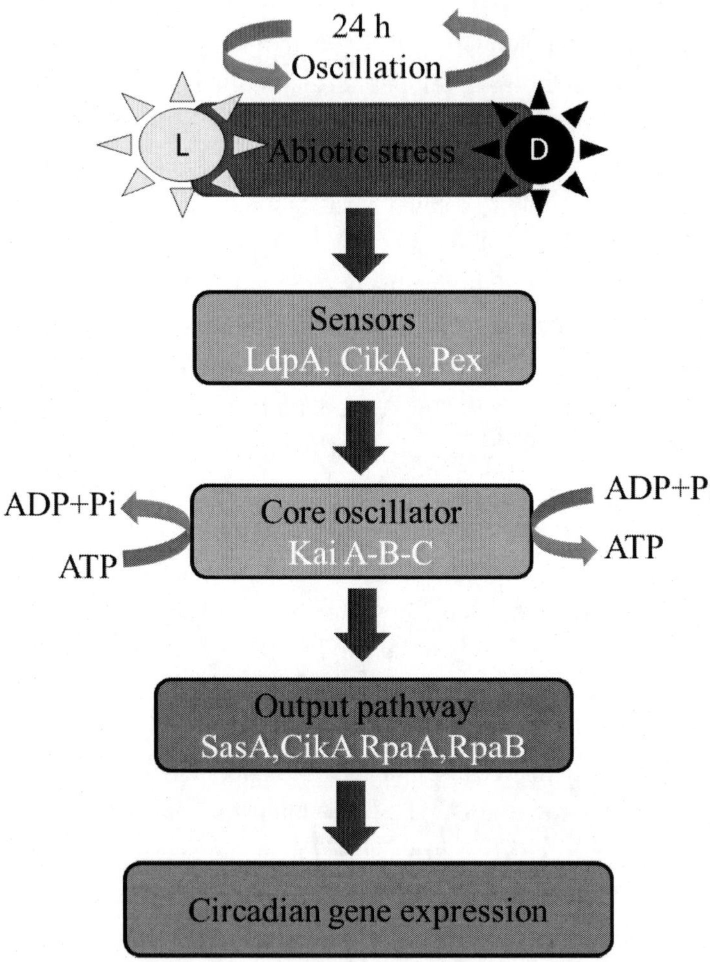

Figure 2. Putative pathways of circadian oscillation in cyanobacteria under environmental fluctuation or abiotic stress. For more detail see the references (Cohen and Golden 2015). L: Light, D: Dark.

Structurally, each Kai protein is interacted with individual monomer and form super nanocomplex for regulation of gene expression, cell division timing, and chromosome compaction (Xu et al. 2004; Cohen and Golden 2015). However, deactivation of any of the Kai gene leads to the non-functionality of the clock and feedback inhibition of KaiC or KaiA proteins (Dong et al. 2010a). More interestingly, endogenous clock proteins KaiA, KaiB and KaiC in cyanobacteria have almost insignificant similarity in sequences of eukaryotic clock proteins (Clodong et al. 2007). Experimentally, it has been found that nearly all genes are expressed in *Synechococcus elongatus* under rhythmic oscillation (Liu et al. 1995; Vijayan et al. 2011). Clock regulatory gene expression profile is generally divided into two classes where Class I typified as *kaiBC* genes and Class II exhibiting peak by purF promoter in timekeeping cycle (Cohen and Golden 2015). There are two promoters that drive the expression of genes during dusk (p*kaiA*) and dwan (p*kaiBC*) for rhythmic expression of genes in circadian rhythms (Ishiura et al. 1998).

STRUCTURE AND FUNCTION OF KAI C PROTEIN

KaiC (Circadian clock protein B) has 518 amino acid residues with molecular weight of 57.537 kDa (Iwase et al. 2004). KaiC exhibited activity of ATPase, autokinase, and autophosphatase activity during the timekeeping process (Dong et al. 2010a). In autokinase activity, KaiC is autophophorylated with constant rate circadian oscillation and also determine the diurnal period of circadian rhythm (Terauchi et al. 2007). The autokinase and autophosphatase activity in KaiC regulate the nano super complex of KaiABC for circadian oscillation (Rust et al. 2007; Tseng et al. 2017) (Figure 3).

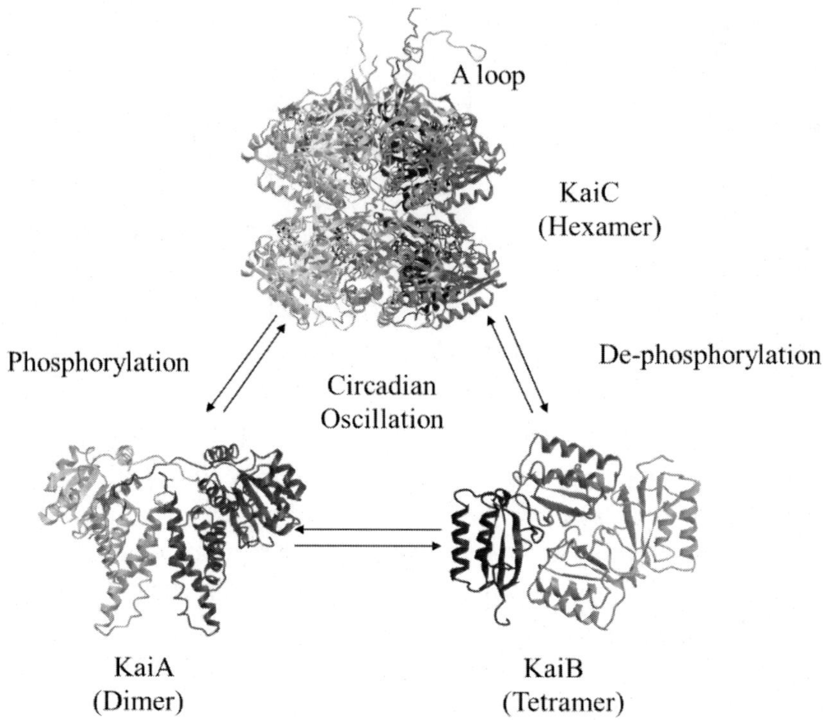

Figure 3. Molecular architecture and interaction of KaiA, KaiB, and KaiC proteins *via* phosphorylation and dephosphorylation mechanism. Structures of KaiA, KaiB, and KaiC have been generated from PDB database with PDB id 2GBL (Pattanayek et al. 2006), 4KSO (Villarreal et al. 2013) and 1R8J (Ye et al. 2004) respectively.

The oscillatory nature of KaiC is very flexible (Nishiwaki et al. 2004) during the process of cell division and metabolism activity under various environmental conditions (Qin et al. 2010). The single molecule methods of PDDA (phosphoform dependent differential affinity) in KaiA has clearly revealed the oscillatory mechanism in circadian rhythm (Mori et al. 2018). In the evolutionary perspectives, it has been found that *rec-A* gene is an ancestor for *kaiC* gene that might have moved from archaea into cyanobacterial ancestry (Iwasaki and Kondo 2000). Structurally, KaiC primary sequence has two P-loops having ATPase domain namely CI and CII (Egli et al. 2013). The Kai C proteins are flexible to form ring structure by N terminal domain (KaiCI) and C terminal domain (KaiCII) (Egli et al.

2013). The CI domain have 245-260 residues that act as linker for maintaining the conformation between CI and CII whereas CII domain have S shaped loop (483−497 residues) and terminal extended residues from 498−519 protrude from loop that exhibit regulation in structural conformation of domains (Egli et al. 2013). In addition, A hexamer is formed when KaiC bound to ATP (adenosine triphosphate) in solution for interactions with the C I subunit. Therefore, phosphate is transferred by C I whereas C II transfers between hydroxyl group-containing sidechains T432 (T- threonine and S-serine) (Swan et al. 2018) (Figure 4).

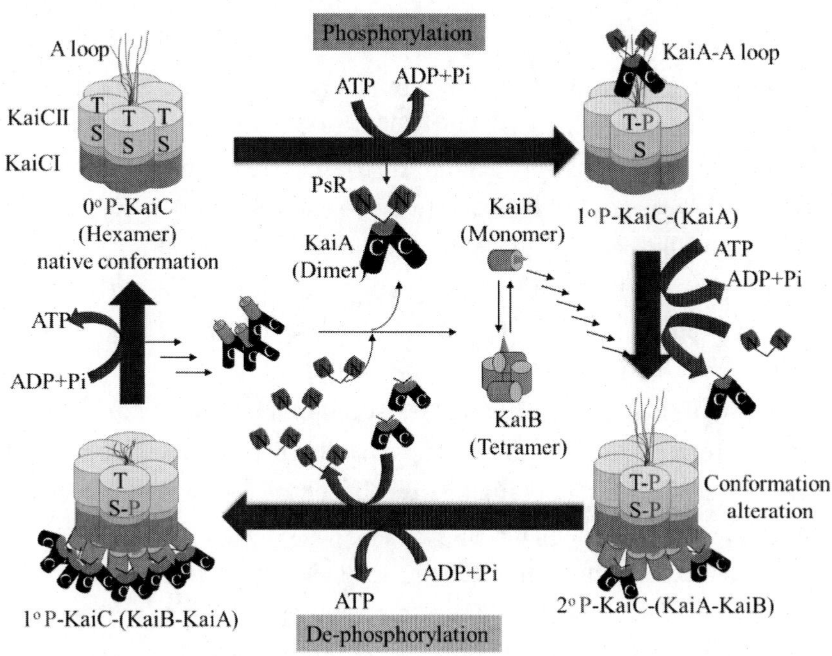

Figure 4. Molecular mechanism of KaiC-KaiBC mediated circadian rhythm in cyanobacteria. For details see the references (Cohen and Golden 2015; Swan et al. 2018; Snijder et al. 2017).

STRUCTURE AND FUNCTION OF KAI B PROTEIN

A KaiB (Circadian clock protein B) consist of 108 amino acid residues with molecular weight of 12.025 kDa (Iwase et al. 2004) has weaken the KaiC phosphorylation during oscillation (Swan et al. 2018). The KaiB belongs to a class of proteins known for metamorphic (reversible fold switching) characteristics (Chang et al. 2015). It has been found that KaiB undergoes major structural rearrangements in tertiary structure from the ground state to fold switched state during the process of circadian timekeeping (Chang et al. 2015). It is a very rare phenomenon that KaiB moves from the ground state to fold switched state and form stable KaiBC complex in the circadian cycle.

There are process of phosphorylation cycle in the KaiC which leads to delayed timing and connecting circadian oscillator for output pathways (Swan et al. 2018). X-ray crystal of KaiB from the cyanobacterium *Thermosynechococcus elongatus* (Iwase et al. 2005) and *Synechococcus elongatus* (Villarreal et al. 2013) has revealed the homo-tetramer structure that are organised as dimer of dimers. However, slight variation in structural conformation in the terminal region of KaiBC complex was observed in wild-type *Synechococcus elongates* by cryo-electron microscopy (Villarreal et al. 2013). In solution, KaiB exists in tetrameric form but recent studies have shown that it binds in monomeric form (Villarreal et al. 2013). For binding of KaiB to KaiC, two conditions should be fulfilled (1) C II domain should be stacked against C I domain to unmask the B-loop and (2) KaiB must be present in fold switched state (Cohen and Golden 2015). Therefore, interaction of KaiB with CII domains and CI ring inhibit the KaiA and promotes dephosphorylation in circadian cycle. In addition, fold switch state of KaiB also plays an important role in the molecular coordination between SasA and CikA during output pathway (Murzin 2008; Taniguchi et al. 2010). The overall fold of the KaiB monomer is a two a-helix/three b-sheet Alpha-beta meander (Figure 3), with helical and sheet secondary structure elements alternating along the peptide Kai.

STRUCTURE AND FUNCTION OF KAI A PROTEINS

KaiA (circadian clock protein A) has a molecular weight of 32.6 kDa with 283 amino acid residues exist in most of the cyanobacteria (Iwase et al. 2004). Moreover, a cyanobacterial group namely *Prochlorococcus* sp. has lost the KaiA proteins during the phase of evolution (Axmann et al. 2009). Interestingly, they have both *kaiB* and *kaiC* gene sequences (Golden and Canales 2003). Structurally, KaiA proteins have dual homodimer structure build up by domain swapping interaction (Clodong et al. 2007; Ye et al. 2004). The structural domain is categorized into pseudo-receiver domain (N terminal) and alpha domain (C terminal). Both domains are interlined together by linker polypeptides and form V shape structure. The cleft of C terminal domain have docking site for CII domain of KaiC proteins (Pattanayek and Egli 2015).

Interestingly, KaiA protein is interacted with KaiC and promotes phosphorylation by structurally rearrangement of CII catalytic core (Kageyama et al. 2006; Kim et al. 2008) and active site replacement for ATP from ADP (Nishiwaki-Ohkawa et al. 2014) whereas A-loop residues are projected towards central cavity of KaiC in unbound condition (Kim et al. 2008; Swan et al. 2018). Thus, A-loops plays an important role in the interaction of CII KaiC and KaiA proteins. In the presence of substrate, KaiA also exhibits nucleotide exchange factor for maintaining the level of phosphorylation on CII (Swan et al. 2018; Nishiwaki-Ohkawa et al. 2014). KaiA also induces the rate of exchange in ATP/ADP ratio on CII for phosphorylation and sustain the energy level in equilibrium condition (Swan et al. 2018). Recently, it has been reported that *kaiBC* gene activity was damped and exhibited low amplitude oscillation in 24 h diurnal rhythm in the *kaiA* inactivated organisms. This indicates that certain other mechanism(s) of circadian rhythm is found in other bacterial species (Kawamoto et al. 2020).

REGULATORY MECHANISMS

The regulatory oscillators TTFL (transcriptional/translational feedback loop) and PTO (post-translational oscillator) play a key role in the regulation of circadian clock. TTFL is mostly regulated in eukaryotic circadian while PTO acts as cyanobacterial oscillator with cyclic proteins KaiA, KaiB, and KaiC (Hurley et al. 2016; Cohen and Golden 2015). In the process of TTFLs, gene product plays negative feedback by repressing the expression when high level is achieved that results in significant changes in circadian oscillation while PTOs are independent of transcription, rather utilizes posttranslational modifications, protein-protein interaction and subcellular localization in cyanobacteria (Dunlap 1999; Mohawk et al. 2012). PTO is regulated by two component system (TCS) of signal transduction. TCSs are widely used in signal transduction of circadian oscillation and environmental induced adaptations in bacteria and archaea (Gao et al. 2007). Histidine residues play a major role with sensor histidine protein kinase in the process of autophosphorylation and transfer the phosphoryl group to N-terminal phosphor-acceptor whereas aspartate (Asp) acts as response regulator leading to conformational changes in the C-terminal effector domain (Stock et al. 2000).

Histidine kinases also recognize multidomain signaling from several response regulators and generate signaling network for light and dark phases in circadian clock (Laub and Goulian 2007; Mitrophanov and Groisman 2008). Circadian oscillator (PTO) transfers the signal to RpaA transcription factor through sensor histidine kinases SasA and CikA for phosphorylation and dephosphorylation (Gutu and O'Shea 2013). Genetically, phosphorylated RpaA (regulator of phycobilisome associated A) is the key protein for controlling the global gene expression in circadian clock (Taniguchi et al. 2007; Takai et al. 2006). However, another sensory regulator RpaB (regulator of phycobilisome associated B) is connected with the RpaA for global regulation. RpaB plays a key role in controlling the gene expression, domain function, cell division kinetics, phosphorylation and helps in rhythmic action during environmental fluctuations (Espinosa et al. 2015). The association of NblS-RpaB is very sensitive to environment which regulates the expression of master regulator RpaA-SasA by signal mediated level of phosphorylation including changes in RpaA that promotes *purA*, *kaiBC* and other related genes (Espinosa et al. 2015). Thus, circadian regulation is modulated by master regulon RpaA and RpaB which affects the expression of clustered genes *kaiABC* in cyanobacteria (Figure 5).

Individually, KaiA and KaiB proteins regulate the phosphorylation of KaiC. KaiA promotes KaiC autokinase activity, whereas, KaiB neutralize the action of KaiA and promotes KaiC dephosphorylation (Kageyama et al. 2006; Ito et al. 2007). The phosphorylation stage of KaiC can control the phase of the clock in several organisms (Rust et al. 2007). Hence, KaiC is considered as molecular timer in circadian rhythm (Terauchi et al. 2007).

Sensory Input Pathway

The key components of input pathway are redox-sensitive CikA (Circadian Input kinase), LdpA (light-dependent period) (Dong and Golden 2008), and Pex (period-extender) proteins (Takai et al. 2006). These components relay the input signal to functional oscillator by sensing the redox potential of plastoquinone (PQ) in varying light intensity (Ivleva et al. 2006; Carneiro et al. 2009; Kim et al. 2012). CikA carries regulatory domain GAF MOTIF near the N-terminus autophosphorylation in both prokaryotes and eukaryotes (Golden and Canales 2003; Mutsuda et al. 2003). CikA also carries pseudo-receiver domain that binds to oxidized form of quinone with KaiA domain from the thylakoid membrane (Wood et al. 2010; Zhang et al. 2006). CikA has ability to senses light/dark period by interacting with quinone molecules and relay the information to central oscillators (Zhang et al. 2006; Ivleva et al. 2006).

It was observed that quick disappearance of quinone molecules in the presence of CikA and KaiA indicates the aggregation *vs* degradation pathways to synchronization of clock with the 24 h day-night cycle (Sehgal 2004; Ivleva et al. 2006; Costantini et al. 2020). The role of *cikA* gene has been clearly demonstrated in experiment with *cikA* null mutant which exhibited shortness of clock in circadian rhythm (Schmitz et al. 2000). The input pathway is also regulated by ferrodoxin family protein LdpA with HcyB domain and specific conserved terminal dual domain (Katayama et al. 2003). Ferrodoxin family have FeS cluster to signify the redox signal transmittance from photosynthesis by variable intensity of light (Katayama et al. 2003).

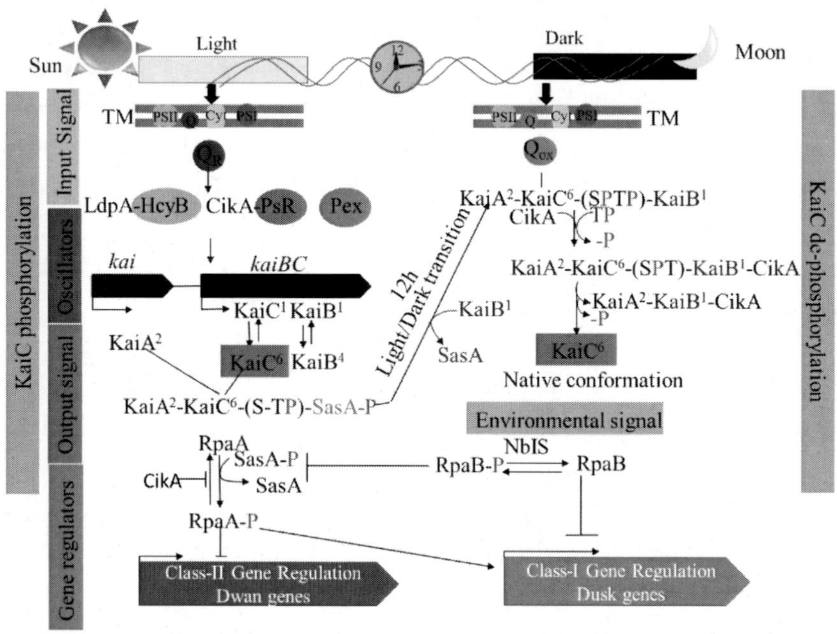

Figure 5. Schematic model of molecular regulation of circadian rhythm in cyanobacteria. Circadian pathways consists of input signal, oscillators, output signal and gene regulators in dawn (day) and dusk (night) environment of circadian rhythm. For details see the references (Swan et al. 2018; Cohen and Golden 2015; Schmelling and Axmann 2018; Espinosa et al. 2015; Egli et al. 2013; Johnson et al. 2008). $KaiA^2$ (KaiA dimer), $KaiB^4$ (KaiB tetramer), $KaiB^1$ (KaiB monomer), $KaiC^6$ (KaiC hexamer), TM: thylakoid membrane, Q_R: quinine reduced, Q_{ox}: quinine oxidized.

Interestingly, LdpA plays a very important role in finding adjustment of circadian clock with external environment with CikA protein and transmit signal to oscillator (Cohen and Golden 2019).

It is found that *ldpA* mutants have the ability to reset and enhance the dark oscillation in low light conditions as compared to wild type *S.*

elongatus (Katayama et al. 2003). A third member of input pathway is Pex (period-extender) which acts as transcriptional repressor of KaiA protein (Kutsuna et al. 1998).

Regulatory Output Pathway

Circadian output pathways are regulated by the downstream transcriptional proteins SasA (*Synechococcus* adaptive sensor A)-RpaA (regulator of phycobilisome-associated) in circadian clock (Takai et al. 2006). This SasA and RpaA regulatory pathways are required to drive whole circadian gene expression and play a key role in regulation of process of cell division and metabolism (Dong et al. 2010b). SasA is consists of N-terminal domain like KaiB which directly interact with KaiC and transfer the phosphate group (His161) to Asp53 of RpaA protein (Iwasaki et al. 2000; Takai et al. 2006).

Moreover, phosphorylated RpaA domain acts as a DNA-binding transcription factor and promotes expression from the kaiBC promoter and repress the expression of purF promoter (Markson et al. 2013). Thus, the interaction of KaiC-SasA-RpaA acts as a positive pathway that are known to regulate the circadian shift in gene expression (Takai et al. 2006). In addition, RpaB is a conserved response regulator that inhibits the phosphorylation of RpaA. More interestingly, RpaB also modulates number of genes in response to oxidative stress, high light, temperature, and osmotic response in cell (Kappell and van Waasbergen 2007; López-Redondo et al. 2010; Yasuda et al. 2020). In the phosphorylated state, RpaB can bind to HLR1 (high light regulatory 1) which are present in promoter *kaiBC* for the regulation of gene expression in constant light (Kappell and van Waasbergen 2007). Thus, both RpaA and RpaB work collectively to modulate the expression of circadian rhythm. This report indicates that not only RpaA and RpaB are competing for binding to targets but also that RpaB can act as a modulator of RpaA activity (Espinosa-Diez et al. 2015). These new findings suggest how crosstalk occurs between circadian (SasA-RpaA) proteins and environmentally regulated pathways (NblS-RpaB).

In another pathway, circadian clock modulates circadian gene expression LabA (low-amplitude and bright) which acts as a negative feedback regulator of KaiC (Taniguchi et al. 2007). Moreover, overexpression of *labA* causes lower gene expression of circadian clock. Both LabA and SasA work alternate side to regulate the clock gene expression in *S. elongates*. Overall, RpaA plays a key role in conveying circadian outputs (Markson et al. 2013). Other than these components, recently a *crm* (circadian rhythm modulator) gene has been identified that also plays an important role in the circadian clock system. The *crm* is a small gene that encodes a 62-amino-acid peptide without any functional domain. This was found in the upstream of *rpaA* which affects the expression of *rpaA* (Cohen and Golden 2015). Crm peptide is proposed to function as an allosteric modulator of RpaA (Boyd et al. 2013).

CONCLUSION

Cyanobacteria are the oldest living prokaryotes that have survived in extreme environmental conditions. Circadian clock is the key feature for survivability of cyanobacteria under various light-dark periods, temperature, oxidation and pressure regimes. Oscillatory clock plays an important role in the regulation of cellular physiology, metabolism, and expression of genes for adaptation in daily environmental fluctuations. Most of the cyanobacteria have key regulatory genes such as *kaiA-B-C* and *rpaA-B* which are involved in time homeostasis and adaptation under various environmental conditions. Recent studies have highlighted our understanding about integration of external signal, core oscillator, and effective output cellular processes. However, many unknown mysteries have still to be discovered to understand the whole mechanism of survivability in habitat specific cyanobacteria. In future, this amazing clock will be a great tool in synthetic biology and industrial biotechnology for the production of valuable metabolites from cyanobacteria.

ACKNOWLEDGMENTS

This work was supported by IoE-Seed Grant, BHU and Core Research Grant, DST-SERB, New Delhi, sanctioned to Vinod K. Kannaujiya.

REFERENCES

Akasaka, K. (2006). Probing conformational fluctuation of proteins by pressure perturbation. *Chem Rev*, *106*(5), 1814-1835.

Ashby, M. K. & Mullineaux, C. W. (1999). Cyanobacterial ycf 27 gene products regulate energy transfer from phycobilisomes to photosystems I and II. *FEMS Microbiol Lett*, *181*(2), 253-260.

Axmann, I. M., Dühring, U., Seeliger, L., Arnold, A., Vanselow, J. T., Kramer, A. & Wilde, A. (2009). Biochemical evidence for a timing mechanism in *Prochlorococcus*. *J Bacteriol*, *191*(17), 5342-5347.

Bell-Pedersen, D., Cassone, V. M., Earnest, D. J., Golden, S. S., Hardin, P. E., Thomas, T. L. & Zoran, M. J. (2005). Circadian rhythms from multiple oscillators: lessons from diverse organisms. *Nat Rev Genet*, *6*(7), 544-556.

Boyd, J. S., Bordowitz, J. R., Bree, A. C. & Golden, S. S. (2013). An allele of the crm gene blocks cyanobacterial circadian rhythms. *Proc Natl Acad Sci USA*, *110*(34), 13950-13955.

Carneiro, R. L., dos Santos, M. E., Pacheco, A. B. & Azevedo, S. M. (2009). Effects of light intensity and light quality on growth and circadian rhythm of saxitoxins production in *Cylindrospermopsis raciborskii* (Cyanobacteria). *J Plankton Res*, *31*(5), 481-488.

Chang, Y. G., Cohen, S. E., Phong, C., Myers, W. K., Kim, Y. I., Tseng, R., Lin, J., Zhang, L., Boyd, J. S. & Lee, Y. (2015). A protein fold switch joins the circadian oscillator to clock output in cyanobacteria. *Science*, *349*(6245), 324-328.

Chen, T. H., Chen T. L., Hung L. M. & Huang, T. C. (1991) .Circadian rhythm in amino acid uptake by *Synechococcus* RF-1. *Plant Physiol*, *97*(1), 55-59.

Clodong, S., Dühring, U., Kronk, L., Wilde, A., Axmann, I., Herzel, H. & Kollmann, M. (2007). Functioning and robustness of a bacterial circadian clock. *Mol Syst Biol.*, *3* (1), 90.

Cohen, S. E. & Golden, S. S. (2015) Circadian rhythms in cyanobacteria. *Microbiol Mol Biol Rev*, *79*(4), 373-385.

Cohen, S. E. & Golden, S. S. (2019). Circadian rhythmicity in prokaryotes. *Encyclopedia of Microbiology* 4e doi:10.1016/B978-0-12-809633-8.20670-5 pp . 1-9.

Costantini, C., Renga, G., Sellitto, F., Borghi, M., Stincardini, C., Pariano, M., Zelante, T., Chiarotti, F., Bartoli, A. & Mosci, P. (2020). Microbes in the era of circadian medicine. *Front Cell Infect Microbiol*, *10*, 30.

Diamond, S., Jun, D., Rubin, B. E. & Golden, S. S. (2015). The circadian oscillator in *Synechococcus* elongatus controls metabolite partitioning during diurnal growth. *Proc Natl Acad Sci USA*, *112*(15), E1916-E1925.

Diamond, S., Rubin, B. E., Shultzaberger, R. K., Chen, Y., Barber, C. D. & Golden, S. S. (2017). Redox crisis underlies conditional light–dark lethality in cyanobacterial mutants that lack the circadian regulator, RpaA. *Proc Natl Acad Sci USA*, *114*(4), E580-E589.

Dong, G. & Golden, S. S. (2008). How a cyanobacterium tells time. *Curr Opin Microbiol*, *11*(6), 541-546.

Dong, G., Kim, Y. I. & Golden, S. S. (2010a). Simplicity and complexity in the cyanobacterial circadian clock mechanism. *Curr Opin Genet Dev*, *20*(6), 619-625.

Dong, G., Yang, Q., Wang, Q., Kim, Y. I., Wood, T. L., Osteryoung, K. W., van Oudenaarden, A. & Golden, S. S. (2010b). Elevated ATPase activity of KaiC applies a circadian checkpoint on cell division in *Synechococcus elongatus*. *Cell*, *140*(4), 529-539.

Dunlap, J. C. (1999) Molecular bases for circadian clocks. *Cell*, *96*(2), 271-290.

Egli, M., Pattanayek, R., Sheehan, J. H., Xu, Y., Mori, T., Smith, J. A. & Johnson, C. H. (2013). Loop–loop interactions regulate KaiA-stimulated KaiC phosphorylation in the cyanobacterial KaiABC circadian clock. *Biochem*, *52*(7), 1208-1220

Espinosa-Diez, C., Miguel, V., Mennerich, D., Kietzmann, T., Sánchez-Pérez, P., Cadenas, S. & Lamas, S. (2015). Antioxidant responses and cellular adjustments to oxidative stress. *Redox Biol 6*, 183-197.

Espinosa, J., Boyd, J. S., Cantos, R., Salinas, P., Golden, S. S. & Contreras, A. (2015). Cross-talk and regulatory interactions between the essential response regulator RpaB and cyanobacterial circadian clock output. *Proc Natl Acad Sci USA*, *112*(7), 2198-2203.

Gao, R., Mack, T. R. & Stock, A. M. (2007). Bacterial response regulators: versatile regulatory strategies from common domains. *Trends Biochem Sci*, *32*(5), 225-234.

Golden, S. S. & Canales, S. R. (2003). Cyanobacterial circadian clocks—timing is everything. *Nat Rev Microbiol*, *1*(3), 191-199.

Golden, S. S., Ishiura, M., Johnson, C. H. & Kondo, T. (1997). Cyanobacterial circadian rhythms. *Annu Rev Plant Biol*, *48*(1), 327-354.

Grobbelaar, N., Huang, T., Lin, H. & Chow, T. (1986). Dinitrogen-fixing endogenous rhythm in Synechococcus RF-1. *FEMS Microbiol Lett*, *37*(2), 173-177.

Gutu, A. & O'Shea, E. K. (2013). Two antagonistic clock-regulated histidine kinases time the activation of circadian gene expression. *Mol cell*, *50*(2), 288-294.

Häder, D. P., Helbling E., Williamson C. & Worrest R. (2011). Effects of UV radiation on aquatic ecosystems and interactions with climate change. *Photochem Photobiol Sci*, *10*(2), 242-260.

Hansen, J., Sato, M., Ruedy, R., Lo, K., Lea, D. W. & Medina-Elizade, M. (2006). Global temperature change. *Proc Natl Acad Sci USA*, *103*(39), 14288-14293.

Holtzendorff, J., Partensky, F., Jacquet, S., Bruyant, F., Marie, D., Garczarek, L., Mary, I., Vaulot, D. & Hess, W. (2001). Diel expression of cell cycle-related genes in synchronized cultures of *Prochlorococcus* sp. strain PCC 9511. *J Bacteriol*, *183*(3), 915-920.

Hörnlein, C., Confurius-Guns, V., Grego, M., Stal, L. J. & Bolhuis, H. (2020). Circadian clock-controlled gene expression in co-cultured, mat-forming cyanobacteria. *Sci Rep*, *10*(1), 1-17.

Huang, R. C. (2018). The discoveries of molecular mechanisms for the circadian rhythm: The 2017 Nobel Prize in Physiology or Medicine. *Biomed J*, *41*(1), 5-8.

Hurley, J. M., Loros, J. J. & Dunlap, J. C. (2016). Circadian oscillators: around the transcription–translation feedback loop and on to output. *Trends Biochem Sci*, *41*(10), 834-846.

Ishiura, M., Kutsuna, S., Aoki, S., Iwasaki, H., Andersson, C. R., Tanabe, A., Golden, S. S., Johnson, C. H. & Kondo, T. (1998). Expression of a gene cluster kaiABC as a circadian feedback process in cyanobacteria. *Science*, *281*(5382), 1519-1523.

Ito-Miwa, K., Furuike, Y., Akiyama, S. & Kondo, T. (2020). Tuning the circadian period of cyanobacteria up to 6.6 days by the single amino acid substitutions in KaiC. *Proc Natl Acad Sci USA*, *117*(34), 20926-20931.

Ito, H., Kageyama, H., Mutsuda, M., Nakajima, M., Oyama, T. & Kondo, T. (2007). Autonomous synchronization of the circadian KaiC phosphorylation rhythm. *Nat Struct Mol Biol*, *14*(11), 1084-1088.

Ivleva, N. B., Bramlett, M. R., Lindahl, P. A. & Golden, S. S. (2005). LdpA: a component of the circadian clock senses redox state of the cell. *EMBO J*, *24*(6), 1202-1210.

Ivleva, N. B., Gao, T., LiWang, A. C. & Golden, S. S. (2006). Quinone sensing by the circadian input kinase of the cyanobacterial circadian clock. *Proc Natl Acad Sci USA*, *103*(46), 17468-17473.

Iwasaki, H. & Kondo, T. (2000). The current state and problems of circadian clock studies in cyanobacteria. *Plant Cell Physiol*, *41*(9), 1013-1020.

Iwasaki, H., Williams, S. B., Kitayama, Y., Ishiura, M., Golden, S. S. & Kondo, T. (2000). A KaiC-interacting sensory histidine kinase, SasA, necessary to sustain robust circadian oscillation in cyanobacteria. *Cell*, *101*(2), 223-233.

Iwase, R., Imada. K., Hayashi, F., Uzumaki, T., Morishita, M., Onai, K., Furukawa, Y., Namba, K. & Ishiura, M. (2005). Functionally important substructures of circadian clock protein KaiB in a unique tetramer complex. *J Biol Chem*, *280* (52), 43141-43149.

Iwase, R., Imada, K., Hayashi, F., Uzumaki, T., Namba, K. & Ishiura, M. (2004). Crystallization and preliminary crystallographic analysis of the circadian clock protein KaiB from the thermophilic cyanobacterium *Thermosynechococcus elongatus* BP-1. *Acta Crystallogr D Biol Crystallogr*, *60*(4), 727-729.

Jacquet, S., Partensky, F., Marie, D., Casotti, R. & Vaulot, D. (2001). Cell cycle regulation by light in Prochlorococcus strains. *Appl Environ Microbiol*, *67*(2), 782-790.

Johnson, C. H. & Egli, M. (2014). Metabolic compensation and circadian resilience in prokaryotic cyanobacteria. *Appl Environ Microbiol*, *83*, 221-247.

Johnson, C. H., Mori, T. & Xu, Y. (2008). A cyanobacterial circadian clockwork. *Curr Biol*, *18*(17), R816-R825.

Johnson, C. H., Stewart, P. L. & Egli, M. (2011). The cyanobacterial circadian system: from biophysics to bioevolution. *Annu Rev Biophys*, *40*, 143-167.

Kageyama, H., Nishiwaki T., Nakajima, M., Iwasaki, H., Oyama, T. & Kondo, T. (2006). Cyanobacterial circadian pacemaker: Kai protein complex dynamics in the KaiC phosphorylation cycle *in vitro*. *Mol Cell*, *23* (2), 161-171.

Kannaujiya, V. K., Richa, & Sinha, R. P. (2014). Peroxide scavenging potential of ultraviolet-B-absorbing mycosporine-like amino acids isolated from a marine red alga *Bryocladia* sp. *Front Environ Sci*, *2*, 26.

Kannaujiya, V. K. & Sinha, R. P. (2015). Impacts of varying light regimes on phycobiliproteins of *Nostoc* sp. HKAR-2 and *Nostoc* sp. HKAR-11 isolated from diverse habitats. *Protoplasma*, *252*(6), 1551-1561.

Kannaujiya, V. K. & Sinha, R. P. (2017). Impacts of diurnal variation of ultraviolet-B and photosynthetically active radiation on phycobiliproteins of the hot-spring cyanobacterium *Nostoc* sp. strain HKAR-2. *Protoplasma*, *254*(1), 423-433.

Kappell, A. D. & van Waasbergen, L. G. (2007). The response regulator RpaB binds the high light regulatory 1 sequence upstream of the high-light-inducible hliB gene from the cyanobacterium *Synechocystis* PCC 6803. *Arch Microbiol*, *187* (4), 337-342

Katayama, M., Kondo, T., Xiong, J. & Golden, S. S. (2003). ldpA encodes an iron-sulfur protein involved in light-dependent modulation of the circadian period in the cyanobacterium *Synechococcus elongatus* PCC 7942. *Am Soc Microbiol*, 1415-1422.

Kawamoto, N., Ito, H., Tokuda, I. T. & Iwasaki, H. (2020). Damped circadian oscillation in the absence of KaiA in *Synechococcus*. *Nat Commun*, *11*(1), 1-12.

Kawasaki, K. & Iwasaki, H. (2020). Involvement of glycogen metabolism in circadian control of UV resistance in cyanobacteria. *PLoS Genetics*, *16*(11), e1009230.

Kim, P., Kaur, M., Jang, H. I. & Kim, Y. I. (2020). The circadian clock—A molecular tool for survival in cyanobacteria. *Life*, *10*(12), 365.

Kim, Y. I., Dong, G., Carruthers, C. W., Golden, S. S. & LiWang, A. (2008). The day/night switch in KaiC, a central oscillator component of the circadian clock of cyanobacteria. *Proc Natl Acad Sci USA*, *105*(35), 12825-12830.

Kim, Y. I., Vinyard, D. J., Ananyev, G. M., Dismukes, G. C. & Golden, S. S. (2012). Oxidized quinones signal onset of darkness directly to the cyanobacterial circadian oscillator. *Proc Natl Acad Sci USA*, *109*(44), 17765-17769.

Kitahara, R., Oyama, K., Kawamura, T., Mitsuhashi, K., Kitazawa, S., Yasunaga, K., Sagara, N., Fujimoto, M. & Terauchi, K. (2019). Pressure accelerates the circadian clock of cyanobacteria. *Sci Rep*, *9*(1), 1-8.

Kolowrat, C., Partensky, F., Mella-Flores, D., Le Corguillé, G., Boutte, C., Blot, N., Ratin, M., Ferréol, M., Lecomte, X. & Gourvil, P. (2010a).

Ultraviolet stress delays chromosome replication in light/dark synchronized cells of the marine cyanobacterium *Prochlorococcus marinus* PCC9511. *BMC Microbiol*, *10*(1), 1-24.

Kolowrat, C., Partensky, F., Mella-Flores, D., Le Corguillé, G., Boutte, C., Blot, N., Ratin, M., Ferréol, M, Lecomte, X., Gourvil, P., Lennon, J. F., Kehoe, D. M. & Garczarek, L. (2010b). Ultraviolet stress delays chromosome replication in light/dark synchronized cells of the marine cyanobacterium *Prochlorococcus marinus* PCC9511. *BMC Microbiol*, *10*(1), 204.

Kondo, T. & Ishiura, M. (2000). The circadian clock of cyanobacteria. *Bioessays*, *22*(1), 10-15.

Kondo, T., Strayer, C, A., Kulkarni, R. D., Taylor, W., Ishiura, M., Golden, S. S. & Johnson, C. H. (1993). Circadian rhythms in prokaryotes: luciferase as a reporter of circadian gene expression in cyanobacteria. *Proc Natl Acad Sci USA*, *90*(12), 5672-5676.

Kono, M. & Terashima, I. (2014). Long-term and short-term responses of the photosynthetic electron transport to fluctuating light. *J Photochem Photobiol B*, *137*, 89-99.

Kumar, D., Kannaujiya, V. K., Pathak, J., Sundaram, S. & Sinha, R. P. (2018). Composition and functional property of photosynthetic pigments under circadian rhythm in the cyanobacterium *Spirulina platensis*. *Protoplasma*, *255*(3), 885-898.

Kurosawa, G., Aihara, K. & Iwasa, Y. (2006). A model for the circadian rhythm of cyanobacteria that maintains oscillation without gene expression. *Biophys J*, *91*(6), 2015-2023.

Kutsuna, S., Kondo, T., Aoki, S. & Ishiura, M. (1998). A period-extender gene, pex, that extends the period of the circadian clock in the cyanobacterium *Synechococcus* sp. strain PCC 7942. *J Bacteriol* *180*(8), 2167-2174.

Lai, A. G., Doherty, C. J., Mueller-Roeber, B., Kay, S. A., Schippers, J. H. & Dijkwel, P. P. (2012). CIRCADIAN clock-associated 1 regulates ROS homeostasis and oxidative stress responses. *Proc Natl Acad Sci USA*, *109*(42), 17129-17134.

Laub, M. T. & Goulian, M. (2007). Specificity in two-component signal transduction pathways. *Annu Rev Genet*, *41*, 121-145.

Liao, Y. & Rust, M. J. (2021). The circadian clock ensures successful DNA replication in cyanobacteria. *Proc Natl Acad Sci USA*, *118*(20), e2022516118.

Liu, Y., Tsinoremas, N. F., Johnson, C. H., Lebedeva, N. V., Golden, S. S., Ishiura, M. & Kondo, T. (1995). Circadian orchestration of gene expression in cyanobacteria. *Genes Dev*, *9*(12), 1469-1478.

López-Redondo, M. L., Moronta, F., Salinas, P., Espinosa, J., Cantos, R., Dixon, R., Marina, A. & Contreras, A. (2010). Environmental control of phosphorylation pathways in a branched two-component system. *Mol Microbiol*, *78*(2), 475-489.

Lorne, J., Scheffer, J., Lee, A., Painter, M. & Miao, V. P. (2000). Genes controlling circadian rhythm are widely distributed in cyanobacteria. *FEMS Microbiol Lett*, *189*(2), 129-133.

Maniscalco, M., Nannen, J., Sodi, V., Silver, G., Lowrey, P. & Bidle, K. (2014). Light-dependent expression of four cryptic archaeal circadian gene homologs. *Front Microbiol*, *5*, 79.

Markson, J. S., Piechura, J. R., Puszynska, A. M. & O'Shea, E. K. (2013). Circadian control of global gene expression by the cyanobacterial master regulator RpaA. *Cell*, *155*(6), 1396-1408.

Masahiro, N., Hirofumi, A. & Takeshi, M. (1993). Cloning of a sensory-kinase-encoding gene that belongs to the two-component regulatory family from the cyanobacterium Synechococcus sp. PCC7942. *Gene*, *131*(1), 119-124.

Mitrophanov, A. Y. & Groisman, E. A. (2008). Signal integration in bacterial two-component regulatory systems. *Genes Dev*, *22*(19), 2601-2611.

Mohawk, J. A., Green, C. B. & Takahashi, J. S. (2012). Central and peripheral circadian clocks in mammals. *Annu Rev Neurosci*, *35*, 445-462.

Mori, T., Binder, B. & Johnson, C. H. (1996). Circadian gating of cell division in cyanobacteria growing with average doubling times of less than 24 hours. *Proc Natl Acad Sci USA*, *93*(19), 10183-10188.

Mori, T., Sugiyama, S., Byrne, M., Johnson, C. H., Uchihashi, T. & Ando, T. (2018). Revealing circadian mechanisms of integration and resilience by visualizing clock proteins working in real time. *Nat Commun*, *9*(1), 1-13.

Murayama, Y., Kori, H., Oshima, C., Kondo, T., Iwasaki, H. & Ito, H. (2017). Low temperature nullifies the circadian clock in cyanobacteria through Hopf bifurcation. *Proc Natl Acad Sci USA*, *114*(22), 5641-5646.

Murzin, A. G. (2008). Metamorphic proteins. *Science*, *320*(5884), 1725-1726.

Mutsuda, M., Michel, K. P., Zhang, X., Montgomery, B. L. & Golden, S. S. (2003). Biochemical properties of CikA, an unusual phytochrome-like histidine protein kinase that resets the circadian clock in *Synechococcus elongatus* PCC 7942. *J Biol Chem*, *278*(21), 19102-19110.

Nakajima, M., Imai, K., Ito, H., Nishiwaki, T., Murayama, Y., Iwasaki, H., Oyama, T. & Kondo, T. (2005). Reconstitution of circadian oscillation of cyanobacterial KaiC phosphorylation *in vitro*. *Science*, *308*(5720), 414-415.

Nishiwaki-Ohkawa, T., Kitayama, Y., Ochiai, E. & Kondo, T. (2014). Exchange of ADP with ATP in the CII ATPase domain promotes autophosphorylation of cyanobacterial clock protein KaiC. *Proc Natl Acad Sci USA*, *111*(12), 4455-4460.

Nishiwaki, T., Satomi, Y., Nakajima M., Lee, C., Kiyohara, R., Kageyama, H., Kitayama, Y., Temamoto, M., Yamaguchi, A. & Hijikata, A. (2004). Role of KaiC phosphorylation in the circadian clock system of *Synechococcus elongatus* PCC 7942. *Proc Natl Acad Sci*, *101*(38), 13927-13932.

Paerl, H. W. & Huisman, J. (2008). Blooms like it hot. *Science*, *320*(5872), 57-58.

Pathak, J., Ahmed, H., Singh, D. K., Singh, P. R., Kumar, D., Kannaujiya, V. K., Singh, S. P. & Sinha, R. P. (2019). Oxidative Stress and Antioxidant Defense in Plants Exposed to Ultraviolet Radiation. In: Hasanuzzaman M, Fotopoulos V, Nahar K, Fujita M, editors. *Reactive*

Oxygen, Nitrogen and Sulfur Species in Plants: Production, Metabolism, Signaling and Defense Mechanisms: John Wiley & Sons Ltd, 371-420.

Pathak, J., Rajneesh, R., Sonker, A. S., Kannaujiya, V. K. & Sinha, R. P. (2016). Cyanobacterial extracellular polysaccharide sheath pigment, scytonemin: A novel multipurpose pharmacophore. In: *Se-Kwon Kim*, editor. *Marine Glycobiology*: *Principles and Applications*: Taylor and Francis Group, CRC Press, Boca Raton, Florida, USA, CRC Press, 343-358.

Pattanayek, R. & Egli, M. (2015). Protein–protein interactions in the cyanobacterial circadian clock: structure of KaiA dimer in complex with C-terminal KaiC peptides at 2.8 Å resolution. *Biochem*, *54*(30), 4575-4578.

Pattanayek, R., Williams, D. R., Pattanayek, S., Xu, Y., Mori, T., Johnson, C. H., Stewart, P. L. & Egli, M. (2006). Analysis of KaiA–KaiC protein interactions in the cyano-bacterial circadian clock using hybrid structural methods. *EMBO J*, 25(9), 2017-2028.

Puszynska, A. M. & O'Shea, E. K. (2017). Switching of metabolic programs in response to light availability is an essential function of the cyanobacterial circadian output pathway. *elife*, *6*, e23210.

Qin, X., Byrne, M., Xu, Y., Mori, T. & Johnson, C. H. (2010). Coupling of a core post-translational pacemaker to a slave transcription/ translation feedback loop in a circadian system. *PLoS Biol*, *8*(6), e1000394.

Rajneesh Pathak, J., Kannaujiya, V. K., Singh, S. P. & Sinha, R. P. (2017). Codon usage analysis of photolyase encoding genes of cyanobacteria inhabiting diverse habitats. *3 Biotech*, *7*(3), 192.

Roche, J., Caro, J. A., Norberto, D. R., Barthe, P., Roumestand, C., Schlessman, J. L., Garcia, A. E. & Royer, C. A. (2012). Cavities determine the pressure unfolding of proteins. *Proc Natl Acad Sci USA*, *109*(18), 6945-6950.

Rust, M. J., Golden, S. S. & O'Shea, E. K. (2011). Light-driven changes in energy metabolism directly entrain the cyanobacterial circadian oscillator. *Science*, *331*(6014), 220-223.

Rust, M. J., Markson, J. S., Lane, W. S., Fisher, D. S. & O'Shea, E. K. (2007). Ordered phosphorylation governs oscillation of a three-protein circadian clock. *Science*, *318*(5851), 809-812.

Schmelling, N. M. & Axmann, I. M. (2018). Computational modelling unravels the precise clockwork of cyanobacteria. *Interface Focus*, *8*(6), 20180038.

Schmitz, O., Katayama, M., Williams, S. B., Kondo, T. & Golden, S. S. (2000). CikA, a bacteriophytochrome that resets the cyanobacterial circadian clock. *Science*, *289*(5480), 765-768.

Sehgal, A. (2004). *Molecular biology of circadian rhythms.* John Wiley & Sons, p. 283.

Smith, R. M. & Williams, S. B (2006). Circadian rhythms in gene transcription imparted by chromosome compaction in the cyanobacterium *Synechococcus elongatus*. *Proc Natl Acad Sci USA*, *103*(22), 8564-8569.

Snijder, J., Schuller, J. M., Wiegard, A., Lössl, P., Schmelling, N., Axmann, I. M., Plitzko, J. M., Förster, F. & Heck, A. J. (2017). Structures of the cyanobacterial circadian oscillator frozen in a fully assembled state. *Science*, *355*(6330), 1181-1184.

Stangherlin, A. & Reddy, A. B. (2013). Regulation of circadian clocks by redox homeostasis. *J Biol Chem*, *288*(37), 26505-26511.

Stock, A. M., Robinson, V. L. & Goudreau, P. N. (2000). Two-component signal transduction. *Annu Rev Biochem*, *69*(1), 183-215.

Swan, J. A., Golden, S. S., LiWang, A. & Partch, C. L. (2018). Structure, function, and mechanism of the core circadian clock in cyanobacteria. *J Biol Chem*, *293*(14), 5026-5034.

Takai, N., Ikeuchi, S., Manabe, K. & Kutsuna, S. (2006). Expression of the circadian clock-related gene pex in cyanobacteria increases in darkness and is required to delay the clock. *J Biol Rhythms*, *21*(4), 235-244.

Tanaka, K., Shimakawa, G. & Nakanishi, S. (2020). Time-of-day-dependent responses of cyanobacterial cellular viability against oxidative stress. *Sci Rep*, *10*(1), 20029.

Taniguchi, Y., Katayama, M., Ito, R., Takai, N., Kondo, T. & Oyama, T. (2007). labA: a novel gene required for negative feedback regulation of the cyanobacterial circadian clock protein KaiC. *Genes Dev*, *21*(1), 60-70.

Taniguchi, Y., Takai, N., Katayama, M., Kondo, T. & Oyama, T. (2010). Three major output pathways from the KaiABC-based oscillator cooperate to generate robust circadian kaiBC expression in cyanobacteria. *Proc Natl Acad Sci USA*, *107*(7), 3263-3268.

Taton, A., Erikson, C., Yang, Y., Rubin, B. E., Rifkin, S. A., Golden, J. W. & Golden, S. S. (2020). The circadian clock and darkness control natural competence in cyanobacteria. *Nat Commun*, *11*(1), 1-11.

Terauchi, K., Kitayama, Y., Nishiwaki, T., Miwa, K., Murayama, Y., Oyama, T. & Kondo, T. (2007). ATPase activity of KaiC determines the basic timing for circadian clock of cyanobacteria. *Proc Natl Acad Sci USA*, *104*(41), 16377-16381.

Tseng, R., Goularte, N. F., Chavan, A., Luu, J., Cohen, S. E., Chang, Y. G., Heisler, J., Li, S., Michael, A. K. & Tripathi, S. (2017). Structural basis of the day-night transition in a bacterial circadian clock. *Science*, *355*(6330), 1174-1180.

Vijayan, V., Jain, I. H. & O'Shea, E. K. (2011). A high resolution map of a cyanobacterial transcriptome. *Genome Biol*, *12*(5), 1-18.

Villarreal, S. A., Pattanayek, R., Williams, D. R., Mori, T., Qin, X., Johnson, C. H., Egli, M. & Stewart, P. L. (2013). CryoEM and molecular dynamics of the circadian KaiB–KaiC complex indicates that KaiB monomers interact with KaiC and block ATP binding clefts. *J Mol Biol*, *425*(18), 3311-3324.

Welkie, D. G., Rubin, B. E., Diamond, S., Hood, R. D., Savage, D. F. & Golden, S. S. (2019). A hard day's night: cyanobacteria in diel cycles. *Trends Microbiol*, *27*(3), 231-242.

Wijffels, R. H., Kruse, O. & Hellingwerf, K. J. (2013). Potential of industrial biotechnology with cyanobacteria and eukaryotic microalgae. *Curr Opin Biotechnol*, *24*(3), 405-413.

Wijnen, H. & Young, M. W. (2006). Interplay of circadian clocks and metabolic rhythms. *Annu Rev Genet*, *40*, 409-448.

Williamson, M. P. & Kitahara, R. (2019). Characterization of low-lying excited states of proteins by high-pressure NMR. *Biochim Biophys Acta*, *1867*(3), 350-358.

Woelfle, M. A. & Johnson, C. H. (2009). The adaptive value of the circadian clock system in cyanobacteria. In: *Bacterial Circadian Programs*. Springer, 205-221.

Wood, T. L., Bridwell-Rabb, J., Kim, Y. I., Gao, T., Chang, Y. G, LiWang, A., Barondeau, D. P. & Golden, S. S. (2010). The KaiA protein of the cyanobacterial circadian oscillator is modulated by a redox-active cofactor. *Proc Natl Acad Sci USA*, *107*(13), 5804-5809.

Xu, Y., Ma, P., Shah, P., Rokas, A., Liu, Y.& Johnson, C. H. (2013). Non-optimal codon usage is a mechanism to achieve circadian clock conditionality. *Nature*, *495*(7439), 116-120.

Xu, Y., Mori, T., Pattanayek, R., Pattanayek, S., Egli, M. & Johnson, C. H. (2004). Identification of key phosphorylation sites in the circadian clock protein KaiC by crystallographic and mutagenetic analyses. *Proc Natl Acad Sci USA*, *101*(38), 13933-13938.

Yasuda, A., Inami, D. & Hanaoka, M. (2020). RpaB, an essential response regulator for high-light stress, is extensively involved in transcriptional regulation under light-intensity upshift conditions in *Synechococcus elongatus* PCC7942. *J Gen Appl Microbiol, 66(2),* 73-79.

Ye, S., Vakonakis, I., Ioerger, T. R., LiWang, A. C. & Sacchettini, J. C. (2004). Crystal structure of circadian clock protein KaiA from *Synechococcus elongatus*. *J Biol Chem*, *279*(19), 20511-20518.

Zhang, X., Dong, G. & Golden, S. S. (2006). The pseudo-receiver domain of CikA regulates the cyanobacterial circadian input pathway. *Mol Microbiol*, *60*(3), 658-668.

In: Circadian Rhythms and Their Importance ISBN: 978-1-68507-547-7
Editor: Rajeshwar P. Sinha

Chapter 4

PLANT CIRCADIAN RHYTHM: A BIOLOGICAL CLOCK AS DEVELOPMENTAL AND METABOLIC REGULATOR

Nidhi Rai*[1]*, Sabitri Kumari*[1]*, Pajeb Saha*[1]*, Apoorva*[1]*, Sanjay Kumar Rai*[2]*, Ram Prasad Meena*[3] *and Shashi Pandey-Rai*[1,]**

[1]Centre of Advanced Study in Botany, Institute of Science, Banaras Hindu University, Varanasi, Uttar Pradesh, India

[2]Dr. Rajendra Prasad Central Agricultural University, PUSA, Samastipur, Bihar, India

[3]Department of Computer Science, IIT, BHU, Varanasi, Uttar Pradesh, India

ABSTRACT

Plants have an internal biological system that receives differential environmental fluctuations/stimuli such as temperature and light

* Corresponding Author's E-mail: shashi.bhubotany@gmail.com.

controlling circadian rhythm for maintenance of growth and developmental processes. These biological rhythms are regulated by the interaction of certain external signals and internal receptors. In plants, it is complex networking within transcription factors that functions in feedback loops. These light-induced phototropic controls are mediated by photoreceptors like phytochromes, cryptochrome, phototropin and master genes/regulators for floral development. These responses are genetic in nature and have master clock genes which further regulates many copies of the master transcription factor that are responsible for regulating/switch-on many important genes of metabolism by binding with the promoter region of target genes. The diurnal behavior in plants has been observed because of the existence of a feedback loop and a phosphorylation-dephosphorylation cycle. The light and temperatures positively regulate the induction of various genes along with a set of polycomb gene. Many long non-coding RNAs, micro-RNA and RNA-dependent polymerases are indispensable parts of the diurnal cycle in plants. Environmental signals are involved in activating clock genes, and clock repressor circuits work to alienate and degrade these extrinsic gene activation pathways. Most transcription factors are cyclic and these subclasses can regulate clock parameters. Transcriptional regulators and associated chromatids that control transcriptional regulation are only one step in a multistep regulatory network. Post-translational relaxation, nuclear-cytoplasmic dissociation, RNA splicing and proteolytic functions participate in the stimulation. Homogenization of all these activities leads to the generation and sustainable facilitation of the robust rhythm and response to the diurnal variations of the environment. The purpose of this chapter is to explain the physiological and molecular mechanisms of the circadian clocks of plants, including biochemistry, and to demonstrate the function/role of the circadian clock in metabolic, physiological processes and plant behavior.

Keywords: circadian rhythm, diurnal behavior, phytochrome, polynuclear genes, transcriptional regulators

INTRODUCTION

The rotation of the Earth around its own axis causes day and night and is responsible for the creation of two powerful cycles i.e., the light and temperature cycles in the environment. It is obvious that living organism's physiology, biochemistry and behavior are affected and regulated by the environmental oscillation of day and night thus generating the diurnal rhythms. It occurs in the presence of environmental cues/signals that have the capacity to rearrange the clock (phase, time and temperature cycles). In order to adjust the continuously varying environmental conditions plant develops many strategies for their adaptation and future existence. In relation to diurnal rhythms, plants have developed a time-dependent mechanism which controls the biological events in plants with the help of environmental cues (light and temperature) i.e., the circadian clock. An internal clock system is used by the plants for the anticipation of seasonal variations. Circadian rhythms can be fully understood by three basic parameters: period, phase and amplitude. The period is the time it takes to complete a cycle and is measured as the time between two uninterrupted peaks i.e., maximum (peak) periods and minimum (peak) periods.

This period usually lasts around 24 hs. A phase is related to the rest of the cycle and can be any point in the defined cycle. The most demarcated phase is the upper and lower positions. Other components, such as amplitude, are measured such as the distance between peaks and valleys. The amplitude of the circadian clock can change, but the period is usually constant. The biological clock is a complex and fragile key regulator of plant cellular and physiological processes, but on an overall basis the specific controller that led to its origin and evolution is light. That is, plants need light to function for photosynthesis throughout the day plants had no choice except to get stuck with light and dark cycles. Thereby the usage of solar power plants spate for carbon fixation and ends at the night. The problem would not be solved by moving around. Accordingly, they adjust their physiological mechanisms as well as developmental processes. Plants biological clocks which are under the control of environmental cycles also support their adaptivity. The plant biological clock also directs reproductive development period so that other events related to this can be facilitated like flowering, nectar synthesis, pollination, leaf development/movement, plant growth, stomatal gaseous exchange, secondary metabolism, enzyme function and photosynthesis etc. (Fenske and Imaizumi 2016) as shown in Figure 1.

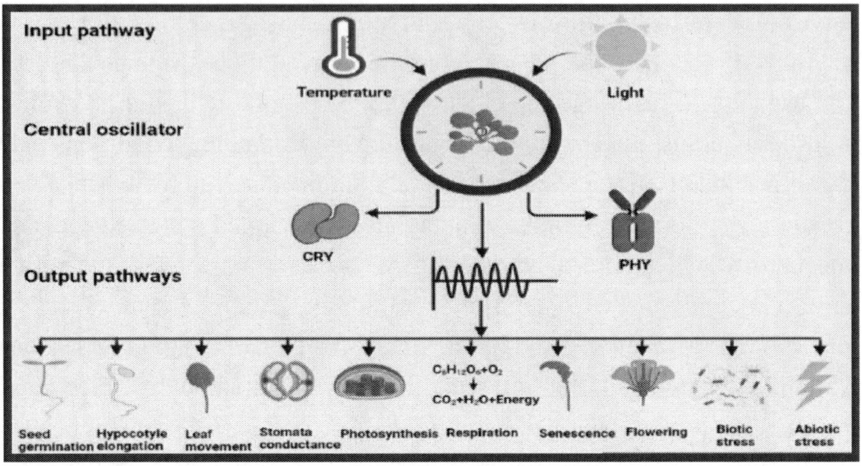

Figure 1. Generalized schematic representation of plant circadian clock showing relationship between the circadian clock and environmental stimulus and various plant activities/responses under its regulation. Abbreviations: CRY (cryptochrome); PHY (phytochrome) (Adapted and modified from Srivastava et al., 2019).

Further, the light quality is also an important factor for the long day and short day plants in context with the photoperiodic flowering. Therefore, the biological clock is very vital for flower development. This can be well understood with the help of impairment in the clock functioning. Studies on plant biological clock with the impaired functional circadian clock have resulted in the inference that a systematically organized functioning of the biological clock gives plant a better evolutionary advantage. It is noticed that an impaired circadian clock can give rise to late or early flowering with less viable phenotypes under specific photoperiods. Plants that are working according to light and temperature have been found to accumulate high chlorophyll and also increased photosynthetic capacity as well as high biomass accumulation (Linde et al. 2017). It is also experimentally proven that an optimal working biological clock helps plants in the degradation of starch at the night by efficient utilization of carbon which helps plants to continue growing during the night time.

The biological clock also acts as an internal molecular timekeeper which plays role in sensing the environmental periodic changes (daily or seasonal variations) along with daily or seasonal fluctuations on each day (Mizuno et al. 2014). The fitness and physiology of an organism are obtained through the efficient use of the cellular system only after synchronization of the circadian clock of the body with the physiological, metabolic and developmental processes in plants (Greenham and McClung 2015). Circadian oscillators or the components of the biological clock stabilizes the performance of plants in accordance to environmental cycles and provide them an internal sequential skeleton. Plants, like other eukaryotic and prokaryotic organisms, have also regulated under light/dark cycles (photoperiods) by establishing the biological clock mechanism, which generates harmonized signals/rhythms in controlling many aspects of life processes such as metabolism, physiology, and behavior.

The circadian rhythm of plants follows a 24 h cycle, the day/night cycle, similar to that of other organisms on the planet. Periods other than 24 h are only observed under special conditions or in *in-vitro* conditions. Periodic rhythms are a subclass of circadian rhythms with cycles, defined under 24 -h cycles of light and dark. With this characteristic, in 1959 Franzhalberg coined the term biology from the Latin words "Circa" and "die." The main distinguishing feature of circadian rhythms is that they are endogenously generated and self-sustaining. Another key attribute is light–dark environmental signals along with temperature which is responsible for the induction of rhythm according to the environmental cues. The circadian clock can also be entrained by temperature fluctuation, unlike other biological processes which is known as a phenomenon of temperature compensation. This is believed to be one aspect of the overall mechanism that protects the clock from changes at cellular and/or organelles metabolism.

This chapter is framed such, as it apprehends the historical journey to understand the plant circadian rhythms. The molecular pillars of the circadian clock, entrainment factors, and genetic bases along with regulation of the circadian rhythm in plants also gives comprehensive literature on the various roles of the circadian clock in plants development and fitness.

Plant Circadian Clock: Historical Perspective

Diurnal clocks were observed a century ago, but it was not realized that they could be endogenous. These biological clocks are present in all plants which control important processes such as transcription, mitosis, meiosis and photosynthesis at different cycles of day and night. In 1729, BC Androsthene first described the daily movement in the leaves of *Tamarindus indicus* plants. After that the French astronomer Jean-Jacques d' Ortous de Mairan, (1729) pointed out in detail about the intrinsic origin of *Mimosa pudica's* circadian rhythm processes relating it to sleep rhythm. The origin of circadian rhythms has been independently proven by various workers. Further in 1880, the inheritance of circadian rhythms was proposed by Charles and Francis Darwin, due to exposure to the developing diurnal cycle.

Recently, several workers have demonstrated the conserved architecture of biological clock with various independent origins (Dunlap 1999; Young & Kay 2001; McClung 2013). Further, (Harmer 2009) demonstrated 24 h rhythmic patterns and its regulation by gene expression/transcriptional and protein synthesis/translational feedback loops. The interaction of clock with changing environment needs to be adjusted daily by the plants as they are sessile in nature (Johnson et al. 2003). Beside this the clock has evolved in plants having more complex, flexible and robust nature (Rand et al. 2004; Tsai et al. 2008). Circadian rhythms have been also largely investigated in vascular plants for various behaviors of plants such as stomatal opening, growth, metabolism, leaf movement, floral induction and in response to various stresses (Greenham & McClung 2015). Correspondingly, large numbers of genes have been identified having major role in circadian control (Covington et al. 2008; Michael et al. 2008). Disruption of the biological clock has also been demonstrated by Yerushalmi & Green (2009) in relation to plant fitness implicating its importance. The recently well characterized circadian clock model of the *Arabidopsis* clearly demonstrated the whole network along with interlocked regulatory transcriptional feedback loops (Pokhilko et al. 2013; Fogelmark & Troein 2014; De Caluwé et al. 2016).

The main components of this model are represented by putative set of single MYB domain containing clock associated transcription factors known as PSEUDO-RESPONSE REGULATORs (PRRs: PRR1, PRR3, PRR5, PRR7 and PRR9), and several other plant key regulators/genes having regulatory role such as CO (CONSTANS).

THE ORGANIZATION OF PLANT CIRCADIAN CLOCK AT THE MOLECULAR LEVEL

Biological clocks are circadian periodicity setters that synchronize plants developmental process with the external environmental factors *via* combining temperature and light stimulus. The rotation of the Earth on its own axis generates predictable diurnal variations in response to daylight and temperature and functions endogenously to bring together inner physiological behavior to go with the already known environmental circumstances. The plant biological clock system unswervingly governs an array of yield pathways, which includes processes such as anabolism and catabolism, hormone interaction, hypocotyls and plant development, flower development, photosynthesis and responses against biotic and abiotic stresses. It gives an upper hand in survival advantage to plants that match with 24 h circadian system. Internal periodic 24-h rhythms are formed *via* a sequence of interrelated transcriptional-translational feedback loops, collectively known as the central oscillator.

The main oscillator is made up of many mutual feedback loops in the plants and has many roles for e.g., the determination of oscillator function is in the hands of input as well as output pathways by central oscillators in a feedback loop mechanism. Input signaling have an important role in the regulation and functioning of the clock associated components and affects clock output results, but the input signals are in direct circadian control *via* a gating system. However, the gate system is concerned with forming the receptiveness of the clock to assure inputs in a particular time of day. The plant processes which are "gated" by the circadian clock are stomatal closing and opening, hormone response, and cold sensitivity. The gene level studies have given light to the molecular pillars of the circadian clock contains morning genes, evening genes, evening complex, pseudo response regulator family and other clock associated genes as depicted in Figure 2.

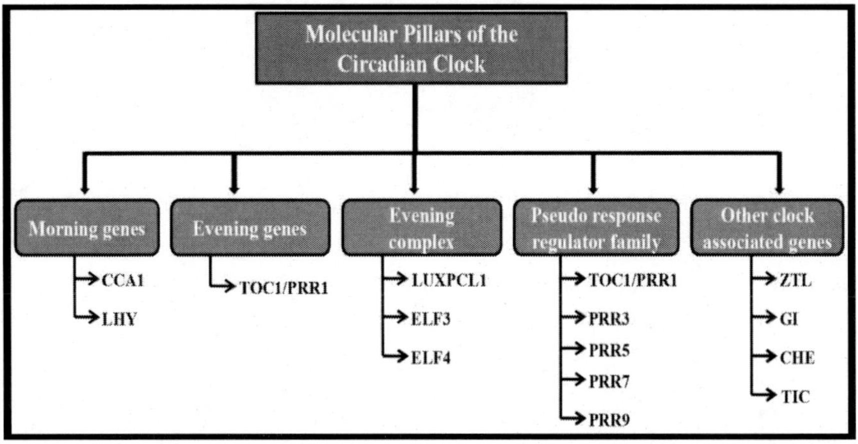

Figure 2. The pictorial representation of circadian genes evolved during various processes at the molecular level. Abbreviations: CCA1 (Circadian Clock Associated 1); LHY (Late Elongated Hypocotyl); TOC1 (Timing of Cab Expression 1); PRRs (Pseudo-Response Regulators); LUX (LUX Arrhythmo); ELF (Early Flowering); ZTL (Zeitlupe); GI (Gigantea); CHE (CCA1 Hiking Expedition 1); TIC (Time for Coffee).

Generally circadian clock has three basic components (1) An oscillator for induction of rhythms (2) transduction pathway for input-signal in response to environmental cues (3) transduction pathways for output signals to regulate biological processes such as gene expression, ion homeostasis, leaf/flower movements etc.

The Mechanism of Plant Circadian Clock

The biological clock is an internal timekeeper for plants as it helps in establishing a connection between yearly and daytime variations in the surrounding with their life processes. The endogenous clock plays a vital role in regulating plants biological activities like growth, development and reproduction etc. Plant circadian clock is also a master controller of the plant gene expression in response to external environmental conditions. Large number of molecular investigations on the circadian clock have relieved that it consists of genes that interact with the sequences of transcriptional and posttranscriptional feedback loops thereby regulating rhythmic expression of genes.

The major identified and characterized core circadian clock genes having interlocked feedback regulatory mechanism are Circadian Clock Associated 1 (CCA1), Pseudo Response Regulators (PRRs), Timing of Cab Expression 1 (TOC1), LUX Arrhythmo (LUX), Early Flowering (ELF), Gigantea (GI), Late Elongated Hypocotyl (LHY) Light-Regulated WD (LWD), Reveille (RVE), and Night Light-Inducible and Clock-Regulated (LNK) showing circadian regulation in response to environmental stimuli (Figure 3). The fluctuations in light and temperature signals also results in the entrainment process as input signals in the identification of time cues by the multiple pathways having differential role on the plant oscillator. The main entraining signal is light (early morning). The plants sense the light through the receptors commonly known as photoreceptors such as light-oxygen-voltage (LOV) sensors, phytochromes (for red/far-red), cryptochromes, phototropins, ZTL/ADO/FKF1/LKP2 (for blue/UV-A) and UVR8 (for UV-B light) family proteins involved in plant circadian clock for sensing the solar spectrum of light in plants for induction of rhythmic responses. The communication that occurs between PFR forms of PHY A and PHY B along with other cellular components have vital role for transferring the external environmental signals to the nucleus for generating whole circadian responses. The mechanism underlying the function of the biological clock is highly conserved phenomenon that ensures the internal time set of approximately twenty hs (24 hr). A number of regulatory transcriptional and translational feedback loops have been reported that defines the essential structural design of the circadian clock in the model plant *Arabidopsis thaliana*.

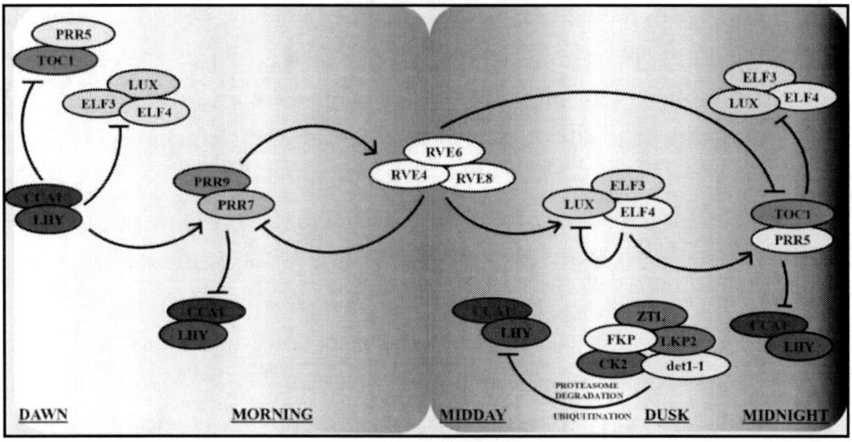

Figure 3. Diagrammatic representation of circadian clock genes as a whole day mechanism. The color background shows the morning to evening transition. Light induces transcription of LHY and CCA1 that inhibit the expression of evening complex (ELF3, ELF4, LUX) and TOC1, PRR5. As day proceeds activator RVFs and LNKs activate the expression of the evening complex. PRR5 and TOC1 express during the evening and together with other PRRs inhibit expression of LHY and CAA1. Abbreviations: PRRs (Pseudo-Response Regulators); TOC1 (Timing of Cab Expression 1); LUX (LUX Arrhythmo): ELF (Early Flowering); CCA1 (Circadian Clock Associated 1); LHY (Late Elongated Hypocotyl); PRR (Pseudo Response Regulator); RVE (Reveille); ZTL (Zeitlup); CK2 (casein kinase II); LKP2 (Lov Kelch Protein2) (adopted and modified from Gill et al. 2019).

There are two roles of MYB- domain transcription factors, Circadian Clock Associated 1, Late Elongated Hypocotyl, a Pseudo-Response Regulator also called as Timing of Cab Expression 1 is the part of the central regulatory network. After translation, CCA1 and LHY work together to inhibit the expression of the *TOC1* gene in the morning. At the evening time, CCA1 and TOC1 level decreases and functions to suppress CCA1 and LHY transcription. One more addition to this network is formed by PRR9, PRR7 and PRR5 function all day sequence with PRR9 in the early hs of the day. The PRR gene family members are reluctant and homologs to TOC1 as they inhibit the transcription of LHY & CCA1 genes. In fact, lab researchers reported that PRR9 & PRR7 may be initiated byCCA1 & LHY clock genes. But new data reveals that LHY is the repressor of these two genes, PRR5. Besides this, they are also restricted by the TOC1 gene and PRR9 is suppressed by an evening components system. The EC consists of two nuclear proteins the Phytoclock1, Early Flowering 3 and Early Flowering 4.

The evening complex in directly regulates the expression of morning genes but LHY and CCA1 have a suppressive role in the EC components i.e., reciprocally regulated by each other. Even the LHY and CCA1 can bind to their own promoter and suppress their expression. In present times, a new feedback loop system was explored surrounding dawn sets of genes, REVEILLE which functions the same as CCA1 and LHY. RVE8 functions in the induction of the transcriptional activity of afternoon and evening genes such as PRR5, TOC1, LUX, and ELF4. RVE8 is further repressed by PRR9, PRR7, and PRR5 whereas RVE6 and RVE4 seem to have little diffused roles with RVE8. Another known morning-expressed genes family that functions with the circadian clock is composed of the NIGHT LIGHT INDUCIBLE AND CLOCK-REGULATED 1, 2, 3 and 4 components. These genes communicate to RVEs in order to co-activate or antagonize RVE8. There are reports that CCA1, LHY, and TOC1 suppress GIGANTEA (GI) function, resulting in the initiation of CCA1 & LHY functions (Adams et al. 2015). The GIGANTEA is not a main component of the circadian clock but it acts as a connecting link between the clock and plant developmental processes (Mishra and Panigrahi 2015). The above-mentioned feedback loops and transcriptional and post-transcriptional networking of the biological clock are linked to other mechanisms as well for e.g., epigenetic, transcriptional, and translational regulation (Romanowski and Yanovsky 2015).

Further interaction between nucleus and chloroplast genes are important for the appropriate functioning of the circadian clock. Additionally, like animals the plants also have tissues to operate particular clock system (Takahashi et al. 2015). The whole interlinking and structure of the circadian clock need a more clear digging. In recent years many more clock genes are added to the clock genes. In addition to this, the specific transcriptional regulation, post transcriptional modification, alternative splicing, chromatin modification and protein turnover further refine the circadian rhythm.

REGULATION OF THE PLANT CIRCADIAN CLOCK

The plant circadian clock regulation can be subdivided into four distinct categories which include morning, daytime, afternoon, and evening components that are under control of various transcription factors and environmental cues. The key transcription factors involved in regulation of morning components in response to sun light are CCA1 (Circadian Clock Associated1) and LHY (Late Elongated Hypocotyl). The expression of both gene trancripts and proteins has been found in abundance during the morning time at the time of sunrise (Lu et al., 2009). Further, these proteins control and regulate the expression of the gene TOC1 (Timing of Cab Expression) which is known as chief component of the evening phase that is also associated with the PRR1 (Pseudo-Response Regulator 1). There is direct link between the CCA1 and LHY proteins in controlling the TOC1 promoters in evening element (EE). Further, the CCA1 and LHY additionally acts as repressors for expression of evening genes such as EFL3 and EFL4 (Early Flowering 3 and 4) genes. Besides acting as repressor, the CCA1 and LHY also work as activators for regulating the activity genes in the day time such as PRR9 and PRR7. The PRR9 and PRR7 genes, along with other PRRs (PRR1 and PRR5), also involved in controlling the fundamental processes related to circadian rhythm. The CCA1 and LHY genes, regulated the induction of PRR9 gene followed by the other PRRs genes during entire day time till dawn and increases the levels of TOC1 transcripts/proteins during dark/night time (Farre and Liu 2013).

Recently, it has been also shown that the PRR7 gene expression is strongly being regulated by light, also under the control of the sugar levels induced by photosynthetic activities. Therefore, during sunrise, after induction of the photosynthesis activities there is increases in glucose/sugars synthesis that also activates transcription factors and its expression. Contrastingly, the genes of the REVEILLE family (*RVE4, RVE8* and *RVE6*) act in opposite direction as shown in Figure 4 during afternoon. Further, RVE is also responsible for the induction of the expression of TOC1 gene during evening time. Therefore, the evening phase of low light conditions is mainly governed by the TOC1 and PRR5 TFs (Hsu and Harmer 2014). By observing all mechanisms starting from morning to evening, it can be inferred that the plant circadian clock involves the action of various transcription factors induced by the level of sunlight and photosynthetic products in a feedback mechanism, in which the actions of genes/transcription factor are either as suppressor or activator in the circadian cycle. Both transcriptional, as well as post-transcriptional regulators are vital for the circadian clock functioning.

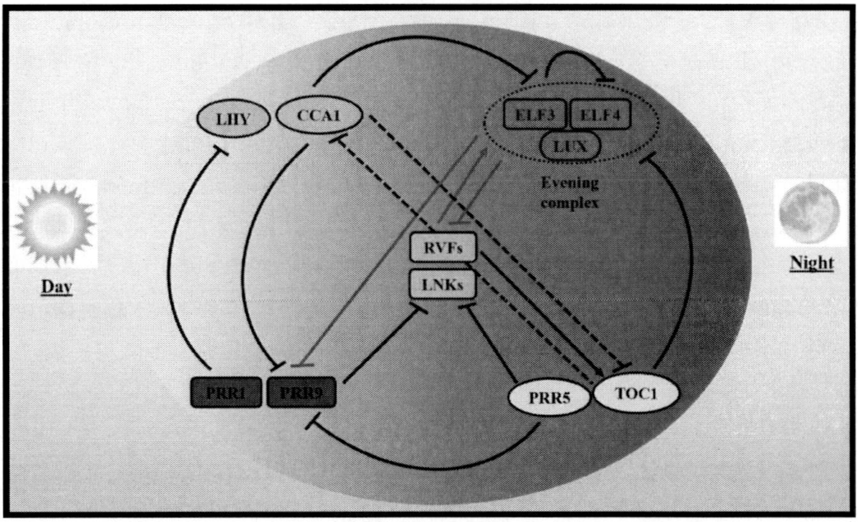

Figure 4. Diagram showing interlinked transcriptional-translational feedback loop. Background is showing dawn to dusk transition. CCA1 and LHY positively regulate PRR9 and PRR7 whereas negatively regulates PRR5, TOC1, and evening complex in the morning. During midday PRR9 and PRR7 enhance the expression of RVE6, RVE4, and RVE8 and negatively regulate LHY and CCA1. RVE4, RVE6, and RVE8 positively regulate the evening complex in the evening. Evening complex enhances the expression of TOC1 and PRR5 during the night. While CCA1 and LHY degraded by proteasomal degradation via ubiquitination.
Abbreviation: LHY (Late Elongated Hypocotyl); CCA1 (Circadian Clock Associated 1); PRRs (Pseudo Response Regulators); TOC1 (Timing of Cab Expression 1); RVFs (Reveille); LNKs (Night Light-Inducible and Clock-Regulated); LUX (LUX Arrhythmos); ELF (Early Flowering) (adopted and modified from Srivastava et al. 2019).

Beside these genes and transcription factors the action of casein kinase II (CK2) phosphorylates, CCA1 and LHY, and b3 subunit (CKB3) also involved in circadian functioning. It is now well established that CK2-mediated phosphorylation of CCA1 is an essential process. Recently, posttranslational modification and identification of Spindly (N-acetyl glucosamine transferase), Lov Kelch Protein2 (LKP2), Zeitlupe (ZTL), and FLAVIN binding Kelch Repeat F-Box (FKF) having PAS/LOV domains along with Kelch repeats and F-boxes, have been also identified and characterized having their potential role in controlling *Arabidopsis* circadian clock. The LOV domains are like phototropin, present in FKF have a restricted role in photoperiodism. However, ZTL and LKP2 have controlling effect on the clock. Further, the proteasome-mediated degradation of ZTL during day time and its stability at dusk demonstrated the clear-cut interaction of proteasomal degeneration paths with poly-ubiquitination by F-box proteins. Interestingly, the higher expression of ZTL is also involved in photoperiod shortening in dose-dependent manner. Overall, it can be concluded that the transcript level of TOC1 has been fully under the control of transcriptional repressors such as CCA1 and LHY and degradation of ZTL protein, which is known as a key determinant and master regulator of the circadian period.

Plant Circadian Clock and Chromatin Modifications

The plant DNA packaging and chromatin structure have direct linkage with the regulation of genes/transcription factors expression in response environmental cues by modulating accessibility/activity of transcriptional regulatory proteins. Multiple reversible and chemical DNA modifications such as methylation of DNA, histone methylation and histone acetylation have regulatory impact on chromatin function and activity. Large number of evidence reveals that structural change in chromatin have modulatory role in circadian functions. The molecular mechanisms and components responsible for the histone methylation are still in the binging stage. The accumulation of H3K4me3 at the promoter region of LHY, CCA1, and TOC1 have positive role on rhythmic genes and their transcripts *via* histone methyltransferase SDG2/ATXR3 and histone H3K36me2 have negative correlation with these genes (Table 1). Recently, CCA1 and LHY have been also found to reduce H3ac and H3K4Me2 levels by inducing histone modification complex at *TOC 1* locus having LDL1/2 and HDA6 genes. Further, theJMJ30/JMJD5 (Jumonji C domain–containing histone demethylase), is also known for controlling circadian processes along with demethylase of H3K9me3 by core oscillators CCA1 and LHY having repressive activity. By the modulation in the levels of H3K9me3, the JMJ30/JMJD5 has positive impact on expression of CCA1 and LHY.

Interestingly, induction through TOC1 has control on histone acetylation thereby activating chromatin structures for transcription. It has been also observed that H3K9/K14Ac and H3K4Me3 are also associated with transcriptional start sites (TSSs) of circadian clock regulating genes CCA1, LHY and TOC1. Further, several chromatin remodeling factor like HAG3, HDA2/6, SUVH3, and JMJ28 are also known for activating/ regulating circadian oscillation. Nevertheless, deciphering complexity in the circadian regulatory network is the need which is currently required for mechanistic regulation of circadian rhythm and it relationship with histone modification. The interactive interplay between circadian regulators and chromatin components along with the understanding of the molecular network remain to be answered.

Circadian Clock as Regulator of Cellular Processes

The circadian clock is not only active and functional in multicellular organisms but also found to control several biological processes in unicellular organisms. The circadian check points in cell cycle of the cyanobacterium *Synechococcus elongatus* have been deciphered using a time-lapse imaging which is also observed in *Cyanidioschyzon merolae,* a unicellular red algae. This circadian gating related to cell division has been also observed in eukaryotic organisms. The various functions at cellular levels such as stomatal opening, photosynthesis, crassulacean acid metabolism, photorespiration, starch degradation and sugar metabolism are under control of circadian rhythm. The expression and function of genes that are associated with the light-harvesting complex have regulatory impact of the circadian clock such as accumulation level of the chlorophyll a/b binding protein (CAB), ribulose 1,5-bisphosphate carboxylase/ oxygenase activase (RCA), phosphorylation of the D1 photosystem II reaction center protein, phosphoenolpyruvate carboxylase have been also reported under regulation of the circadian clock indicating photosynthetic process in unicellular and multicellular organisms is strictly regulated by various processes at transcriptional and translational levels to provide plant fitness.

Table 1. List of some core circadian clock genes and their role in several key processes/functions

Circadian oscillators	Processes/Functions	References
CCA1/LHY	Biotic stress, flowering, stomata activity, regulation of carbon partitioning and starch degradation (through CaK), DNA methylation and histone acetylation, LHY regulate plant height (Glycine max)	Karapetyana and Donga 2018; Pokhilko et al. 2015; Perales et al. 2007; Baerenfaller et al. 2017; Cheng et al. 2017
TOC1	Circadian clock component, abiotic stress, regulate expression of ABA signaling genes, histone demethylation	Hecht et al. 2005; Legnaioli et al. 2009; Hung et al. 2018
PRR3/7	Control of photoperiodic Flowering, biennial growth habit	Dally et al. 2014; Pin et al. 2012
PRR5/7/9	Triple mutant shows increased cold tolerance, involve in phospho-relay signal transduction	Nakamichi et al. 2009; Liew et el. 2009
ELF3	Circadian clock function, flowering timing, light response, regulating leaf senescence (via ORE1), response to salt stress	Liew et al. 2009; Weller et al. 2012; Sakuraba et al. 2014; Sakuraba et al. 2017
ELF4	Circadian clock, flowering time, evening-expressed gene	Liew et al. 2009; Weller et al. 2012; Galeou et al. 2012
LUX	Flowering time regulation	Weller et al. 2012
GI	Flowering time, growth regulators, circadian clock function, controlling the cycle of glucose metabolism, photoperiod response	Hayama et al. 2003; Watanabe et al. 2011; Haydon et al. 2013; Watanabe et al. 2011
REV4/6/8	Mutant cause extreme long period phenotype	Hsu et al. 2013
ZTL	Control of flowering induction and photoreceptor	Xue et al. 2012

The Clock rhythmic studies in unicellular/single-cells such as in blue green algae (cyanobacteria) by Yang et al., 2010 are helpful in the development of quantitative single-cell models relating *Arabidopsis* circadian clock. These single-cell circadian models would also require significant molecular investigations and computational resources for developing series of datasets for unraveling mechanisms to create avenue of researches in computational perspective.

Thus, the circadian clock genes involve in many intracellular and intercellular processes and functions in plants such as in defense mechanisms (biotic and abiotic stress), chromatin modifications, metabolisms (glucose and starch), hormonal signaling, plant growth, leaf senescence, flowering time regulation etc. as given in Table 1 with their correspond genes.

Circadian Clock Regulators at Tissue/Organ Level Controlling Plant Developmental Processes

The circadian clock in plants has been known for regulating various growth and developmental activities throughout the life cycle starting from germination to flowering. At the earliest stage the photoperiodic control of plant development at the stage of seed germination has been reported by many workers in different plant species. Indeed, imbibitions of seeds, gas exchange, hypocotyl elongation and abscisic acid and gibberellin-related genes expression are also reported under control of the circadian rhythm (Penfield and Hall 2009). The rhythmic pattern and growth of hypocotyls have been regulated by function of major two transcription factors, Phytochrome-Interacting Factor 4 (PIF4) and PIF5. For basic helix loop helix formation in response to external and internal cues, the PIF4 and PIF5 make a signaling hub (Stewart et al. 2011). Other transcription factors are also involved in regulating PIF4 and PIF5 such as *Early flowering* ELF3, ELF4 and LUX act as repressor in the early evening by which the oscillator modulates the light resister. It has been also found that the interaction activated Phytochrome B (PHYB) causes degradation of PIF4 and PIF5 under light conditions. Thus, it can be concluded that the function/activity of PIF4 and PIF5 is under tight control of various regulators at an early stage of plant development.

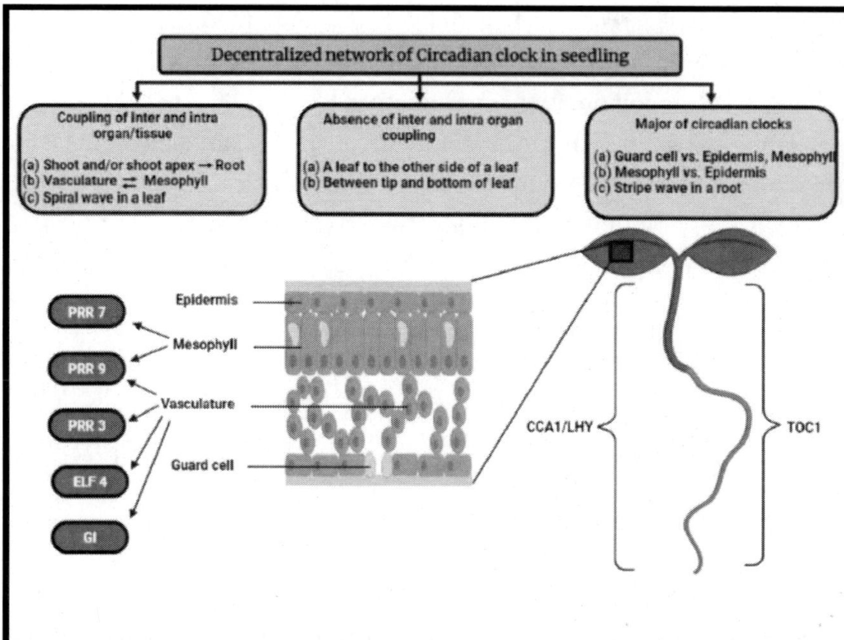

Figure 5. Tissue/organ specific circadian clock and their communication with spatial expression of circadian clock genes; CCA1 (Circadian Clock Associated 1), LHY (Late Elongated Hypocotyl), TOC1 (Timing of Cab Expression 1), PRR3/7/9 (Pseudo Response Regulators), GI (Gigantea), ELF1 (Early Flowering 1).

Furthermore, it has been reported that the function of ELF3 is also required for leaf initiation, leaf growth and movements. In the recent study of sunflower floral movement depicted that heliotropism in this plant is driven by the complex and coordinate mechanism of circadian clock and light-signaling pathways as CONSTANS (CO), a key transcription factor is also involved in its strict regulation. The light signaling mediated by various photoreceptors is also involved in regulation and stability of CO protein *via* complexation with Constitutive Photomorphogenic 1 (COP1) and Suppressor of PHYA/SPA1 during the dark conditions. Further, the coordinated action of light signaling pathways and circadian clock allows induction/accumulation of CO protein under high/long duration light conditions. Recently, the significance of vasculature-specific circadian clock functioning have been established for understanding of photoperiodic flowering. To date, it has been demonstrated that in plants organ/tissue-specific circadian clocks and its coupling have hierarchical networks and regulatory elements of circadian clocks. Advancement in various techniques, such as tissue-isolation, Luciferase Assay (TSLA), and disruption of clock functions at cellular/tissue/organ level along with microscopic observations, laser capture micro dissection will surely open new avenues to decipher circadian clocks at an organ/tissue/cellular levels.

FUNCTION OF PLANT BIOLOGICAL CLOCK UNDER STRESS SIGNALS

The biological clock system in plants also participates in take part in combating various biotic as well as abiotic stresses for e.g., sensitivity towards temperature and light along with maintenance of circadian rhythm which ultimately results in the plant overall development and growth. The external signaling of the surrounding and internal communication in plants happens with the help of the circadian clock *via* switching/turning on the rhythm timing. Earlier it was believed that the biological clock functions include only three modules i.e., input path, central oscillator system and output path but recent researches demonstrated that circadian rhythm have a complex-convoluted feedback-loop system which works in a coordinated manner involving large number of transcription factors at transcriptional and translational level.

Role of Plant Circadian Clock under Abiotic Stresses

Plant circadian clock not only has its coordination with environmental cues but also prepares the plant to acclimatize against environmental extremities by assigning the clock genes for regulating balanced rhythmic expressions. The analytical transcriptomics studies on *Arabidopsis*, rice, barley, and soybean have hinted that maximum abiotic stress-responsive genes are in circadian regulation. The ABA pathway is one of many prominent stress pathways in plants. It is in proximity to the circadian clock leads to many stresses response execution. TOC1 is a central oscillator gene that gets activated by ABA signaling, in turn TOC1expression regulates the circadian rhythm as well as ABA receptor-like gene (ABAR) which facilitates the drought response in the plant system.

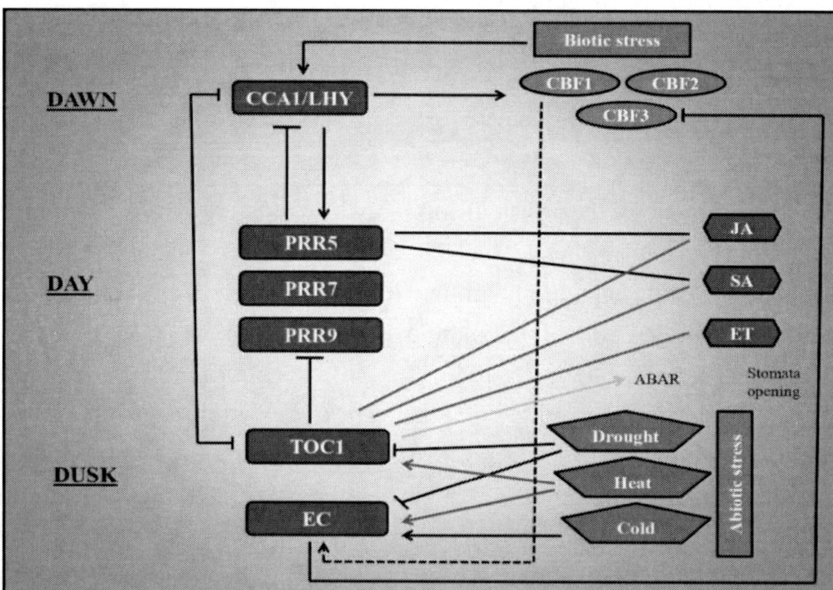

Figure 6. Role of circadian clock in the regulation of biotic and abiotic stresses. Abiotic stress such as drought, cold, and heat has a high effect on plant immunity. Many genes and mechanisms belong to defense system including JA and SA is circadian regulator after biotic stress. Abbreviations: CCA1 (Circadian Clock Associated 1); LHY (Late Elongated Hypocotyl); PRRs (Pseudo Response Regulators); TOC1 (Timing of Cab Expression 1); EC (Evening complex); CBFs (C binding repeat factor 1); ABAR (ABA receptor-like gene); JA (Jasmonic acid); SA (Salicylic acid); ET (Ethylene) (Adopted and modified from Bhattacharya et al., 2017).

Many other genes for e.g., cold-responsive genes (COR15B, and CORA, salt responsive as well as drought-related genes like early response to dehydration ERD10& ERD7), works with the circadian clock. The transcription factors C binding repeat factor 1 and dehydration responsive element binding 1 (CBF1 and DREB1) to induce the cold-responsive gene. Furthermore, this helps to provide resistance to salt, cold and salt stress. Other circadian regulated genes are WNK-like genes, which are found in plants e.g., *Oryza sativa* WNK gens coordinates with the circadian clock and is responsible for the differential expression of the plant in various abiotic stress at different time points. A chill induced co-activator MYB1 of rice helps in rhythmic expression while combating different abiotic stress in soybean, the clock genes (PRR7, PRR5 TOC1, SUB1, 1 and ABAR) governs the water lodging and water deficit conditions via alternative splicing processing (Srivastava et al. 2019). The biological clock and abiotic stress relation can be well understood with the help of the below given Figure 6.

Plant Biological Clock in Relation with Biological Stress Factors

Plants are organisms that lack locomotion. Hence, they have developed mechanisms in order to combat environmental stress. The circadian clock plays an important role in the proper function of plants in biotic danger circumstances via governing the plants innate immunity. There are two immune responses by plants, one PAMP pathogen-associated molecular patterns triggered immunity PTI and another ETI effecter triggered immunity. PTI recognizes the pathogen chitin, flagella, etc. in short it is the basic level of defense response by the plants. But ETI is induced by r proteins which identify the pathogen effectors molecules and provide gene-mediated resistance to plants. It is well reported that many clock genes like TOC1, LHY, and CHE, CAA1 play role in plant defense against biotic stress *via* light and dark cycles as shown in Figure 6.

The JA, SA and ethylene signaling functions in the regulation of plant biotic response. In *Pryrecularia syringe* infection there is an accumulation of a CO_2 and more susceptible phenotype when over-expression of clock genes CAA1 AND LHY was performed. The above report showed the direct involvement of clock gene. Both SA and JA gens work in accordance with circadian timing e.g., in the night time pathogen invasion is fought when SA peaks while in the daytime when JA is at peak fungal pathogens are combated. It is documented that the defense-related gene in plants for e.g., Resistance to *Phakopsora pachyrhizi* 4, alkyrin, Leucine-rich repeat kinase etc have promoters for the clock genes for e.g., CCA1 and LHY mutually interfere basic level and gene-related defense system in *Arabidopsis* against *Pseudomonas syringae* and *Hyaloperonospora* arabidopsidis. Around 40 per cent of wound response-related genes are under circadian clock regulation. Thus, it is clear that there is a direct role of the circadian system in plants defense strategies. Figure 6 explains the crosstalk between the circadian and biotic stress response in plants along with some genes in Table 1.

Circadian Clock in the Plant Development and Other Processes

Plant developmental processes like photosynthesis, germination, flowering, senescence etc are under the direct control of the circadian mechanism (Figure 1). Many aspects of plant growth such as seed germination, hypocotyl elongation, and rhythmic growth/movement are showing dependence on the action of t basic helix-loop-helix transcription factors and Phytochrome Interacting Factor (PIF4 and PIF5). The clock genes are also involved in ABA and GA signaling that is major players of seed germination and dormancy and phytochrome interacting factors (PIF5 and PIF4) have significant control on hypocotyls elongation. Although, the expression of these two (PIF4 and PIF 5) are under the control of *ELF4, ELF3* and *LUX* genes during dusk time. However, at the time of early night, the *ELF3* inhibits *PIF4* and limiting its accumulation at late night. Further, the clock gene ELF3 also involved in leaf initiation and development. The transition of vegetative phage to the flowering phase is also under the control of expression of various genes regulated by the circadian clock.

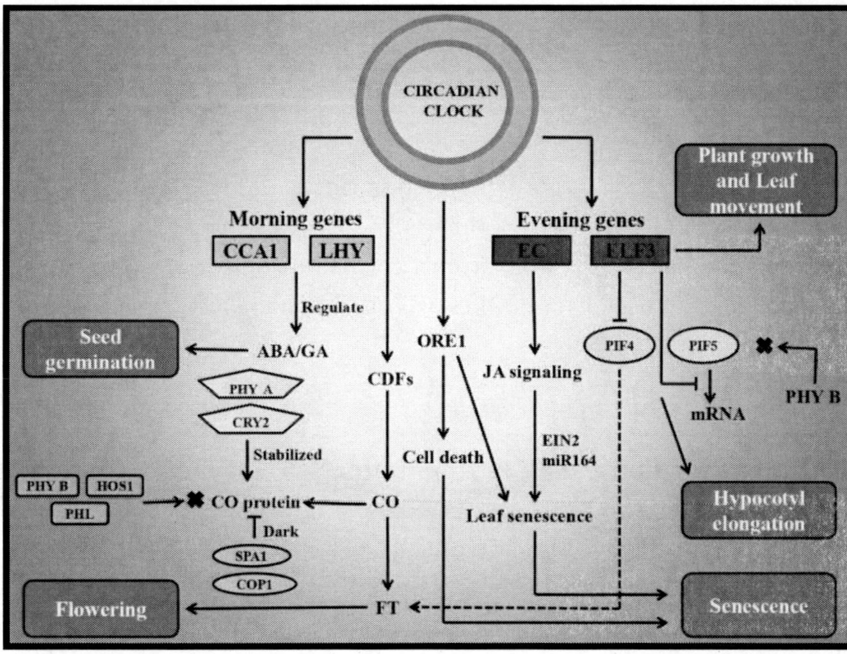

Figure 7. This illustration shows various developmental processes regulated through circadian clock; seed germination, hypocotyl elongation, plant growth, leaf movement, flowering and in senescence. Hypocotyl growth is targeted by light signaling pathways including circadian clock regulation. Plant growth and leaf movement are associated with ELF3. CO is responsible for flowering. Circadian clock regulates of Gibberellic acid (GA) and Abscisic acid (ABA) which are necessary for seed germination. ORE1 is well known for senescence. Pointed arrows are showing activation, blunt arrows showing repression while cross sign represents degradation. Abbreviations: CCA1 (Circadian Clock Associated 1); Lhy (Late Elongated Hypocotyl); EC (Evening complex); ELF3 (Early Flowering3); ABA (Abscisic acid); GA (Gibberlic acid); PHYA (Phytochrome A); CRY2 (Cryptochrome2); CO (Constans); SPA1 (Suppressor of Phya); COP1 (Constitutive Photomorphogenic 1); EIN2 (Ethylene-insensitive protein2); PIFs (Phytochrome Interacting Factors); CDFs (Cycling DOF Factors); PHY B (Phytochrome B); ORE1 (ORESARA1); FT (Flowering Locus T)

The duration of light i.e., photoperiodism and various environmental factors such as temperature also contribute to flowering via activation of circadian clock genes. The CONSTANS (CO) gene is majorly known for the flowering transition by activating the expression/transcription of a florigen-encoding Flowering Locus T (FT) gene for inducing photoperiodic flowering (Srivastava et al. 2019). Previous findings have already demonstrated the function and presence of *phyA, CRY2, COP1, SPA1*, and *CO* in the companion cells of phloem for controlling photoperiodic flowering.

Additionally, the circadian clock genes have regulatory function in flower opening and successful pollination *via* regulating the timing of petal opening when potential pollinators are present or most active. Furthermore, release of some aromatic volatile compounds from flowers and synthesis of nectar for pollinator's attraction are also under control of circadian clock.

Role of Circadian Clock in Regulation of Secondary Metabolism

Plant metabolism is an integral part for determining the external phenotype and whole life cycle. The interaction between genetic, development and environment inputs define the growth, metabolisms and performance of plants. Various researchers have deciphered the circadian control of metabolism which is widely characterized in model plants to demonstrate the metabolic shifts in response to circadian regulators. The changes and fluctuations in environmental cues are well known for shifting the primary metabolism toward the synthesis of secondary metabolites. The biosynthesis of these secondary metabolites utilizes plant primary metabolism products such as amino acids and glucose etc. as substrates/cofactor/ligands in inducing the enzymatic reactions. For instance, the circadian clock has a major impact on regulation of starch turnover and mobilization during light and dark conditions along with deli regulation of amino acid and organic acid metabolism.

Further, recent studies have also revealed that the starch utilization and photosynthetic products are under regulation of circadian clock and alteration in sugar status due to photosynthetic evolution modulates functioning of circadian clock. It can be inferred that the circadian clock directly regulates primary metabolism requires for plant development, growth and shifting of primary metabolism towards the synthesis of secondary metabolites affecting crop productivity and quality through metabolite changes in response to environmental stimuli.

CONCLUSION

The circadian clock is a well-networked endogenous complex acting as a timekeeper under the control of day-to-day environmental changes and seasonal variations. In response to the exogenous stimulus, the various constituents of the circadian clock have been regulated at both transcription and post-transcriptional levels by involving a large number of transcription factors and genes. In recent years, understanding towards plant circadian clocks has opened an enormous data, predominantly in the identification and characterization of molecular components, function and regulatory participation. Since the biological clock controls the majority of plant life processes throughout the life span of the plant, the knowledge can be utilized in building plants adaptive fitness.

However, molecular studies are helpful in deciphering the basic mechanism but a thorough understanding related to its functioning is still needed. The recent technological advancements such as chromatin immune precipitation (ChIP), ChIP-sequencing and be utilized to unravel the direct connections between the main circadian clock system with output signaling pathways. Genome editing in circadian clock genes can be useful in improving crop yield and combing with forward and reverse quantitative/qualitative genetics the plant behavior can be explored in regulation in abiotic and biotic stresses as well as in food crop domestication.

REFERENCES

Adams S, Manfield I, Stockley P, Carré IA (2015) Revised morning loops of the *Arabidopsis* circadian clock based on analyses of direct regulatory interactions. *PLoS One* 10(12): e0143943.

Bhattacharya A, Khanale V, Char B (2017) Plant circadian rhythm in stress signaling. *Indian J Plant Physiol* 22(2): 147-155.

Bray MS, Shaw CA, Moore MW, Garcia RA, Zanquetta MM, Durgan DJ, Jeong WJ, Tsai JY, Bugger H, Zhang D, Rohrwasser A (2008) Disruption of the circadian clock within the cardiomyocyte influences myocardial contractile function, metabolism, and gene expression. *Am J Physiol Heart Circ Physiol* 294(2): H1036-H1047.

Covington MF, Maloof JN, Straume M, Kay SA, Harmer SL (2008) Global transcriptome analysis reveals circadian regulation of key pathways in plant growth and development. *Genome Biol* 9(8): 1-8.

De Caluwé J, Xiao Q, Hermans C, Verbruggen N, Leloup JC, Gonze D (2016) A compact model for the complex plant circadian clock. *Front Plant Sci* 7: 74.

Farré EM, Liu T (2013) The PRR family of transcriptional regulators reflects the complexity and evolution of plant circadian clocks. *Curr Opin Plant Boil* 16(5): 621-629.

Fenske MP, Imaizumi T (2016) Circadian rhythms in floral scent emission. *Front Plant Sci* 7: 462.

Fogelmark K, Troein C (2014) Rethinking transcriptional activation in the *Arabidopsis* circadian clock. *PLoS Comput Biol* 10(7): e1003705.

Gil KE, Park CM (2019) Thermal adaptation and plasticity of the plant circadian clock. *New Phytol* 221(3): 1215-1222.

Greenham K, McClung CR (2015) Integrating circadian dynamics with physiological processes in plants. *Nat Rev Genet* 16(10): 598-610.

Harmer SL (2009) The circadian system in higher plants. *Annu Rev Plant Biol* 60: 357-377.

Hirschie Johnson C, Elliott JA, Foster R (2003) Entrainment of circadian programs. *Chronobiol Int* 20(5): 741-774.

Hsu PY, Harmer SL (2014) Wheels within wheels: the plant circadian system. *Trends Plant Sci* 19(4): 240-249.

Dunlap JC (1999) Molecular Bases for Circadian Clocks. *Cell* 96(2): 271-290.

Johnson CH, Stewart PL, Egli M (2011) The cyanobacterial circadian system: from biophysics to bioevolution. *Annu Rev Biophys* 40: 143-167.

Linde AM, Eklund DM, Kubota A, Pederson ER, Holm K, Gyllenstrand N, Lagercrantz U (2017) Early evolution of the land plant circadian clock. *New Phytol* 216(2): 576-590.

Lu SX, Knowles SM, Andronis C, Ong MS, Tobin EM (2009) Circadian clock associated1 and late elongated hypocotyl function synergistically in the circadian clock of *Arabidopsis*. *Plant Physiol* 150(2): 834-843.

McClung CA (2013) How might circadian rhythms control mood? Let me count the ways. *Biol Psychiatry* 74(4): 242-249.

Michael TP, Breton G, Hazen SP, Priest H, Mockler TC, Kay SA, Chory J (2008) A morning-specific phytohormone gene expression program underlying rhythmic plant growth. *PLoS Boil* 6(9): e225.

Mishra P, Panigrahi KC (2015) Gigantea–an emerging story. *Front Plant Sci* 6: 8.

Mizuno T, Nomoto Y, Oka H, Kitayama M, Takeuchi A, Tsubouchi M, Yamashino T (2014) Ambient temperature signal feeds into the circadian clock transcriptional circuitry through the EC night-time repressor in *Arabidopsis thaliana*. *Plant Cell Physiol* 55(5): 958-976.

Penfield S, Hall A (2009) A role for multiple circadian clock genes in the response to signals that break seed dormancy in *Arabidopsis*. *Plant Cell* 21(6): 1722-32.

Pokhilko A, Bou-Torrent J, Pulido P, Rodríguez-Concepción M, Ebenhöh O (2015) Mathematical modelling of the diurnal regulation of the MEP pathway in *Arabidopsis*. *New Phytologist* (3): 1075-1085.

Rand DA, Shulgin BV, Salazar D, Millar AJ.(2004) Design principles underlying circadian clocks. *J R Soc Interface* 1(1): 119-130.

Romanowski A, Yanovsky MJ (2015) Circadian rhythms and post-transcriptional regulation in higher plants. *Front Plant Sci* 6: 437.

Srivastava D, Shamim M, Kumar M., Mishra A, Maurya R, Sharma D, Singh KN (2019) Role of circadian rhythm in plant system: An update from development to stress response. *Environ Exp Bot* 162: 256-271.

Takahashi JS (2015) Molecular components of the circadian clock in mammals. *Diabetes Obes Metab* 17: 6-11.

Yang Q, Pando BF, Dong G, Golden SS, van Oudenaarden A (2010) Circadian gating of the cell cycle revealed in single cyanobacterial cells. *Science* 327(5972): 1522-1526.

Yerushalmi S, Green RM (2009) Evidence for the adaptive significance of circadian rhythms. *Ecol Lett* 12(9): 970-981.

Young MW, Kay SA (2001) Time zones: comparative genetics of circadian clocks. *Nat Rev Genet* 2(9): 702-715.

In: Circadian Rhythms and Their Importance ISBN: 978-1-68507-547-7
Editor: Rajeshwar P. Sinha

Chapter 5

Circadian Clock: Photic Regulation and Its Disruption on Health

Vivek Verma*[*], *Ruchika Kumari*[*]
***and Muniyandi Singaravel*[†]**
Chronobiology Lab, Department of Zoology, Institute of Science,
Banaras Hindu University, Varanasi, Uttar Pradesh, India

Abstract

Circadian variation in organism activity follows a 24 h oscillatory pattern which is strongly directed by the external changing light-dark (LD) cycle. The circadian rhythm is under the control of a pair of bi-lobed nuclei situated in the anterior hypothalamic area referred to as the suprachiasmatic nucleus (SCN). A non-image-forming retinal input of the external LD cue synchronizes the endogenously originated rhythm of SCN which then translates into effective behavioral and physiological responses thereby maintaining normal body homeostasis. The internal master clock, SCN synchronizes organisms' rhythm such as core body temperature, heart rate, endocrine secretions, metabolism, cognitive

[*] These authors contributed equally.
[†] Corresponding Author's E-mail: m.singaravel@bhu.ac.in.

functions, gene expression and sleep/wake cycle etc. according to the external 12:12 h LD cycle. The crucial period of light and dark of the 12:12 h LD cycle acts as an important controlling factor in clock regulation and maintenance of a plethora of biological processes whose disruption leads to several negative health consequences. To achieve body homeostasis, the master clock establishes several neural and humoral cross-talk with several peripheral clocks for the time-dependent release of several bioactive molecules including gene expression. The disruption of the core molecular machinery of the clock due to changing natural LD cycle leads to circadian rhythm disturbances hierarchically and subsequently causing several negative health consequences such as metabolic syndrome, cardiovascular diseases, mental illness, de-regulation of reproductive function, cancer and many more.

Keywords: circadian rhythm, suprachiasmatic nucleus, photic cues, circadian clock, shift work, jet lag

Introduction

Most behavioral and physiological functions such as body temperature, hormone rhythm, mood and sleep/wake or rest/activity cycle exhibit daily rhythm and are programmed by an endogenous clock. The circadian rhythm has a periodicity of approximately 24 h and is governed by a master pacemaker, the suprachiasmatic nucleus (SCN) located in the anterior hypothalamus (Weaver 1998; Mohawk et al. 2012). SCN is a bilateral paired structure that consists of a cluster of densely packed neurons of about 20,000 in number and is capable of self-oscillation and runs a rhythm of more or less 24 h cycle referred to as free-run. The autonomous endogenous oscillatory pattern of the master clock is synchronized to the external environment to maintain 24 h rhythm of electrical activity and neurotransmitter release (Gillette and Reppert 1987). Light acts as a potent *zeitgeber* (time cue) that synchronizes the clock according to the pattern of the light-dark (LD) cycle by a process of entrainment (Brainard and Hanifin 2005; Moore 1996).

Apart from light, non-photic cues such as food, exercise and other chemicals (e.g., caffeine) can also act as important synchronizers of the circadian clock (Aschoff et al. 1971; Klerman et al. 1998; Winget et al. 1989). External light information is relayed to SCN through the retinohypothalamic tract (RHT), arising from the innermost layer of the retina, which communicates the light information by secretion of different neurotransmitters such as glutamate, aspartate, pituitary adenylate cyclase-activating peptide (PACAP) and substance P (Chen et al. 1999; Ebling 1996; Fahrenkrug 2006; Hannibal 2002, 2006).

These neurotransmitters act as a messenger to control the expression of several clock genes within SCN which in turn coordinates to different downstream peripheral target tissues, thus influencing the behavior and physiology according to environmental periodicity induced by axial rotation of the Earth (Reppert and Weaver 2002). A peripheral target-mediated release of corticotrophin releasing hormone and adrenocorticotropic hormone (ACTH) is mediated by vasopressin projections from SCN to dorsomedial hypothalamus and subsequently to the paraventricular nucleus (Kalsbeek et al. 1996) is an example of SCN directed target tissue function.

The advancement in modern technology promotes the non-traditional work schedule which reduced the dependency of social and work activities with normal environment LD cycle. Circadian misalignment is associated with the change in timing of the normal LD cycle which results in a shift in the phase of circadian rhythm and leads to several short- and long-term consequences. This shift in the LD cycle provides conflicting signals that disrupt the synchronized temporal relationship of the SCN with the normal LD event (Hastings et al. 2003). The non-traditional work schedule comprises a wide variety of recurring or fixed night and morning shifts including frequent transmeridian travels (e.g., jet lag) leading to misalignment of the social time from the biological clock (Wittmann et al. 2006). The disrupted active schedule and perturbation in the normal sleep/wake cycle end up in anxiety, depression and altered behavioral response (Akerstedt and Wright 2009; Wirz-Justice 2006).

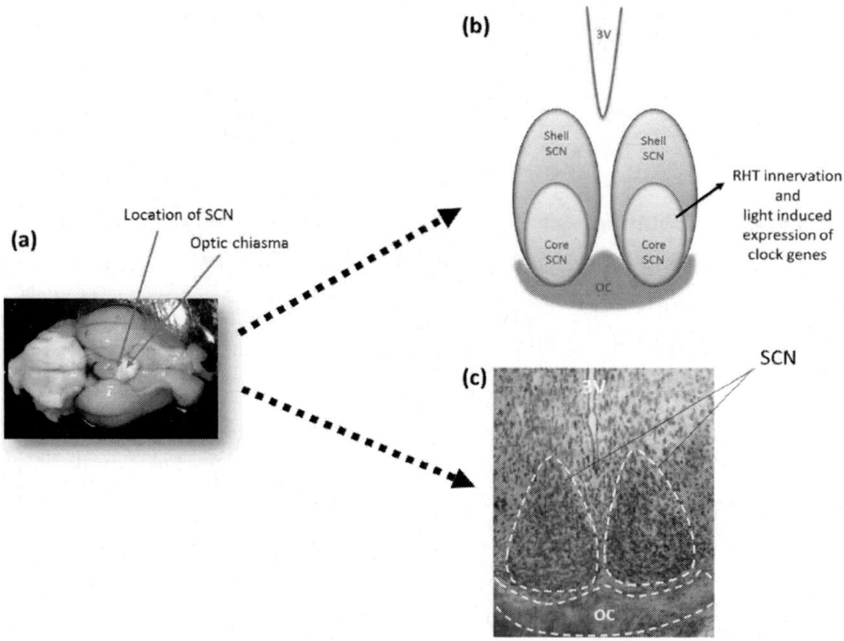

Figure 1. Location and structure of the suprachiasmatic nucleus. Ventral view of a rodent brain showing optic chiasma and the location of the suprachiasmatic nucleus (a), the topological organization of suprachiasmatic nucleus includes the core or ventrolateral area and the shell or dorsomedial area depicted in the schematic diagram (b), the cresyl violet stain of coronal section of brain showing dense nuclei of a pair of suprachiasmatic nuclei positioned over the optic chiasma (OC) (c).

The loss of circadian synchrony among different peripheral tissues with that of the central clock results in loss of body homeostasis and well-being (Davidson et al. 2009; Waterhouse et al. 2007). Apart from the behavioral alteration, circadian misalignment leads to fluctuations in the circadian rhythm of hormone and metabolic response as well as mood. Disturbed circadian rhythm has been suggested to be a strong promoter of obesity and several other metabolic syndromes which may further lead to cardiovascular diseases, gastric disorder or cancer that eventually results in increased mortality in male and female shift workers (Tenkanen et al. 1998; Li et al. 2011; Kubo et al. 2006; Filipski et al. 2009; Davis et al. 2001). Apart from that, circadian misalignment by artificial light exposure distorts the normal functioning of brain regions involved in mood regulation which end up in numerous psychiatric disorders (Bedrosian and Nelson 2017).

Molecular Feedback Mechanism in SCN

SCN generated rhythm is the network of molecular feedback loops that positively and negatively regulates the transcription of core clock genes to drive the oscillation in clock genes expression with a period of 24 h (Lowrey and Takahashi 2004; Okamura et al. 2002; Reppert and Weaver 2001; Balsalobre et al. 2000; Shearman et al. 2000). Core clock genes are referred to as genes whose proteins are necessary for the generation and regulation of circadian rhythms within individual cells of the organisms (Takahashi 2004). Such autonomously generated SCN rhythms follow the pattern of transcription – translation mechanism of a set of clock genes which includes *Period* genes (*Per1, Per2, Per3*), *Cryptochrome* (*Cry1 & 2*), *Bmal1* and *Clock* (circadian locomotor output cycles kaput) to evoke behavioral, physiological and molecular response in an intrinsic time-dependent manner (Nagoshi et al. 2004; Gekakis et al. 1998). The product of these genes interacts with each other and their expression is controlled or regulated by a negative feedback mechanism which provides a basis to generate circadian rhythmicity (Sato et al. 2006). The positive limb of the transcription-translation loop is formed by CLOCK and BMAL1 while PER and CRY formed the negative limb of the transcription-translation feedback loop (Schroeder and Colwell 2013). CLOCK and BMAL1 belong to members of the basic helix-loop-helix Period-Arnt-Single-minded family.

Additionally, the C terminal domain of CLOCK protein has histone acetyltransferase activity and gets activated upon binding with BMAL1 (Doi et al. 2006). Under activated condition CLOCK/BMAL1 heterodimer binds to promoter region (E box) of *Per* and *Cry* and facilitates its expression (Gekakis et al. 1998; Kume et al. 1999; Zheng et al. 2001). During the day, light-induced expression of *Clock* and *Bmal1* and their product interact with each other to form a heterodimer and cause transcription and translation of *Per* and *Cry* gene. Prolonged expression results in the accumulation of PER and CRY protein in the cytoplasm during the afternoon which then translocates back to the nucleus and inhibits their expression by inhibiting transcription activator CLOCK/BMAL1 heterodimer (Sangoram et al. 1998; Griffin et al. 1999). The cyclic transcription activity is achieved by chromatin remodeling done by a group of clock-specific and ubiquitous histone-modifying proteins that rhythmically acetylate and deacetylate H3 and H4 histone proteins at multiple sites on clock genes (Ripperger and Schibler 2006; Etchegaray et al. 2003).

Another feedback loop is initiated by the CLOCK/BMAL1 heterodimer which acts in conjunction with the above-described loop. The CLOCK/BMAL1 heterodimer activates the transcription of retinoic acid-related orphan nuclear receptors, *Rev-erba* and *Rora* by binding with E-box (Preitner et al. 2002; Sato et al. 2004; Guillaumond et al. 2005; Triqueneaux et al. 2004; Akashi and Takumi 2005). The REV-ERB and ROR are then able to compete for binding to retinoic acid-related orphan receptor response element (RORE) present within the promoter region of *Bmal1*. ROR and REV-ERB positively/negatively influence the transcription of *Bmal1* where ROR activates the transcription of *Bmal1* and REV-ERB suppresses its transcription (Guillaumond et al. 2005; Preitner et al. 2002). The members of ROR (α, β, and γ) and REV-ERB (α and β) are known to regulate the transcription of *Bmal1* by binding to RORE.

The resetting of the molecular clock which generates 24 h rhythm requires post-translational modifications such as phosphorylation and ubiquitination of negative limb protein i.e., PER and CRY. The stability and degradation rate of PER and CRY are critical to set the period of the mammalian clock. The *casein kinase 1 epsilon* (*CK1ε*) and *casein kinase 1 delta* (*CK1δ*) are known to phosphorylate PER which then subsequently targeted for ubiquitination by βTγCP and degradation by 26S proteasome (Camacho et al. 2001; Eide et al. 2005; Shirogane et al. 2005). The tau mutant in the hamster which results from a mutation in *CK1ε* and *CK1δ* displays a free-running rhythm with a period of 20 h (Ralph and Menaker 1988). The proteasomal degradation of PER and CRY is required to remove the repression to start the next cycle again (Dardente and Cermakian 2007; Ko and Takahashi 2006).

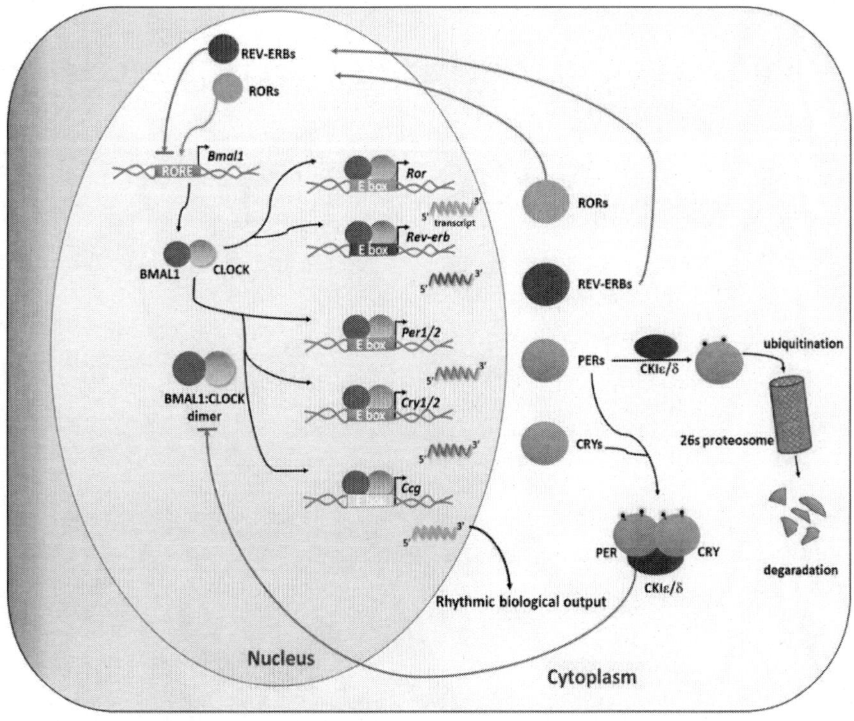

Figure 2. Schematic diagram of the network of molecular transcriptional-translational feedback loop of the mammalian circadian clock.

CLOCK: BMAL1 heterodimer which constitutes the positive arm of feedback loop binds and initiates the transcription of target genes containing E-box cis regulatory enhancer sequence including *Per1/2/3* and *Cry1/2*. The heterodimer of PER and CRY in the cytoplasm get phosphorylated by CK1ε/δ and translocate back to the nucleus and suppresses its transcription by inhibiting CLOCK: BMAL1 complex and thus forms the negative arm of the feedback loop. BMAL1 production was further regulated by ROR and REV-ERB. Binding of ROR and REV-ERB to RORE present in *Bmal1* promoter activates and repress the transcription of *Bmal1*, respectively. In addition to the regulation of transcription of core clock genes, these circadian modulators also control the expression of various clock-controlled genes (Ccgs) which includes several key regulators of cell cycle and metabolism.

PHOTIC INPUTS TO SUPRACHIASMATIC NUCLEUS

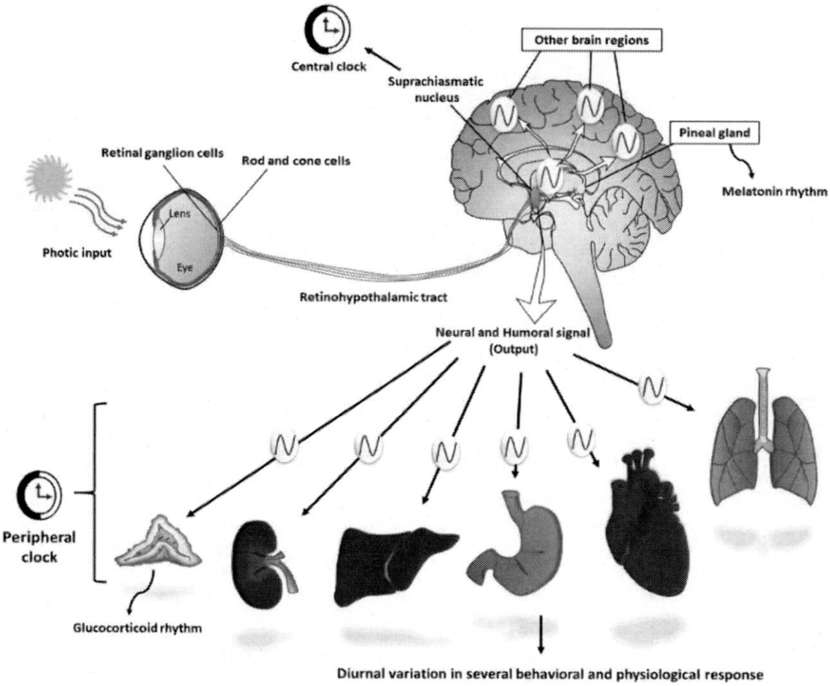

Figure 3. The schematic representation of Photic synchronization of the central and peripheral clock. Light information from the external recurrent light-dark cycle is transmitted to the suprachiasmatic nucleus via ganglions of the retinohypothalamic tract. The temporal information within the central clock relayed to peripheral tissue through neural (autonomic) or humoral (circulating or hormonal) pathways for the rhythmic expression of clock genes. Perfect alignment between central and peripheral clock with external light-dark cycle elicits behavioral and physiological response in a time-dependent manner.

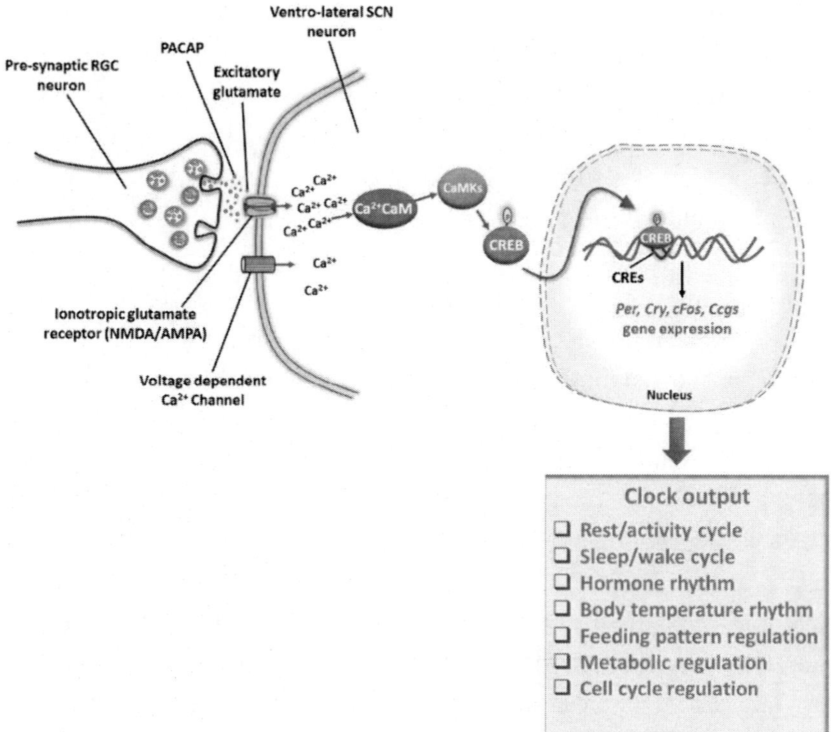

Figure 4. A schematic diagram showing light-induced signaling pathway between the axonal end of retinohypothalamic tract and neurons of core or ventral part of SCN. Light-induced release of glutamate is the main signal for photic entrainment of the central clock. Binding of glutamate to the NMDA or AMPA receptor present on the membrane of neurons of core SCN cause influx of Ca2+ ion. The increase in Ca2+ level activates downstream signaling pathways leading to the activation of several kinases. The phosphorylation and activation of nuclear transcription factor CREB by an enzyme CAMKII, an isoform of a Ca2+/calmodulin (Ca2+CaM) dependent protein kinase. The nuclear translocation of phosphorylated CREB allow its binding to the CRE region present on the promoter region of the target genes and induce their expression.

Light is the most potent environmental cue for the entrainment of endogenous circadian pacemaker that allows it to synchronize most of the behavioral and physiological processes with the 24 h environment cycle (Roenneberg and Foster 1997).

In mammals, RHT appears to play a major role in communicating photic information from the external environment to the hypothalamic SCN (Hattar et al. 2006). The classical report demonstrates that the destruction of RHT innervation abolishes circadian entrainment to the LD cycle (Johnson et al. 1988). The RHT which arises from retinal ganglion cells innervates most of its axons to the core region of SCN also referred to as the ventrolateral area which is characterized by neurons that express vasoactive intestinal polypeptide and gastrin-releasing peptide. In addition to that, the smaller proportion of RHT axons reach the SCN shell also termed as the dorsomedial region which contains arginine vasopressin (AVP) and calretinin neurons (Moga and Moore 1997; Morin 2007; Yan et al. 2007). The photoreceptor cells of the retina include the rod, cones and the most importantly melanopsin-containing retinal ganglion cells which relay photic information to SCN for circadian-related functions (Hattar et al. 2003; Panda et al. 2003). Melanopsin-containing retinal ganglion cells also referred to as intrinsically photosensitive retinal ganglion cells (ipRGCs) having a dual role as photoreceptors and retinal ganglion cells (Hattar et al. 2002). The extrinsic response of light input from rod or cone cells is also communicated to ipRGCs for light-induced circadian response mediated by pigment melanopsin (Wong et al. 2007; Panda et al. 2002). In absence of rod and cone function, the ipRGCs are sufficient to induce a circadian response to light in blind humans with intact inner retina (Freedman et al. 1999; Zaidi et al. 2007).

The principal neurotransmitter through which RHT transmits light information from the retina to SCN is glutamate, an excitatory amino acid (Castel et al. 1993; Ebling 1996). Another neuropeptide the PACAP is secreted along with glutamate at synapse which modifies the action of glutamate (Hannibal et al. 2000). Upon synaptic release of glutamate and PACAP depolarizes the membrane of SCN neurons by binding with N-methyl-D-aspartate (NMDA) and α-amino-3-hydroxy-5-methyl-4-isoxazolepropionic acid (AMPA) receptor (Schurov et al. 1999). Upon neurotransmitter binding, activation of voltage-sensitive calcium ion channel cause an influx of calcium ion (Gillette and Tischkau 1999; Meijer and Schwartz 2003). The rise in calcium level activates a group of kinases like calcium/calmodulin-dependent protein kinase (CaMK), mitogen-activated protein kinase and protein kinase A which then phosphorylates and activates nuclear transcription factor cyclic-AMP response element binding protein (CREB). CREB along with CREB binding protein (CBP) binds to the cyclic-AMP response element (CRE) region of clock genes and induces their transcription (Shearman et al. 1997; Carlezon et al. 2005). Nocturnal light pulse mediated neuronal firing is associated with CREB phosphorylation (Ser133 and Ser142) and *Per* gene expression (Schwartz et al. 2005; Shigeyoshi et al. 1997; Reppert and weaver 2001). Light pulse during night cause signaling in the ventrolateral region of SCN which includes expression of *cFos* (immediate early gene), *Per1* and *Per2* gene. However, dorsomedial SCN does not get stimulated upon nocturnal light pulse but involve in the circadian pattern of those gene expressions (Hamada et al. 2004; Yan and Okamura 2002; Yan and Silver 2004).

Circadian Rhythm Disruption on Melatonin and Glucocorticoid Rhythm

Melatonin, N-acetyl-5-methoxy tryptamine, a hormone synthesized and secreted primarily by the pineal gland and also by the retina and some other regions like the gastrointestinal system, bone marrow, blood platelets, skin etc. Circadian rhythmicity of melatonin release is generated by SCN in response to LD cycle with its peak level during the night (02:00 – 04:00) and low during the daytime (12:00 – 18:00) (Wurtman et al. 1963; Reiter 1993). SCN controlled melatonin rhythm is mediated by a multi-synaptic pathway that arises from SCN and transmitted sequentially to paraventricular nuclei (PVN), intermediolateral nucleus of the spinal cord, superior cervical ganglia that terminate in the release of noradrenaline from fibers of postganglionic superior cervical ganglia in the pineal gland (Klein and Moore 1979; Borjigin et al. 1999). The glutaminergic synaptic output from SCN to autonomic PVN neurons stimulates pineal melatonin synthesis during the night whereas during daytime blocking of glutaminergic inputs to PVN occur via GABAergic synaptic input from the SCN terminal (Hermes et al. 1996; Buijs et al. 2003; Perreau-Lenz et al. 2004). The rhythm of melatonin varies greatly by abrupt changes in the normal LD cycle and eventually disrupts melatonin influenced several biological functions.

The exposure of light produces a phase-shift effect on melatonin rhythm and acutely inhibits its synthesis (Shanahan and Czeisler 1991). Administration of exogenous melatonin promotes sleep in humans and reduces sleep onset latency during daytime (Hughes and Badia 1997; Rajaratnam et al. 2004; Wyatt et al. 2006) and thus, melatonin plays an important role in the regulation of sleep/wake cycle. A significant reduction in melatonin release after light pulse during the night has been seen in rats and other organisms including humans (Illnerova et al. 1979; Lewy et al. 1980). Both the nocturnal and diurnal rodents are known to have a similar inhibitory effect of nocturnal light exposure on melatonin secretion (Trinder et al. 1996; Aoki et al. 2001; Kumari et al. 2021). Nocturnal light-mediated reduction in melatonin level is dependent on both intensity and wavelength of light. Light of shorter wavelength of range 460-480nm (blue light) effectively reduces melatonin release by pineal (Brainard et al. 2001; Thapan et al. 2001). Since external light information to SCN is relayed through the retina, degeneration of the retina abolishes melatonin rhythm and its nocturnal suppression (Stephan and Zucker 1972). Apart from SCN neurons, the melatonin receptor is also located on several peripheral tissues such as the adrenal gland, kidney, heart, gonads etc. (Slominski et al. 2012). The phase shift effect of melatonin is due to the presence of its receptor in SCN and thus, melatonin resets the oscillatory pattern of the central clock upon nocturnal light exposure (Boivin et al. 1996; Zeitzer et al. 2000).

Glucocorticoid mainly includes cortisol in primates and corticosterone in rodents, secreted by adrenocortical steroidogenic cells that follow circadian rhythmicity whose secretion is governed by the Hypothalamic-pituitary-adrenal axis (HPA). SCN controls PVN to secrete corticotrophin releasing hormone which in turn stimulates the anterior pituitary to release ACTH.

ACTH then stimulates the adrenal cortex for the production and secretion of glucocorticoid (Whitnall et al. 1985; Chrousos 1998; Jacobson 2005). Thus, the SCN-HPA axis forms the basis for the circadian release of glucocorticoid with its peak level during the active phase of an organism i.e., the circadian peak of glucocorticoid occurs during early morning in diurnal and an early night in nocturnal animals (Perlow et al. 1981; Cheifetz 1971; Klemcke et al. 1989; Van Cauter 1990). The AVP, a hormone produced by SCN neurons put a negative effect on glucocorticoid release by the adrenal cortex (Buijs et al. 1999). In general, the synthesis of glucocorticoid occurs as an ultradian pulse with a frequency of about 1-2 per hour, due to alternating activation and inhibition of the HPA axis (Holaday et al. 1977; Lightman et al. 2008). Apart from stress regulation, the widespread effects of glucocorticoid on the body include lipid metabolism, gluconeogenesis, reproduction, cardiovascular function, inflammation, immune and cognitive modulation (Leliavski et al. 2015; Sapolsky et al. 2000).

Glucocorticoid act on target peripheral tissue by binding to glucocorticoid receptor and thus helps in maintaining normal physiological functions. Any desynchrony associated with altered LD cycle leads to internal misalignment with the central clock which eliminates the normal biological rhythm of glucocorticoid leading to several physiological and metabolic imbalance in the body (So et al. 2009; Lamia et al. 2011). In hamster, exposure to 5 lux intensities of nocturnal light flattened the endogenous cortisol rhythm (Bedrosian et al. 2013a). Furthermore, an increased level of circulating corticosterone has been reported in mice and grass rats exposed to dim light at night (Fonken et al. 2012; Martynhak et al. 2011). Light-induced corticosterone response is not observed in SCN-lesioned mice which show the involvement of SCN in maintaining and modulating glucocorticoid release from the adrenal gland (Ishida et al. 2005). Disruption of sleep-in shift/night work leads to baseline elevation of cortisol level and eventually causes hypertension and hyperglycemia (Niu et al. 2011; Dc Bacquer et al. 2009).

Influence of Circadian Rhythm Disruption on Mental Health

The disruption of circadian rhythm by the prolonged accumulation of irregular LD cycles is the major hallmark of many psychological disorders characterized by sleep disruption. It is known to be critical for the diagnosis of several mood-related disorders including major depressive disorder and bipolar disorder (Hickie et al. 2013; Rosenwasser and Wirz-Justice 1997). Nowadays with the increase in industrialization and modernization majority of the population follow night shift schedule to fulfill their need which distorts their circadian homeostasis. Untimed lighting condition is associated with several mood-related disorders like depression and mania. One such condition can be observed in seasonal affective disorder where short and long photoperiod associated with winter and summer, respectively, evoke mental disorders characterized by anxiety, cognitive impairment and anhedonic response (Lambert et al. 2015). These photoperiod conditions disrupt the normal sleep/wake cycle and in due course cause disruption in circadian rhythmicity in endocrine functions, metabolism, physiological functions etc. (Buijs et al. 2003). Such desynchrony can also be seen in workers engaged in different shift/night work schedule and also in travelers of the frequent transmeridian journey and thus, plays a crucial role in the progression of major depressive disorders (Scott et al. 1997; Inder et al. 2016).

Optimal functioning of the body requires coordinated regulation of SCN with its peripheral clock from biochemical to behavioral level however, the detailed mechanism regarding how nighttime activity brings deleterious effect on mood is still unclear. Increased depression-like behavior has been reported in mice devoid of melatonin (Brown et al. 1987). Since light during the night does not affect the sleep of the nocturnal animals as this period is considered as an active period for them (Borniger et al. 2013) but for diurnal rodents, the light at night disrupts sleep and responsible for impaired mood and cognition as evidence from the study done in *Avicanthis niloticus* (Fonken et al. 2012). Similarly, another study where the light at night of 5 lux similar to electronic gadgets (computers, e-readers and television) that humans are usually exposed are provided to female Siberian hamsters showed impaired locomotor activity pattern, decreased BDNF expression with reduced CA1 dendritic spine density and depression-like behavior (Bedrosian et al. 2013b). A recent study has shown the deleterious impact of nocturnal light exposure on the mood of both a nocturnal and a diurnal rodent where prolonged constant light exposure leads to anhedonia-like behavior (inability to feel pleasure).

Furthermore, in response to constant light illumination, these rodents have shown a reduction in antioxidant enzyme system and enhancement in proinflammatory cytokine in the hippocampus (Kumari et al. 2021). Impaired circadian pattern of neurotransmission activity of certain monoamine neurotransmitters like serotonin, dopamine and epinephrine, associated with nocturnal light exposure results in depressive symptoms (Hampp et al. 2008). The influence of circadian rhythm over limbic brain regions, monoamine neurotransmitters and the HPA axis and its disruption by untimed or irregular light exposure becomes the possible reason that contributes to increasing the occurrence of mood-related disorders (Nader et al. 2010; Li et al. 2013; Lambert et al. 2015).

CIRCADIAN RHYTHM DISRUPTION ON METABOLISM

Daily regulation and maintenance of body metabolism require circadian clock dependent oscillation of energy metabolites levels in the body which includes feeding time, energy storage and breakdown (Dibner and Schibler 2015; Hastings et al. 2007). In addition to the sleep/wake and feeding/fasting cycle, circadian oscillation of the clock also influences rhythmic change in the downstream molecular and physiological pathway that involves nutrition and metabolism (Ramsey and Bass 2009). Direct regulation of metabolic pathways by circadian clock occurs by controlling the expression of enzymes that regulate the rate-limiting steps of the metabolic pathway while indirect regulation involves clock-controlled influence on the release of endocrine factors which influences metabolically active tissues in a time-of-day dependent manner (Brown et al. 1979; Nakahata et al. 2009). The evidence to support the role of circadian control of metabolic process came from genetic mutations in the core components of the circadian clock leading to severe metabolic consequences. For example, obesity and subsequent progression to metabolic syndromes such as hyperlipidemia, hyperleptinemia and visceral adiposity have been observed in mice harboring a mutation in the clock genes which encode transcription factor that affects persistence and period of circadian rhythm (Marcheva, et al. 2010; Turek et al. 2005). Similarly, a *Per2*$^{-/-}$ mice display abnormal feeding pattern and gain more weight than control mice when fed with a high-fat diet (Yang et al. 2009).

Time-dependent oscillation in the glucose metabolism observed in both humans and rodent models. Maintenance of proper glucose homeostasis requires coordinated regulation of exogenous and endogenous glucose input and their disposal in which the circadian clock plays a major role by sustaining a balance between gluconeogenesis and glycogen turnover (Sollberger 1964). The circadian pattern of insulin secretion by beta cells of the pancreas and insulin sensitivity is also responsible for maintaining glucose homeostasis. Impaired insulin secretion and pancreatic cell proliferation were observed in *Clock* mutant mice which subsequently lead to hyperglycemia and hypoinsulinemia condition (Marcheva et al. 2010). The level of blood glucose and its transporter i.e., GLUT 2, GLUT 5 and sodium-glucose cotransporter in the intestine varies throughout the day with peak level during the initial phase of the active period and thus, maximum absorption of glucose occur during the middle of the active period (Balakrishnan et al. 2008; Pan and Hussain 2009; Iwashina et al. 2011).

Ablation of SCN and degeneration of autonomic tract connecting SCN to the liver, completely abolish glucose rhythm and insulin sensitivity (la Fleur et al. 2001). Alpha-amylase secreted by the salivary (parotid) gland and pancreas exhibit diurnal variation in its activity and is responsible for the hydrolysis of starch (polysaccharides) into simpler di- or monosaccharides. Since, carbohydrate preference is associated with the initial phase of the day and thus, the activity of alpha-amylase starts

increasing with the start of the light period with peak level found to be in the middle of the light phase irrespective of diurnality or nocturnality of an organism (Leibowitz et al. 1989; Girard-Globa et al. 1980; Bellavia et al. 1992). These all processes are anticipated with increase expression of GLUT expression in the intestinal epithelial cells (la Fleur et al. 2001; Pan and Hussain 2009). Timing of clock genes expression in the gut (intestine) is anticipated with the time of feeding in food entrained animals. The biological action of the circadian clock and food consumption is in accordance to get enough energy to fulfill body needs and to avoid excessive weight loss.

Circadian rhythm in protein and enzyme metabolism plays a crucial role to carry out whole biological functions in the time-dependent module. Albumin, a serum protein synthesized and secreted by the liver follows 24 h rhythms with a peak during the morning period i.e., between 9:00-10:00 (Hardeland et al. 1973). Clock protein dependent regulation of transport protein in intestine involves gene and protein expression on intestinal lumen with a peak during night period (active phase) and is independent of time of meal consumed (Pan and Hussain 2009). Amino acids which are utilized for various purposes by the cell include de novo nucleic acid synthesis and incorporation into a polypeptide chain during translational process exhibits circadian rhythm. The level of amino acid in blood shows circadian fluctuation with highest during noon to night and lowest during the midnight to early morning (Feigin et al. 1967). The plasma level of tyrosine, tryptophan and phenylalanine shows the highest-level during day time and lowest during the night (Wurtman et al. 1968).

Most of the enzymes in the liver exhibit change in activity in a circadian fashion, influence by the external LD cycle and feeding time. Time-dependent change in the activity of several liver enzymes such as isocitrate dehydrogenase, alcohol dehydrogenase, glutathione reductase, glyoxylate reductase, L- alanine aminotransferase, glutamate oxaloacetate transaminase, pyruvate decarboxylase, fructose-1-phosphate aldolase, fructose diphosphate aldolase, fructose-1, 6-diphosphatase, fatty acid synthase, with enhanced activity associated with the start of dark phase or the initiation of activity in nocturnal animal whereas brain enzymes such as phosphogluco isomerase, adenosine triphosphatase, creatine phosphokinase, pyruvate kinase, adenylate kinase, malate dehydrogenase, shows peak activity during beginning of light or inactive period of nocturnal animal (North et al. 1981). N-acetyltransferase (NAT), an essential pineal enzyme responsible for melatonin synthesis shows circadian rhythm with higher activity during the night and low during day time. Light pulse at night leads to a significant decrease in the activity of NAT which ultimately lowers melatonin levels (Vollrath and Huesgen 1988; Hamm and Menaker 1980).

Proper functioning of core clock genes is crucial for the regulation of gene expression required for lipid metabolism, adipogenesis in mature adipocytes and adipose leptin production (Kalsbeek et al. 2001). SCN ablation distorts circadian leptin secretion, suggesting the direct control of the central clock on leptin expression (Simon et al. 1998; Ahima et al. 1998). *Per2 and Bmal1* mutant mice show abnormal lipid metabolism characterized by increase adiposity with disrupted diurnal feeding behavior (Lamia et al. 2008; Yang et al. 2009). Sleep disruption in homozygous *Clock* delta mutant mice becomes obese and hyperphagic and ultimately leads to hypertriglyceridemia condition (Turek et al. 2005).

Evidence from epidemiological studies suggests a possible correlation between metabolic syndrome and shift work accompanied by light exposure at night (Karlsson et al. 2001). Disruption of circadian rhythm is associated with a risk factor for promoting cardiovascular-related disorder accompanying several metabolic syndromes (Ashby and Louis 2019; Antunes et al. 2010). Mice exhibited loss of internal synchrony when exposed to 20 h LD cycle which resulted in negative consequences on physiology including high insulin and leptin level, altered body temperature, weight gain and obesity (Karatsoreos et al. 2011). Modernization associated with a changed lifestyle leads to an increase in the prevalence of type I and type II diabetes mellitus worldwide. Numerous reports indicate chronodisruption in form of the night shift, rotating shift or regular day/night shift leads to an increase in the rate of diabetes when compared to individuals not involved in shift work (Gan et al. 2015; Pan et al. 2011). Often, night/shift work accompanied with sleep disruption accounts for the increased incidence of obesity and type II diabetes mellitus (Suwazono et al. 2009; Nedeltcheva and Scheer 2014). Rat exposed to constant light become more susceptible to beta-cell dysfunction and develops type II diabetes, the most common form of diabetes.

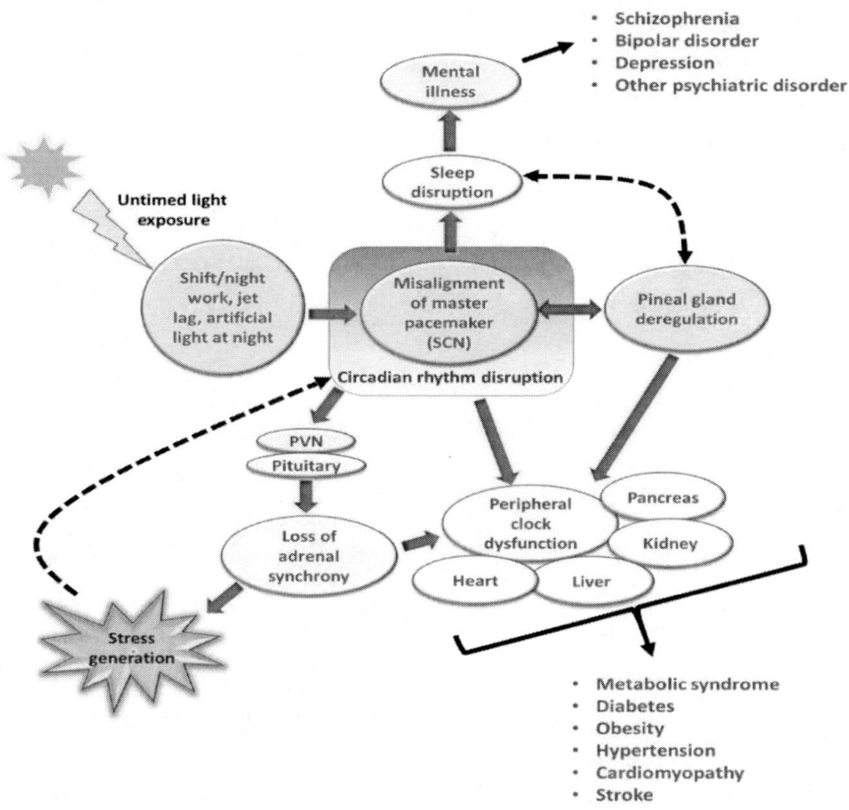

Figure 5. The diagram representing pathological conditions associated with circadian rhythm disruption induced by alterations in the external light-dark cycle. The light-induced disruption of the central clock and its interactions with the peripheral clock as well as other brain regions leading to abnormal endocrine rhythm, altered mood associated behaviors and metabolic syndrome.

Diminished glucose-stimulated insulin secretion with disturbed islet circadian clock also seen in rats exposed to 10 weeks of constant light (Qian et al. 2013). Women engaged in an extended period of night shift (three-night shift/month) showed an increased risk of type II diabetes (Pan et al. 2011). Another study on nurses of rotational night shift showed impaired beta-cell responsivity for insulin synthesis and secretion in response to postprandial glucose concentration (Sharma et al. 2017). Workers associated with the night shift schedule show a high level of cholesterol and triglyceride, low level of HDL cholesterol and increased body mass index and thus become more prone to cardiovascular diseases (Di Lorenzo et al. 2003; Buchvold et al. 2018; Ha and Park 2005). Mice exposed to constant illumination of bright light exhibit glucose intolerance and increased body weight compared with control mice exposed to standard LD cycle (Fonken et al. 2010). Thus, light in the photoperiod skeleton of the 12:12 h LD cycle is essential for proper synchronization of the central and peripheral clock and this stable phase establishment is required for maintaining metabolic balance in our body.

CONCLUSION

Mammals possess an endogenous timing system in the form of a bilobed structure, the SCN which resides in the anterior hypothalamic region of the brain. SCN allows the organism to synchronize the behavioral and physiological rhythm according to the external environmental LD cycle. The molecular oscillation of clock genes (*Per* and *Cry*) within SCN in form of transcriptional-translational feedback loop brings about rhythmic changes in the downstream molecular cascade and thus, mediates the circadian rhythm of multiple behavioral and physiological parameters. The two important endocrine manifestations of circadian rhythm include the rhythm of melatonin and glucocorticoids which show a recurrent change in its level according to the LD cycle. The ipRGCs of the inner layer of the retina relays external light information to SCN for proper alignment of the central clock and peripheral clock to the external LD cycle. The altered temporal phase relationship of endogenous biological clock with external LD cycle leads to disruption of stabilized circadian rhythm leading to several health consequences. The use of artificial light at night and the increased demand of modern society encourages humans to follow 24 h work schedule. An altered lifestyle and traditional work schedule are associated with untimed light exposure, frequent transmeridian travel, jet lag and shift/night work. Shift/night work and light exposure at night has been shown to produce disrupted circadian rhythm characterized by abnormal sleep/wake or rest/activity cycle and altered hormone rhythm, ultimately affecting the mental health and metabolic processes. The current information provides the interactive relationship of mental and metabolic consequences with that of hormone rhythm. Detail knowledge is required to get the exact molecular mechanism responsible for such health consequences and to establish methods that reduce detrimental effects of light to provide better health and well beings in modern societies.

ACKNOWLEDGMENTS

Financial support by the UGC, New Delhi (No.6-2/2017/IC) to MS is greatly acknowledged. CSIR JRF & SRF to VV (File No. 09/013/(0717)/2017-EMR-I) and CSIR-UGC JRF & SRF to RK (UGC Ref. No. 992/(SC) (CSIR-UGC NET DEC. 2016) is acknowledged.

REFERENCES

Ahima RS, Prabakaran D, Flier JS (1998) Postnatal leptin surge and regulation of circadian rhythm of leptin by feeding. Implications for energy homeostasis and neuroendocrine function. *J Clin Invest* 101(5): 1020-1027.

Akashi M, Takumi T (2005) The orphan nuclear receptor ROR alpha regulates circadian transcription of the mammalian core-clock Bmal1. *Nat Struct Mol Biol* 12(5): 441-448.

Akerstedt T, Wright KP Jr (2009) Sleep Loss and Fatigue in Shift Work and Shift Work Disorder. *Sleep Med Clin* 4(2): 257-271.

Antunes LC, Levandovski R, Dantas G, Caumo W, Hidalgo MP (2010) Obesity and shift work: chronobiological aspects. *Nutr Res Rev* 23(1): 155-168.

Aoki H, Ozeki Y, Yamada N (2001) Hypersensitivity of melatonin suppression in response to light in patients with delayed sleep phase syndrome. *Chronobiol Int* 18(2): 263-271.

Aschoff J, Fatranska M, Giedke H, Doerr P, Stamm D, Wisser H (1971) Human circadian rhythms in continuous darkness: entrainment by social cues. *Science* 171(3967): 213-215.

Ashby T, Louis M (2019) Circadian misalignment and cardiovascular risk. *Cardiovasc Innov Appl* 3(4): 435-440.

Balakrishnan A, Stearns AT, Rounds J, Irani J, Giuffrida M, Rhoads DB, Ashley SW, Tavakkolizadeh A (2008) Diurnal rhythmicity in glucose uptake is mediated by temporal periodicity in the expression of the sodium-glucose cotransporter (SGLT1). *Surgery* 143(6): 813-818.

Balsalobre A, Marcacci L, Schibler U (2000) Multiple signaling pathways elicit circadian gene expression in cultured Rat-1 fibroblasts. *Curr Biol* 10(20): 1291-1294.

Bedrosian TA, Galan A, Vaughn CA, Weil ZM, Nelson RJ (2013a) Light at night alters daily patterns of cortisol and clock proteins in female Siberian hamsters. *J Neuroendocrinol* 25(6): 590-596.

Bedrosian TA, Nelson RJ (2017) Timing of light exposure affects mood and brain circuits. *Transl Psychiatry* 7(1): e1017.

Bedrosian TA, Weil ZM, Nelson RJ (2013b) Chronic dim light at night provokes reversible depression-like phenotype: possible role for TNF. *Mol Psychiatry* 18(8): 930-936.

Bellavia SL, Sanz EG, Sereno R, Vermouth NT (1992) Alpha-amylase circadian rhythm of young rat parotid gland: an endogenous rhythm with maternal coordination. *Arch Oral Biol* 37(5): 429-433.

Boivin DB, Duffy JF, Kronauer RE, Czeisler CA (1996) Dose-response relationships for resetting of human circadian clock by light. *Nature* 379(6565): 540-542.

Borjigin J, Li X, Snyder SH (1999) The pineal gland and melatonin: molecular and pharmacologic regulation. *Annu Rev Pharmacol Toxicol* 39: 53-65.

Borniger JC, Weil ZM, Zhang N, Nelson RJ (2013) Dim light at night does not disrupt timing or quality of sleep in mice. *Chronobiol Int* 30(8): 1016-1023.

Brainard GC, Hanifin JP (2005) Photons, clocks, and consciousness. *J Biol Rhythms* 20(4): 314-325.

Brainard GC, Hanifin JP, Greeson JM, Byrne B, Glickman G, Gerner E, Rollag MD (2001) Action spectrum for melatonin regulation in humans: evidence for a novel circadian photoreceptor. *J Neurosci* 21(16): 6405-6412.

Brown MS, Goldstein JL, Dietschy JM (1979) Active and inactive forms of 3-hydroxy-3-methylglutaryl coenzyme A reductase in the liver of the rat. Comparison with the rate of cholesterol synthesis in different physiological states. *J Biol Chem* 254(12): 5144-5149.

Brown RP, Kocsis JH, Caroff S, Amsterdam J, Winokur A, Stokes P, Frazer A (1987) Depressed mood and reality disturbance correlate with decreased nocturnal melatonin in depressed patients. *Acta Psychiatr Scand* 76(3): 272-275.

Buchvold HV, Pallesen S, Waage S, Bjorvatn B (2018) Shift work schedule and night work load: Effects on body mass index - a four-year longitudinal study. *Scand J Work Environ Health* 44(3): 251-257.

Buijs RM, van Eden CG, Goncharuk VD, Kalsbeek A (2003) The biological clock tunes the organs of the body: timing by hormones and the autonomic nervous system. *J Endocrinol* 177(1): 17-26.

Buijs RM, Wortel J, Van Heerikhuize JJ, Feenstra MG, Ter Horst GJ, Romijn HJ, Kalsbeek A (1999) Anatomical and functional demonstration of a multisynaptic suprachiasmatic nucleus adrenal (cortex) pathway. *Eur J Neurosci* 11(5): 1535-1544.

Camacho F, Cilio M, Guo Y, Virshup DM, Patel K, Khorkova O, Styren S, Morse B, Yao Z, Keesler GA (2001) Human casein kinase Idelta phosphorylation of human circadian clock proteins period 1 and 2. *FEBS Lett* 489(2-3): 159-165.

Carlezon WA Jr, Duman RS, Nestler EJ (2005) The many faces of CREB. *Trends Neurosci* 28(8): 436-445.

Castel M, Belenky M, Cohen S, Ottersen OP, Storm-Mathisen J (1993) Glutamate-like immunoreactivity in retinal terminals of the mouse suprachiasmatic nucleus. *Eur J Neurosci* 5(4): 368-381.

Cheifetz PN (1971) The daily rhythm of the secretion of corticotrophin and corticosterone in rats and mice. *J Endocrinol* 49(3): xi-xii.

Chen D, Buchanan GF, Ding JM, Hannibal J, Gillette MU (1999) Pituitary adenylyl cyclase-activating peptide: a pivotal modulator of glutamatergic regulation of the suprachiasmatic circadian clock. *Proc Natl Acad Sci USA* 96(23): 13468-13473.

Chrousos GP (1998) Ultradian, circadian, and stress-related hypothalamic-pituitary-adrenal axis activity--a dynamic digital-to-analog modulation. *Endocrinology* 139(2): 437-440.

Dardente H, Cermakian N (2007) Molecular circadian rhythms in central and peripheral clocks in mammals. *Chronobiol Int* 24(2): 195-213.

Davidson AJ, Castanon-Cervantes O, Leise TL, Molyneux PC, Harrington ME (2009) Visualizing jet lag in the mouse suprachiasmatic nucleus and peripheral circadian timing system. *Eur J Neurosci* 29(1): 171-180.

Davis S, Mirick DK, Stevens RG (2001) Night shift work, light at night, and risk of breast cancer. *J Natl Cancer Inst* 93(20): 1557-1562.

De Bacquer D, Van Risseghem M, Clays E, Kittel F, De Backer G, Braeckman L (2009) Rotating shift work and the metabolic syndrome: a prospective study. *Int J Epidemiol* 38(3): 848-854.

Di Lorenzo L, De Pergola G, Zocchetti C, L'Abbate N, Basso A, Pannacciulli N, Cignarelli M, Giorgino R, Soleo L (2003) Effect of shift work on body mass index: results of a study performed in 319 glucose-tolerant men working in a Southern Italian industry. *Int J Obes Relat Metab Disord* 27(11): 1353-1358.

Dibner C, Schibler U (2015) Circadian timing of metabolism in animal models and humans. *J Intern Med* 277(5): 513-527.

Doi M, Hirayama J, Sassone-Corsi P (2006) Circadian regulator clock is a histone acetyltransferase. *Cell* 125(3): 497-508.

Ebling FJ (1996) The role of glutamate in the photic regulation of the suprachiasmatic nucleus. *Prog Neurobiol* 50(2-3): 109-132.

Eide EJ, Woolf MF, Kang H, Woolf P, Hurst W, Camacho F, Vielhaber EL, Giovanni A, Virshup DM (2005) Control of mammalian circadian rhythm by CKIepsilon-regulated proteasome-mediated PER2 degradation. *Mol Cell Biol* 25(7): 2795-2807.

Etchegaray JP, Lee C, Wade PA, Reppert SM (2003) Rhythmic histone acetylation underlies transcription in the mammalian circadian clock. *Nature* 421(6919): 177-182.

Fahrenkrug J (2006) PACAP--a multifacetted neuropeptide. *Chronobiol Int* 23(1-2): 53-61.

Feigin RD, Klainer AS, Beisel WR (1967) Circadian periodicity of blood amino-acids in adult men. *Nature* 215: 512–514.

Filipski E, Subramanian P, Carriere J, Guettier C, Barbason H, Levi F (2009) Circadian disruption accelerates liver carcinogenesis in mice. *Mutat Res* 680(1-2): 95-105.

Fonken LK, Kitsmiller E, Smale L, Nelson RJ (2012) Dim nighttime light impairs cognition and provokes depressive-like responses in a diurnal rodent. *J Biol Rhythms* 27(4): 319-327.

Fonken LK, Workman JL, Walton JC, Weil ZM, Morris JS, Haim A, Nelson RJ (2010) Light at night increases body mass by shifting the time of food intake. *Proc Natl Acad Sci USA* 107(43): 18664-18669.

Freedman MS, Lucas RJ, Soni B, von Schantz M, Munoz M, David-Gray Z, Foster R (1999) Regulation of mammalian circadian behavior by non-rod, non-cone, ocular photoreceptors. *Science* 284(5413): 502-504.

Gan Y, Yang C, Tong X, Sun H, Cong Y, Yin X, Li L, Cao S, Dong X, Gong Y, Shi O, Deng J, Bi H, Lu Z (2015) Shift work and diabetes mellitus: a meta-analysis of observational studies. *Occup Environ Med* 72(1): 72-78.

Gekakis N, Staknis D, Nguyen HB, Davis FC, Wilsbacher LD, King DP, Takahashi JS, Weitz CJ (1998) Role of the CLOCK protein in the mammalian circadian mechanism. *Science* 280(5369): 1564-1569.

Gillette MU, Reppert SM (1987) The hypothalamic suprachiasmatic nuclei: circadian patterns of vasopressin secretion and neuronal activity in vitro. *Brain Res Bull* 19(1): 135-139.

Gillette MU, Tischkau SA (1999) Suprachiasmatic nucleus: the brain's circadian clock. *Recent Prog Horm Res* 54: 33-59.

Girard-Globa A, Bourdel G, Lardeux B (1980) Regulation of protein synthesis and enzyme accumulation in the rat pancreas by amount and timing of dietary protein. *J Nutr* 110(7): 1380-1390.

Griffin EA Jr, Staknis D, Weitz CJ (1999) Light-independent role of CRY1 and CRY2 in the mammalian circadian clock. *Science* 286(5440): 768-771.

Guillaumond F, Dardente H, Giguere V, Cermakian N (2005) Differential control of Bmal1 circadian transcription by REV-ERB and ROR nuclear receptors. *J Biol Rhythms* 20(5): 391-403.

Ha M, Park J (2005) Shiftwork and metabolic risk factors of cardiovascular disease. *J Occup Health* 47(2): 89-95.

Hamada T, Antle MC, Silver R (2004) Temporal and spatial expression patterns of canonical clock genes and clock-controlled genes in the suprachiasmatic nucleus. *Eur J Neurosci* 19(7): 1741-1748.

Hamm HE, Menaker M (1980) Retinal rhythms in chicks: circadian variation in melantonin and serotonin N-acetyltransferase activity. *Proc Natl Acad Sci USA* 77(8): 4998-5002.

Hampp G, Ripperger JA, Houben T, Schmutz I, Blex C, Perreau-Lenz S, Brunk I, Spanagel R, Ahnert-Hilger G, Meijer JH, Albrecht U (2008) Regulation of monoamine oxidase A by circadian-clock components implies clock influence on mood. *Curr Biol* 18(9): 678-683.

Hannibal J (2002) Neurotransmitters of the retino-hypothalamic tract. *Cell Tissue Res* 309(1): 73-88.

Hannibal J (2006) Roles of PACAP-containing retinal ganglion cells in circadian timing. *Int Rev Cytol* 251: 1-39.

Hannibal J, Møller M, Ottersen OP, Fahrenkrug J (2000) PACAP and glutamate are co-stored in the retinohypothalamic tract. *J Comp Neurol* 418(2): 147-155.

Hardeland R, Hohmann D, Rensing L (1973) The rhythmic organization of Rodent Liver a review. *Biol Rhythm Res* 4(2): 89-118.

Hastings M, O'Neill JS, Maywood ES (2007) Circadian clocks: regulators of endocrine and metabolic rhythms. *J Endocrinol* 195(2): 187-198.

Hastings MH, Reddy AB, Maywood ES (2003) A clockwork web: circadian timing in brain and periphery, in health and disease. *Nat Rev Neurosci* 4(8): 649-661.

Hattar S, Kumar M, Park A, Tong P, Tung J, Yau KW, Berson DM (2006) Central projections of melanopsin-expressing retinal ganglion cells in the mouse. *J Comp Neurol* 497(3): 326-349.

Hattar S, Liao HW, Takao M, Berson DM, Yau KW (2002) Melanopsin-containing retinal ganglion cells: architecture, projections, and intrinsic photosensitivity. *Science* 295(5557): 1065-1070.

Hattar S, Lucas RJ, Mrosovsky N, Thompson S, Douglas RH, Hankins MW, Lem J, Biel M, Hofmann F, Foster RG, Yau KW (2003) Melanopsin and rod-cone photoreceptive systems account for all major accessory visual functions in mice. *Nature* 424(6944): 76-81.

Hermes ML, Coderre EM, Buijs RM, Renaud LP (1996) GABA and glutamate mediate rapid neurotransmission from suprachiasmatic nucleus to hypothalamic paraventricular nucleus in rat. *J Physiol* 496(3): 749-757.

Hickie IB, Naismith SL, Robillard R, Scott EM, Hermens DF (2013) Manipulating the sleep-wake cycle and circadian rhythms to improve clinical management of major depression. *BMC Med* 11: 79.

Holaday JW, Martinez HM, Natelson BH (1977) Synchronized ultradian cortisol rhythms in monkeys: persistence during corticotropin infusion. *Science* 198(4312): 56-58.

Hughes RJ, Badia P (1997) Sleep-promoting and hypothermic effects of daytime melatonin administration in humans. *Sleep* 20(2): 124-131.

Illnerova H, Vanecek J, Krecek J, Wetterberg L, Saaf J (1979) Effect of one minute exposure to light at night on rat pineal serotonin N-acetyltransferase and melatonin. *J Neurochem* 32(2): 673-675.

Inder ML, Crowe MT, Porter R (2016) Effect of transmeridian travel and jetlag on mood disorders: evidence and implications. *Aust N Z J Psychiatry* 50(3): 220-227.

Ishida A, Mutoh T, Ueyama T, Bando H, Masubuchi S, Nakahara D, Tsujimoto G, Okamura H (2005) Light activates the adrenal gland: timing of gene expression and glucocorticoid release. *Cell Metab* 2(5): 297-307.

Iwashina I, Mochizuki K, Inamochi Y, Goda T (2011) Clock genes regulate the feeding schedule-dependent diurnal rhythm changes in hexose transporter gene expressions through the binding of BMAL1 to the promoter/enhancer and transcribed regions. *J Nutr Biochem* 22(4): 334-343.

Jacobson L (2005) Hypothalamic-pituitary-adrenocortical axis regulation. *Endocrinol Metab Clin North Am* 34(2): 271-292.

Johnson RF, Moore RY, Morin LP (1988) Loss of entrainment and anatomical plasticity after lesions of the hamster retinohypothalamic tract. *Brain Res* 460(2): 297-313.

Kalsbeek A, Fliers E, Romijn JA, La Fleur SE, Wortel J, Bakker O, Endert E, Buijs RM (2001) The suprachiasmatic nucleus generates the diurnal changes in plasma leptin levels. *Endocrinology* 142(6): 2677-2685.

Kalsbeek A, van der Vliet J, Buijs RM (1996) Decrease of endogenous vasopressin release necessary for expression of the circadian rise in plasma corticosterone: a reverse microdialysis study. *J Neuroendocrinol* 8(4): 299-307.

Karatsoreos IN, Bhagat S, Bloss EB, Morrison JH, McEwen BS (2011) Disruption of circadian clocks has ramifications for metabolism, brain, and behavior. *Proc Natl Acad Sci USA* 108(4): 1657-1662.

Karlsson B, Knutsson A, Lindahl B (2001) Is there an association between shift work and having a metabolic syndrome? Results from a population based study of 27,485 people. *Occup Environ Med* 58(11): 747-752.

Klein DC, Moore RY (1979) Pineal N-acetyltransferase and hydroxyindole-O-methyltransferase: control by the retinohypothalamic tract and the suprachiasmatic nucleus. *Brain Res* 174(2): 245-262.

Klemcke HG, Nienaber JA, Hahn GL (1989) Plasma adrenocorticotropic hormone and cortisol in pigs: effects of time of day on basal and stressor-altered concentrations. *Proc Soc Exp Biol Med* 190(1): 42-53.

Klerman EB, Rimmer DW, Dijk DJ, Kronauer RE, Rizzo JF 3rd, Czeisler CA (1998) Nonphotic entrainment of the human circadian pacemaker. *Am J Physiol* 274(4): R991-R996.

Ko CH, Takahashi JS (2006) Molecular components of the mammalian circadian clock. *Hum Mol Genet* 15: R271-277.

Kubo T, Ozasa K, Mikami K, Wakai K, Fujino Y, Watanabe Y, Miki T, Nakao M, Hayashi K, Suzuki K, Mori M, Washio M, Sakauchi F, Ito Y, Yoshimura T, Tamakoshi A (2006) Prospective cohort study of the risk of prostate cancer among rotating-shift workers: findings from the Japan collaborative cohort study. *Am J Epidemiol* 164(6): 549-555.

Kumari R, Verma V, Kronfeld-Schor N, Singaravel M (2021) Differential response of diurnal and nocturnal mammals to prolonged altered light-dark cycle: A possible role of mood associated endocrine, inflammatory and antioxidant system. *Chronobiol Int* 1-13.

Kume K, Zylka MJ, Sriram S, Shearman LP, Weaver DR, Jin X, Maywood ES, Hastings MH, Reppert SM (1999) mCRY1 and mCRY2 are essential components of the negative limb of the circadian clock feedback loop. *Cell* 98(2): 193-205.

la Fleur SE, Kalsbeek A, Wortel J, Fekkes ML, Buijs RM (2001) A daily rhythm in glucose tolerance: a role for the suprachiasmatic nucleus. *Diabetes* 50(6): 1237-1243.

Lambert KG, Nelson RJ, Jovanovic T, Cerda M (2015) Brains in the city: Neurobiological effects of urbanization. *Neurosci Biobehav Rev* 58: 107-122.

Lamia KA, Papp SJ, Yu RT, Barish GD, Uhlenhaut NH, Jonker JW, Downes M, Evans RM (2011) Cryptochromes mediate rhythmic repression of the glucocorticoid receptor. *Nature* 480(7378): 552-556.

Lamia KA, Storch KF, Weitz CJ (2008) Physiological significance of a peripheral tissue circadian clock. *Proc Natl Acad Sci USA* 105(39): 15172-15177.

Leibowitz SF, Weiss GF, Walsh UA, Viswanath D (1989) Medial hypothalamic serotonin: role in circadian patterns of feeding and macronutrient selection. *Brain Res* 503(1): 132-140.

Leliavski A, Dumbell R, Ott V, Oster H (2015) Adrenal clocks and the role of adrenal hormones in the regulation of circadian physiology. *J Biol Rhythms* 30(1): 20-34.

Lewy AJ, Wehr TA, Goodwin FK, Newsome DA, Markey SP (1980) Light suppresses melatonin secretion in humans. *Science* 210(4475): 1267-1269.

Li JZ, Bunney BG, Meng F, Hagenauer MH, Walsh DM, Vawter MP, Evans SJ, Choudary PV, Cartagena P, Barchas JD, Schatzberg AF, Jones EG, Myers RM, Watson SJ Jr, Akil H, Bunney WE (2013) Circadian patterns of gene expression in the human brain and disruption in major depressive disorder. *Proc Natl Acad Sci USA* 110(24): 9950-9955.

Li Y, Sato Y, Yamaguchi N (2011) Shift work and the risk of metabolic syndrome: a nested case-control study. *Int J Occup Environ Health* 17(2): 154-160.

Lightman SL, Wiles CC, Atkinson HC, Henley DE, Russell GM, Leendertz JA, McKenna MA, Spiga F, Wood SA, Conway-Campbell BL (2008) The significance of glucocorticoid pulsatility. *Eur J Pharmacol* 583(2-3): 255-262.

Lowrey PL, Takahashi JS (2004) Mammalian circadian biology: elucidating genome-wide levels of temporal organization. *Annu Rev Genomics Hum Genet* 5: 407-441.

Marcheva B, Ramsey KM, Buhr ED, Kobayashi Y, Su H, Ko CH, Ivanova G, Omura C, Mo S, Vitaterna MH, Lopez JP, Philipson LH, Bradfield CA, Crosby SD, JeBailey L, Wang X, Takahashi JS, Bass J (2010) Disruption of the clock components CLOCK and BMAL1 leads to hypoinsulinaemia and diabetes. *Nature* 466(7306): 627-631.

Martynhak BJ, Correia D, Morais LH, Araujo P, Andersen ML, Lima MM, Louzada, FM, Andreatini R (2011) Neonatal exposure to constant light prevents anhedonia-like behavior induced by constant light exposure in adulthood. *Behav Brain Res* 222(1):10-14.

Meijer JH, Schwartz WJ (2003) In search of the pathways for light-induced pacemaker resetting in the suprachiasmatic nucleus. *J Biol Rhythms* 18(3): 235-249.

Moga MM, Moore RY (1997) Organization of neural inputs to the suprachiasmatic nucleus in the rat. *J Comp Neurol* 389(3): 508-534.

Mohawk JA, Green CB, Takahashi JS (2012) Central and peripheral circadian clocks in mammals. *Annu Rev Neurosci* 35: 445-462.

Moore RY (1996) Entrainment pathways and the functional organization of the circadian system. *Prog Brain Res* 111: 103-119.

Morin LP (2007) SCN organization reconsidered. *J Biol Rhythms* 22(1): 3-13.

Nader N, Chrousos GP, Kino T (2010) Interactions of the circadian clock system and the HPA axis. *Trends Endocrinol Metab* 21(5): 277-286.

Nagoshi E, Saini C, Bauer C, Laroche T, Naef F, Schibler U (2004) Circadian gene expression in individual fibroblasts: cell-autonomous and self-sustained oscillators pass time to daughter cells. *Cell* 119(5): 693-705.

Nakahata Y, Sahar S, Astarita G, Kaluzova M, Sassone-Corsi P (2009) Circadian control of the NAD+ salvage pathway by CLOCK-SIRT1. *Science* 324(5927): 654-657.

Nedeltcheva AV, Scheer FA (2014) Metabolic effects of sleep disruption, links to obesity and diabetes. *Curr Opin Endocrinol Diabetes Obes* 21(4): 293-298.

Niu SF, Chung MH, Chen CH, Hegney D, O'Brien A, Chou KR (2011) The effect of shift rotation on employee cortisol profile, sleep quality, fatigue, and attention level: a systematic review. *J Nurs Res* 19(1): 68-81.

North C, Feuers RJ, Scheving LE, Pauly JE, Tsai TH, Casciano DA (1981) Circadian organization of thirteen liver and six brain enzymes of the mouse. *Am J Anat* 162(3): 183-199.

Okamura H, Yamaguchi S, Yagita K (2002) Molecular machinery of the circadian clock in mammals. *Cell Tissue Res* 309(1): 47-56.

Pan A, Schernhammer ES, Sun Q, Hu FB (2011) Rotating night shift work and risk of type 2 diabetes: two prospective cohort studies in women. *PLoS Med* 8(12): e1001141.

Pan X, Hussain MM (2009) Clock is important for food and circadian regulation of macronutrient absorption in mice. *J Lipid Res* 50(9): 1800-1813.

Panda S, Provencio I, Tu DC, Pires SS, Rollag MD, Castrucci AM, Pletcher MT, Sato TK, Wiltshire T, Andahazy M, Kay SA, Van Gelder RN, Hogenesch JB (2003) Melanopsin is required for non-image-forming photic responses in blind mice. *Science* 301(5632): 525-527.

Panda S, Sato TK, Castrucci AM, Rollag MD, DeGrip WJ, Hogenesch JB, Provencio I, Kay SA (2002) Melanopsin (Opn4) requirement for normal light-induced circadian phase shifting. *Science* 298(5601): 2213-2216.

Perlow MJ, Reppert SM, Boyar RM, Klein DC (1981) Daily rhythms in cortisol and melatonin in primate cerebrospinal fluid. Effects of constant light and dark. *J Neuroendocrinol* 32(4): 193-196.

Perreau-Lenz S, Kalsbeek A, Pevet P, Buijs RM (2004) Glutamatergic clock output stimulates melatonin synthesis at night. *Eur J Neurosci* 19(2): 318-324.

Preitner N, Damiola F, Lopez-Molina L, Zakany J, Duboule D, Albrecht U, Schibler U (2002) The orphan nuclear receptor REV-ERBalpha controls circadian transcription within the positive limb of the mammalian circadian oscillator. *Cell* 110(2): 251-260.

Qian J, Block GD, Colwell CS, Matveyenko AV (2013) Consequences of exposure to light at night on the pancreatic islet circadian clock and function in rats. *Diabetes* 62(10): 3469-3478.

Rajaratnam SM, Middleton B, Stone BM, Arendt J, Dijk DJ (2004) Melatonin advances the circadian timing of EEG sleep and directly facilitates sleep without altering its duration in extended sleep opportunities in humans. *J Physiol* 561(1): 339-351.

Ralph MR, Menaker M (1988) A mutation of the circadian system in golden hamsters. *Science* 241(4870): 1225-1227.

Ramsey KM, Bass J (2009) Obeying the clock yields benefits for metabolism. *Proc Natl Acad Sci USA* 106(11): 4069-4070.

Reiter RJ (1993) The melatonin rhythm: both a clock and a calendar. *Experientia* 49(8): 654-664.

Reppert SM, Weaver DR (2001) Molecular analysis of mammalian circadian rhythms. *Annu Rev Physiol* 63: 647-676.

Reppert SM, Weaver DR (2002) Coordination of circadian timing in mammals. *Nature* 418(6901): 935-941.

Ripperger JA, Schibler U (2006) Rhythmic CLOCK-BMAL1 binding to multiple E-box motifs drives circadian Dbp transcription and chromatin transitions. *Nat Genet* 38(3): 369-374.

Roenneberg T, Foster RG (1997) Twilight times: light and the circadian system. *Photochem Photobiol* 66(5): 549-561.

Rosenwasser AM, Wirz-Justice A (1997) Circadian rhythms and depression: Clinical and experimental models. In: Redfern PH, Lemmer B, editors. *Physiology and Pharmacology of Biological Rhythms*: Berlin, Springer, 457-486.

Sangoram AM, Saez L, Antoch MP, Gekakis N, Staknis D, Whiteley A, Fruechte EM, Vitaterna MH, Shimomura K, King DP, Young MW, Weitz CJ, Takahashi JS (1998) Mammalian circadian autoregulatory loop: a timeless ortholog and mPer1 interact and negatively regulate clock-bmal1-induced transcription. *Neuron* 21(5): 1101-1113.

Sapolsky RM, Romero LM, Munck AU (2000) How do glucocorticoids influence stress responses? Integrating permissive, suppressive, stimulatory, and preparative actions. *Endocr Rev* 21(1): 55-89.

Sato TK, Panda S, Miraglia LJ, Reyes TM, Rudic RD, McNamara P, Naik KA, FitzGerald GA, Kay SA, Hogenesch JB (2004) A functional genomics strategy reveals Rora as a component of the mammalian circadian clock. *Neuron* 43(4): 527-537.

Sato TK, Yamada RG, Ukai H, Baggs JE, Miraglia LJ, Kobayashi TJ, Welsh DK, Kay SA, Ueda HR, Hogenesch JB (2006) Feedback repression is required for mammalian circadian clock function. *Nat Genet* 38(3): 312-319.

Schroeder AM, Colwell CS (2013) How to fix a broken clock. *Trends Pharmacol Sci* 34(11): 605-619.

Schurov IL, McNulty S, Best JD, Sloper PJ, Hastings MH (1999) Glutamatergic induction of CREB phosphorylation and Fos expression in primary cultures of the suprachiasmatic hypothalamus in vitro is mediated by co-ordinate activity of NMDA and non- NMDA receptors. *J Neuroendocrinol* 11: 43–51.

Schwartz WJ, Aronin N, Sassone-Corsi P (2005) Photoinducible and rhythmic ICER-CREM immunoreactivity in the rat suprachiasmatic nucleus. *Neurosci Lett* 385: 87–91.

Scott AJ, Monk TH, Brink LL (1997) Shiftwork as a Risk Factor for Depression: A Pilot Study. *Int J Occup Environ Health* 3(2): 2-9.

Hickie TL, Czeisler CA (1991) Light exposure induces equivalent phase shifts of the endogenous circadian rhythms of circulating plasma melatonin and core body temperature in men. *J Clin Endocrinol Metab* 73(2): 227-235.

Sharma A, Laurenti MC, Dalla Man C, Varghese RT, Cobelli C, Rizza RA, Matveyenko A, Vella A (2017) Glucose metabolism during rotational shift-work in healthcare workers. *Diabetologia* 60(8): 1483-1490.

Shearman LP, Sriram S, Weaver DR, Maywood ES, Chaves I, Zheng B, Kume K, Lee CC, van der Horst GT, Hastings MH, Reppert SM (2000) Interacting molecular loops in the mammalian circadian clock. *Science* 288(5468): 1013-1019.

Shearman LP, Zylka MJ, Weaver DR, Kolakowski LF Jr, Reppert SM (1997) Two period homologs: circadian expression and photic regulation in the suprachiasmatic nuclei. *Neuron* 19(6): 1261-1269.

Shigeyoshi Y, Taguchi K, Yamamoto S, Takekida S, Yan L, Tei H, Moriya T, Shibata S, Loros JJ, Dunlap JC, Okamura H (1997) Light-induced resetting of a mammalian circadian clock is associated with rapid induction of the mPer1 transcript. *Cell* 91(7): 1043-1053.

Shirogane T, Jin J, Ang XL, Harper JW (2005) SCFbeta-TRCP controls clock-dependent transcription via casein kinase 1-dependent degradation of the mammalian period-1 (Per1) protein. *J Biol Chem* 280(29): 26863-26872.

Simon C, Gronfier C, Schlienger JL, Brandenberger G (1998) Circadian and ultradian variations of leptin in normal man under continuous enteral nutrition: relationship to sleep and body temperature. *J Clin Endocrinol Metab* 83(6): 1893-1899.

Slominski RM, Reiter RJ, Schlabritz-Loutsevitch N, Ostrom RS, Slominski AT (2012) Melatonin membrane receptors in peripheral tissues: distribution and functions. *Mol Cell Endocrinol* 351(2): 152-166.

So AY, Bernal TU, Pillsbury ML, Yamamoto KR, Feldman BJ (2009) Glucocorticoid regulation of the circadian clock modulates glucose homeostasis. *Proc Natl Acad Sci USA* 106(41): 17582-17587.

Sollberger A (1964) The Control of Circadian Glycogen Rhythms. *Ann N Y Acad Sci* (117): 519-554.

Stephan FK, Zucker I (1972) Circadian rhythms in drinking behavior and locomotor activity of rats are eliminated by hypothalamic lesions. *Proc Natl Acad Sci USA* 69(6): 1583-1586.

Suwazono Y, Dochi M, Oishi M, Tanaka K, Kobayashi E, Sakata K (2009) Shiftwork and impaired glucose metabolism: a 14-year cohort study on 7104 male workers. *Chronobiol Int* 26(5): 926-941.

Takahashi JS (2004) Finding new clock components: past and future. *J Biol Rhythms* 19(5): 339-347.

Tenkanen L, Sjoblom T, Harma M (1998) Joint effect of shift work and adverse life-style factors on the risk of coronary heart disease. *Scand J Work Environ Health* 24(5): 351-357.

Thapan K, Arendt J, Skene DJ (2001) An action spectrum for melatonin suppression: evidence for a novel non-rod, non-cone photoreceptor system in humans. *J Physiol* 535(1): 261-267.

Trinder J, Armstrong SM, O'Brien C, Luke D, Martin MJ (1996) Inhibition of melatonin secretion onset by low levels of illumination. *J Sleep Res* 5(2): 77-82.

Triqueneaux G, Thenot S, Kakizawa T, Antoch MP, Safi R, Takahashi JS, Delaunay F, Laudet V (2004) The orphan receptor Rev-erbalpha gene is a target of the circadian clock pacemaker. *J Mol Endocrinol* 33(3): 585-608.

Turek FW, Joshu C, Kohsaka A, Lin E, Ivanova G, McDearmon E, Laposky A, Losee-Olson S, Easton A, Jensen DR, Eckel RH, Takahashi JS, Bass J (2005) Obesity and metabolic syndrome in circadian Clock mutant mice. *Science* 308(5724): 1043-1045.

Van Cauter E (1990) Diurnal and ultradian rhythms in human endocrine function: a minireview. *Horm Res* 34(2): 45-53.

Vollrath L, Huesgen A (1988) Response of pineal serotonin N-acetyltransferase activity in male guinea pigs exposed to light pulses at night. *J Neural Transm* 72(1): 55-66.

Waterhouse J, Reilly T, Atkinson G, Edwards B (2007) Jet lag: trends and coping strategies. *Lancet* 369(9567): 1117-1129.

Weaver DR (1998) The suprachiasmatic nucleus: a 25-year retrospective. *J Biol Rhythms* 13(2): 100-112.

Whitnall MH, Mezey E, Gainer H (1985) Co-localization of corticotropin-releasing factor and vasopressin in median eminence neurosecretory vesicles. *Nature* 317(6034): 248-250.

Winget CM, DeRoshia CW, Ogawa KH, Holley DC (1989) Significance of light and social cues in the maintenance of temporal organization in man. *Physiologist* 32(1): S94-S95.

Wirz-Justice A (2006) Biological rhythm disturbances in mood disorders. *Int Clin Psychopharmacol* 21 (1): S11-S15.

Wittmann M, Dinich J, Merrow M, Roenneberg T (2006) Social jetlag: misalignment of biological and social time. *Chronobiol Int* 23(1-2): 497-509.

Wong KY, Dunn FA, Graham DM, Berson DM (2007) Synaptic influences on rat ganglion-cell photoreceptors. *J Physiol* 582(1): 279-296.

Wurtman RJ, Axelrod J, Phillips LS (1963) Melatonin Synthesis in the Pineal Gland: Control by Light. *Science* 142(3595): 1071-1073.

Wurtman RJ, Rose CM, Chou C, Larin FF (1968) Daily rhythms in the concentrations of various amino acids in human plasma. *N Engl J Med* 279(4): 171-175.

Wyatt JK, Dijk DJ, Ritz-de Cecco A, Ronda JM, Czeisler CA (2006) Sleep-facilitating effect of exogenous melatonin in healthy young men and women is circadian-phase dependent. *Sleep* 29(5): 609-618.

Yan L, Karatsoreos I, Lesauter J, Welsh DK, Kay S, Foley D, Silver R (2007) Exploring spatiotemporal organization of SCN circuits. *Cold Spring Harb Symp Quant Biol* 72: 527-541.

Yan L, Okamura H (2002) Gradients in the circadian expression of Per1 and Per2 genes in the rat suprachiasmatic nucleus. *Eur J Neurosci* 15(7): 1153-1162.

Yan L, Silver R (2004) Resetting the brain clock: time course and localization of mPER1 and mPER2 protein expression in suprachiasmatic nuclei during phase shifts. *Eur J Neurosci* 19(4): 1105-1109.

Yang S, Liu A, Weidenhammer A, Cooksey RC, McClain D, Kim MK, Aguilera G, Abel ED, Chung JH (2009) The role of mPer2 clock gene in glucocorticoid and feeding rhythms. *Endocrinology* 150(5): 2153-2160.

Zaidi FH, Hull JT, Peirson SN, Wulff K, Aeschbach D, Gooley JJ, Brainard GC, Gregory-Evans K, Rizzo JF, Czeisler CA, Foster RG, Moseley MJ, Lockley SW (2007) Short-wavelength light sensitivity of circadian, pupillary, and visual awareness in humans lacking an outer retina. *Curr Biol* 17(24): 2122-2128.

Zeitzer JM, Dijk DJ, Kronauer R, Brown E, Czeisler C (2000) Sensitivity of the human circadian pacemaker to nocturnal light: melatonin phase resetting and suppression. *J Physiol* 526 (3): 695-702.

Zheng B, Albrecht U, Kaasik K, Sage M, Lu W, Vaishnav S, Li Q, Sun ZS, Eichele G, Bradley A, Lee CC (2001) Nonredundant roles of the mPer1 and mPer2 genes in the mammalian circadian clock. *Cell* 105(5): 683-694.

In: Circadian Rhythms and Their Importance ISBN: 978-1-68507-547-7
Editor: Rajeshwar P. Sinha

Chapter 6

CIRCADIAN PHASE SHIFTING OF WHEEL RUNNING AND MODULATION OF SPIKE ACTIVITY IN THE RAT SUPRACHIASMATIC NUCLEUS BY NEUROPEPTIDE Y

Alexey N. Inyushkin* and Albina A. Petrova
Department of Human and Animal Physiology,
Samara National Research University, Samara, Russian Federation

ABSTRACT

The suprachiasmatic nucleus (SCN) is the master clock in mammals, generating circadian rhythms, and synchronising peripheral clocks. Neurons in the SCN respond to the different neuropeptides including neuropeptide Y (NPY). In the present study, the effects of NPY on the circadian rhythm of wheel running and on spike activity of the SCN neurones in vitro were investigated. In the in vivo experiments on male Wistar rats, the effects of bilateral intranasal administration of 0.85 μg NPY at six different projected daily time points on the circadian rhythm

* Corresponding Author's E-mail: ainyushkin@hotmail.com.

of wheel running activity were studied. It was found that the administration of NPY at late projected night induced a significant phase advance in the circadian wheel running rhythm whereas administration of NPY at the other time points did not induce significant phase-shifts of the rhythm. NPY did not induce changes in period of the rhythm or in daily gross activity. On hypothalamic slices from male Wistar rats, the effects of 10 nM NPY on extracellularly recorded spike activity of 81 SCN neurones were studied. It was found that the application of NPY induced a decrease in the mean spike frequency accompanied by an increase in spike irregularity (measured by the entropy of the probability distribution of the log interspike intervals) and in spike patterning (measured by the mutual information between the adjacent log interspike intervals) in 43% of the SCN neurones recorded. A possible modulatory effect of NPY on afferent inputs to the SCN neurones from arcuate nucleus (ARCN) was studied using the electrophysiological peristimulus time histogram (PSTH) technique. The statistically significant responses to the stimulation of the ARCN were found in 26 of 54 tested SCN neurons. Eight neurones responded to the stimulation of the ARCN with a short-latency excitation, 6 neurones – with a short-latency inhibition, and in 12 additional neurones, complex responses containing different combinations of excitations and inhibitions were recorded. Application of NPY induced qualitative changes in the responses (e.g., disappearance of initial response or appearing of new response) in 6 neurones initially responded and in one neurone initially not responded to the ARCN stimulation. Additionally, changes in latency and duration of the responses were often induced by NPY. Our results demonstrate that NPY can phase shift the behavioural circadian rhythm, suggesting the direct effects of NPY on spike activity and spike coding in the SCN and modulation of afferent inputs to the SCN.

Keywords: circadian clock, suprachiasmatic nucleus, wheel running, spike activity, neuropeptide Y

Introduction

The suprachiasmatic nucleus (SCN) of the mammalian hypothalamus contains a circadian pacemaker synchronising daily rhythmic processes of the organism with the daily rhythms in the environment (Hastings et al. 2018). Individual cells of the SCN generate circadian rhythms, and interactions between the cells within the SCN network are required to sustain circadian rhythmicity. The SCN contains a variety of neuropeptides (Moore et al. 2002; Irwin and Allen 2010; Ono et al. 2021), that might contribute to the network functions and therefore influence the parameters of circadian rhythms and their entrainment. Photic and non-photic inputs to the circadian clock can alter rhythms, producing a phase shift (Golombek and Rosenstein 2010; Wams et al. 2017). Photic information from the retina reaches the SCN via the retinohypothalamic tract utilising glutamate and pituitary adenylate cyclase-activating polypeptide (PACAP) as primary neurotransmitters (Hannibal 2002). Non-photic input to the SCN arises from the intergeniculate leaflet (IGL) via the geniculohypothalamic tract utilising neuropeptide Y (NPY) as its principal neurotransmitter (Harrington 1997), and from the serotonergic raphe nuclei (Mistlberger et al. 2000). It was shown that in the hamster, NPY can phase advance circadian rhythms when microinjected into the SCN during the subjective day in vivo and in vitro (Huhman and Albers 1994; Harrington and Schak 2000). Furthermore, lesions of the geniculohypothalamic tract disrupted non-photic phase shifting, suggesting an involvement of NPY in non-photic resetting of circadian rhythms (Janik and Mrosovsky 1994; Wickland and Turek 1994).

At least two NPY receptor mRNAs (Y1 and Y5) are expressed in the SCN (Larsen and Kristensen 1998; Yannielli et al. 2004; Kamiji and Inui 2007; Yi et al. 2018). Additionally, a functional expression of Y2 and Y1 receptors at both presynaptic and postsynaptic sites of the SCN neurones was suggested in electrophysiological study (van den Pol et al. 1996). It was shown that phase shifts in response to NPY are mediated by the Y2 receptor, as indicated by the sensitivity to the Y2-selective agonists, and insensitivity to other agonists, as well as by inability of NPY to reset the circadian clock in the Y2−/− mice (Golombek et al. 1996; Huhman et al. 1996; Gribkoff et al. 1998; Soscia and Harrington 2005).

Although previous studies reported the direct effects of NPY on firing of the SCN neurones, the changes in spike activity have not been well characterised. Initial in vitro studies revealed an excitatory or no effect of NPY on the majority of the SCN cells tested (Mason et al. 1987; Shibata and Moore 1988). Later studies in mice and hamsters though showed an inhibition of firing of the SCN cells during NPY application in vitro (Liou and Albers 1991; Besing et al. 2012). It was suggested that Y5 receptor primarily contributes to direct NPY-mediated neuronal inhibition in the SCN (Gribkoff et al. 1998). It was found in studies with whole-cell recordings, that the inhibitory effects of NPY are explained by a membrane hyperpolarisation, a depression of cytosolic Ca^{2+} and a decrease in the frequency of glutamate-mediated mEPSCs (Chen and van den Pol 1996; van den Pol et al. 1996). It is widely accepted, however, that mean spike frequency alone does not adequately describe all aspects of spike coding (Rieke et al. 1999; Bhumbra et al. 2004, 2005). In an earlier study of rhythmical activity in the suprachiasmatic nucleus (Bhumbra et al. 2005), a method that quantified the variability of the interspike interval distribution and spike patterning was employed. The entropy of the log interval distribution was used as a measure of spike irregularity, and the mutual information between the adjacent log interspike intervals was used as a measure of spike patterning (Bhumbra & Dyball 2005). Importantly, the statistical significance of the rhythm of the entropy of the log interval distribution in recordings from suprachiasmatic cells in vivo appeared to be much greater than for mean spike frequency (Bhumbra et al. 2005). In the present study, the spike activity of single units in the SCN recorded extracellularly was analysed using an information approach that quantifies the coding capacity of each spike by the irregularity of firing, and a similar approach was used to quantify changes in firing pattern (Bhumbra & Dyball 2005).

Additionally, to the direct postsynaptic actions on the SCN neurones, NPY might also indirectly affect the excitability of the SCN cells by modulating their excitatory and/or inhibitory synaptic inputs. Hypothalamic ARCN suggested to be an important source of that input for the SCN. ARCN is critical for the regulation of feeding and metabolism (Myers and Olson 2012). It has been shown previously that the reciprocal interaction between the SCN and ARCN is essential for the expression of physiological rhythms (Buijs et al. 2017; Mendez-Hernandez et al. 2020).

The purpose of the present study on male Wistar rats was to investigate the effects of NPY on circadian rhythms *in vivo*, and on spike activity of the SCN cells in vitro. We first examined NPY-induced phase shifts of the circadian rhythm of wheel running, administering NPY at different daily time points. Second, we examined the effects of NPY on spike activity and spike coding of the SCN neurons through extracellular recordings in acute hypothalamic slices. Finally, we investigated the modulatory action of NPY on afferent inputs to the SCN neurones from the ARCN.

METHODS

Recording of Wheel Running Activity

All experimental procedures were conducted in accordance with the guidelines of the European Community Council Directive 86/609/EEC and a project license approved by Ethical Committee of Samara National Research University. Nineteen male Wistar rats weighting 180 – 210 g were used for recording of circadian rhythm in wheel running activity. Animals were housed individually (cage 40 x 38 x 28 cm) at room temperature of 21 ± 1°C with food and water provided ad libitum. Each cage was equipped with running wheel (21 x 9.5 cm). Throughout the whole experiment, voluntary wheel running activity was measured as a number of wheel revolutions (Jud et al. 2005) by an electromagnetic sensor. Signals from the sensor were registered and analysed on a computer.

At the beginning of the experiment, animals were maintained at least two weeks in individual cages at a standard light/dark (LD) cycle (LD 12:12) for adaptation. After this time, animals were transferred to constant dim red light (DD) until the end of the experiment to shield rats from external time cues and to monitor free-running rhythm of locomotor activity (Mrosovsky, 1996; Jud et al., 2005). After 7 to 10 days of the transition to DD, bilateral intranasal administration of 0.85 μg NPY (Sigma-Aldrich, USA, 2 x 10 μl, aqueous solution) was performed at one of six projected daily time points (ZT 2, ZT 6, ZT 10, ZT 14, ZT 18 or ZT 22). NPY was administered trice at the same time point for three successive days. In a control group of rats, the same volume of distilled water was similarly administered. After the last intranasal administration, monitoring of the locomotor activity was lasted for 7 to 10 successive days. Wheel running activity was plotted as double plotted actograms, and the data for 7 days preceding the first intranasal application were compared with the data for 7 days succeeding the last application. The following parameters were determined using custom data acquisition system (Borisov et al., 2020): period length, acrophase, phase shifts and daily overall activity.

The data obtained from wheel running were subjected to statistical analysis. For statistical comparisons of variables before and after NPY applications, Student's paired t-test or Wilcoxon signed rank test (when appropriate) was used. For comparisons between the data obtained at the experiments where NPY applied at each of six projected daily time points, one way ANOVA test followed by post hoc Bonferroni multiple comparison test was used for normally distributed data; otherwise Kruskal-Wallis ANOVA test on Ranks with post hoc Student-Newman-Keuls multiple comparison test was used. The normality of distribution of variables was verified using Shapiro-Wilk test, and homogeneity of variance was assessed using Levene's test. Normally distributed data are presented as the mean ± the standard error of the mean. A P-value < 0.05 was considered statistically significant.

Recording of Spike Activity of the SCN Neurones *In Vitro*

Thirty-four male Wistar rats weighting between 70 and 160 g were used. The animals were housed under a 12:12 h light/dark cycle for at least 14 days before each experiment with food and water provided ad libitum. The lights went on at 9.00 h, the preparation of slices (see below) was carried out between 10.00 and 11.00 h and all recordings were made during subjective daytime, between 13.00 and 21.00 h (i.e., at ZT 04.00–12.00 h). All experimental procedures were conducted in accordance with the guidelines of the European Community Council Directive 86/609/EEC and a project license approved by Ethical Committee of Samara National

Research University. Rats were anesthetised with urethane (1.2 g/kg i.p.) and decapitated. The brains were excised from the skull cavity and transferred to artificial cerebrospinal fluid (aCSF) at 1–3°C. A Vibratome (Vibroslice NVSL, WIP, USA) was used to cut 300-μm sagittal slices through the suprachiasmatic nucleus. After the slices had been cut, they were kept for at least 1 h at 37°C before being transferred to the recording chamber and perfused with aCSF saturated with 95% oxygen and 5% carbon dioxide. The composition of the aCSF was (in mM): NaCl 124, $NaHCO_3$ 25, KCl 3, $CaCl_2$ 1.5, $MgSO_4$ 1, $NaH2PO_4$ 0.5, and glucose 30. Slices were perfused at a constant rate of 1.5 ml/min with a peristaltic pump (Minipuls 3, Gilson, France).

Extracellular recordings were made at 27–30°C using glass electrodes with a tip diameter of approximately 1 μm, filled with aCSF to give a tip resistance of approximately 10 MΩ. Recordings were made under microscopical control from the region of the hypothalamus <500 μm dorsal to the optic tract and <500 μm lateral to the midline. Recorded signals were amplified (2400 A, Dagan, USA), 50-Hz noise was eliminated (Hum Bug, Quest Scientific, Canada), then the signals were passed through an interface device (Micro 1401, CED, UK) to a personal computer. Spike 2 software (Cambridge Electronic Design) was used to collect the data. The raw waveform signal was sampled at 20 kHz and off-line discrimination of spike events was undertaken using custom software (Bhumbra et al., 2004) for subsequent data analysis. After the spontaneous spike activity of each cell had been recorded in aCSF for at least 5 min, NPY was applied to the perfusion solution at a concentration of 10 nM for 10 min. A 5-min excerpt starting 5 min after the beginning of NPY application was taken for analysis of activity in the presence of NPY. The perfusion solution was then replaced with normal aCSF (free of NPY) and, after 10 min, a 5-min period of recording was used for analysis of the activity during the 'washout period' after NPY application. NPY was applied only once to each slice preparation and slices were changed at the end of each NPY exposure. The recording chamber and tubing was washed for at least 30 min between the end of one application and the beginning of recording from the next cell.

Subsequent spike analysis was undertaken as previously described (Bhumbra & Dyball 2005). The mean spike frequency was calculated, the entropy of the log interval distribution was used as a measure of spike irregularity, and the mutual information between the adjacent log interspike intervals was used as a measure of spike patterning. Student's paired t-test or Wilcoxon signed rank test (when appropriate) was used for comparisons of variables before and after NPY application. The normality of data distributions was verified using Shapiro-Wilk test, and homogeneity of variance was assessed using Levene's test. Normally distributed data are presented as the mean ± the standard error of the mean. Responses were considered statistically significant if $P < 0.05$.

Electrophysiological Characterization of Afferent Inputs from the Arcuate Nucleus (ARCN) to the SCN

A side-by-side bipolar stimulating stainless-steel electrode (diameter, 100 μm, tip separation, 200 μm), placed directly on the ventral area of the ARCN, which has been shown to have reciprocal interaction with the SCN (Yi et al. 2006; Buijs et al. 2017; Mendez-Hernandez et al. 2020). The ARCN was stimulated with biphasic bipolar square-wave pulses of 100 x 100 μA. The durations of the positive and immediately following negative phases were identical (0.5 x 0.5 ms). The pulses were applied at a frequency of 1 Hz. Stimulation pulses were generated using a Model 2100 Isolated Pulse Stimulator (A-M Systems, USA).

Extracellular spike recordings from the SCN cells were made as described above during the stimulation of the ARCN. Peristimulus time histograms (PSTHs) were used to determine the parameters of the responses. Interpretation of the effects of a stimulation using a PSTH is based upon the assumption that the pulse evokes an action potential in axons afferent to the recorded cell and that the stimulated area is restricted to the region local to the tip of the stimulating electrode. The cusum procedure (Ellaway 1978) was applied to the PSTHs to determine the sign, latency, and the duration of the responses (Cui et al. 1997). The period of

200 ms immediately before the application of the stimulus pulses served as the control. The latency of a response was defined as the time interval between the application of the stimulus and the start of the response on the PSTH, and the duration of a response was defined as the time interval between the beginning and the end of the response as measured on the PSTH. The statistical significance of each response was tested using Student's paired t-test or Mann-Whitney rank test, and responses were classified as significant if $P < 0.02$ (Cui et al. 1997).

RESULTS

Effects of NPY on the Circadian Rhythm of Wheel Running Activity

At the beginning of the experiments in LD conditions, wheel running activity prevailed during the dark period in all animals. Subsequent transition to DD conditions inducing a free-running activity (Jud et al. 2005; Mrosovsky 1996), did not significantly change a daily pattern of locomotor activity as it still prevailed during the projected dark period (Figure 1). This was confirmed in particular by lack of changes in acrophase of the rhythm, which was 16.37 ± 0.27 h under entrained LD conditions and 16.88 ± 0.24 h under DD free running conditions (the difference is not statistically significant, $P = 0.146$: Student's paired t-test).

The effects of NPY depended on the projected daily time point of its intranasal administration. Statistically significant changes of wheel running activity were revealed in the experiments, where NPY administered at ZT 22 (i.e., at the projected late dark period). Administration of NPY at any other projected daily time point did not alter the wheel running activity parameters.

Figure 1 shows a representative example of a response to intranasal NPY administration at ZT 22. A comparison between the excerpts of the actogram before and after NPY administration suggests that NPY induced a phase advance of wheel running activity period by 2.77 h.

As a result, a significant part of daily locomotor activity progressively shifted from the projected dark period to the late part of the preceding light period. This phase shifting effect of NPY was accompanied by a slight shortening of the period of free running locomotor activity rhythm, from 24.72 h before the application to 24.10 h after NPY application. An increase in overall activity was also induced by NPY: an overall number of wheel revolutions for 7 days before NPY application was 11666, whereas its value for 7 days after the application was 14583.

Statistical analysis of data obtained from the experiments where NPY applied at ZT 22, revealed a significant phase advance of rhythm acrophase (Figure 2). Value of acrophase before the application was 18.58 ± 0.51 h; after NPY application the acrophase was shifted to 15.46 ± 0.84 h ($P = 0.011$: Student's paired t-test), corresponding to 187 ± 64 min phase advance. However, NPY did not induce statistically significant changes of the period ($P = 0.271$: Student's paired t-test), despite tendency to its shortening (from 24.19 ± 0.19 h to 23.84 ± 0.37 h). Overall activity for 7 days also did not alter by NPY ($P = 0.641$: Wilcoxon signed rank test). Median value of revolutions for 7 days before the applications was 2498; after NPY applications median value of revolutions was 1891.

Intranasal administration of NPY at any other time point of the projected daily cycle (ZT 2, ZT 6, ZT 10, ZT 14 or ZT 18) did not produce statistically significant changes of the wheel running activity parameters. Similarly, intranasal administration of distilled water (control) at any time point of the projected daily cycle including ZT 22 did not produce statistically significant changes of the wheel running activity parameters. These results suggest that the effects of NPY on the circadian rhythm of wheel running activity was time-specific, manifesting only when NPY applied at the projected late dark period.

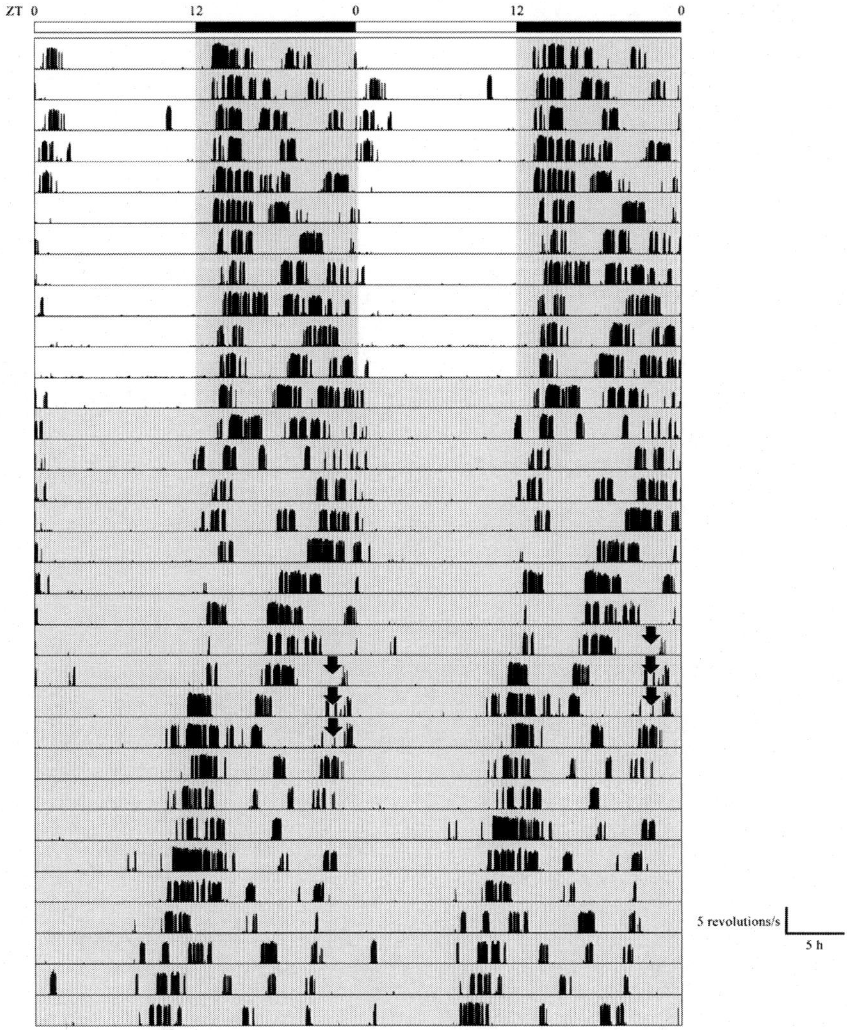

Figure 1. Characteristic response of wheel running activity to 0.85 μg NPY administered intranasally at ZT 22 of the three successive days. Actogram is presented as a double plotted (48 hrs) raster (Jud et al., 2005). Every line contains data from two successive days, so data from the second day is duplicated at the beginning of the next line. Light periods are shown by white background, dark periods are shown by gray background. Administrations of NPY marked by arrows corresponding to ZT 22.

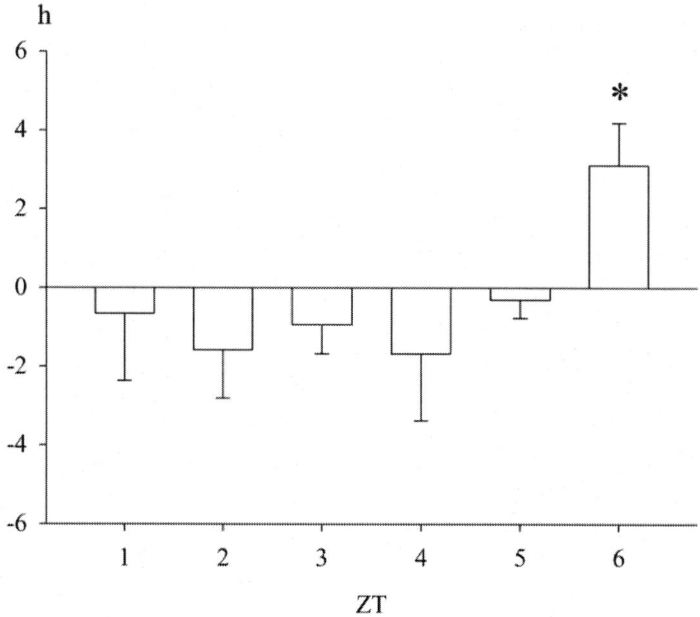

Figure 2. Phase shifts induced by NPY administered at different projected time points of the daily cycle. Averaged data for 7 days are presented as means ± standard errors of means. On Y-axis: positive values – phase advance, negative values – phase delay. *$P < 0.05$: Student's paired t-test.

Effects of NPY on Spike Activity of the SCN Neurons *In Vitro*

A total of 81 single unit extracellular recordings were made from the SCN in hypothalamic slices taken from 34 rats. The application of 10 nM NPY to the bath solution decreased firing rate. A representative response is shown in Figure 3. Ratemeter record of the SCN cell showed a decrease in mean spike frequency, an increase in the entropy of the log interspike interval distribution and in the mutual information between adjacent interspike intervals. This is illustrated by the excerpts of recordings taken before and during the application of NPY.

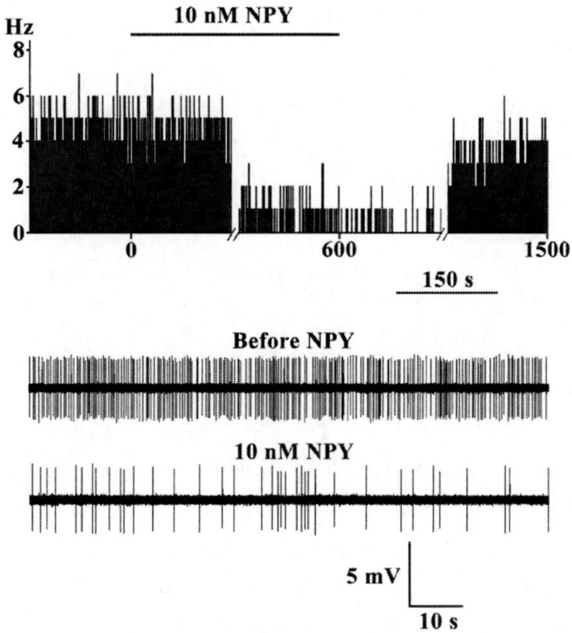

Figure 3. Characteristic example of the response of a suprachiasmatic neurone to application of 10 nM NPY. Ratemeter record presented at the upper part shows that NPY induced a reversible decrease in mean spike frequency. Excerpts of 100 s taken before and during the application of NPY showed an increase in the variability of interspike intervals, and an increase in spike patterning.

Application of NPY to the perfusion solution induced a decrease in the mean spike frequency in 35 of the 81 SCN neurons recorded (43.2%). NPY induced a significant decrease in the mean spike frequency in these neurones from 1.33 to 0.95 s^{-1} ($P < 0.001$: Wilcoxon signed rank test). This effect was accompanied by an increase in median value of the entropy of the log interspike interval distribution from 6.92 to 7.16 bits ($P = 0.005$: Wilcoxon signed rank test) suggesting a greater degree of interspike interval irregularity in the presence of NPY. Spike patterning, quantified by the mutual information between adjacent log interspike intervals, also increased in this group of cells in the presence of NPY ($P = 0.020$, Wilcoxon signed rank test). In most cases, the responses of the SCN neurones to NPY application were reversible since at the end of washout period there was a complete or partial recovery of spike activity to the initial control (Figure 3).

Eight cells (9.9%) demonstrated an increase in firing rate during the application of NPY. In this group of the neurones, median value of the mean spike frequency increased from 0.76 to 1.71 s^{-1} ($P = 0.008$: Wilcoxon signed rank test). By contrast, NPY had no statistically significant effects on the entropy of the log interspike interval distribution ($P = 0.116$: Student's paired t-test) or mutual information between adjacent interspike intervals ($P = 0.250$: Wilcoxon signed rank test) in these neurones. The remaining 38 cells did not respond to the application of NPY. There were no statistically significant changes in mean spike frequency in these neurones: median value was 2.56 s^{-1} before the application and 2.59 s^{-1} in the presence of NPY ($P = 0.331$: Wilcoxon signed rank test). This group of neurons also showed no changes in the entropy of the log interspike interval distribution, which was 6.37 ± 0.16 bits before and 6.34 ± 0.15 bits during the application of NPY ($P = 0.457$: Student's paired t-test); the mutual information between adjacent interspike intervals did not affect by NPY either ($P = 0.922$: Wilcoxon signed rank test).

Grouping all 81 tested neurones together, NPY induced a statistically significant decrease in the mean spike frequency (median value decreased from 1.85 to 1.72 s^{-1}; $P = 0.011$: Wilcoxon signed rank test) along with an increase in the mutual information between adjacent interspike intervals ($P = 0.037$: Wilcoxon signed rank test); however, changes in the entropy of the log interspike interval distribution were not significant ($P = 0.237$: Wilcoxon signed rank test).

Effects of NPY on Afferent Inputs from the ARCN to the SCN *In Vitro*

Fifty-four cells in the SCN were tested for their response to stimulation of the ARCN. Twenty-six of the fifty-four neurones (48.2%) displayed statistically significant orthodromic responses to stimulation ($P < 0.02$: Student's unpaired t-test or Mann-Whitney rank test). The neurones displayed simple excitatory, simple inhibitory or complex responses consisted of combinations of excitations and inhibitions.

Figure 4 shows representative examples of a simple excitatory (A) and a simple inhibitory (B) response of the SCN neurones to stimulation of the ARCN. The excerpt of the extracellular recording in Figure 4A shows that spikes occurred mostly within a short time interval after the stimulation pulses. This is reflected in PSTH constructed on the basis of cumulative data on spike distribution in 325 cycles of 1 Hz stimulation. The PSTH suggests the higher likelihood of a spike occurring within a 200-ms time interval succeeding the stimulus comparing to the control period of 200 ms preceding the stimulus. Such PSTH containing a peak, reflecting a high probability of spikes occurrence shortly after the stimulus, is a characteristic of excitatory response. By contrast, the excerpt of the recording and the corresponding PSTH constructed on the basis of 150 cycles of stimulation, presented in Figure 4B, demonstrate a low probability of spikes occurrence within 80-ms time interval after the stimulus. Such type of PSTH containing a gap directly after the stimulus is a characteristic of inhibitory response.

Before NPY application, eight (14.8%) SCN neurones displayed a simple orthodromic excitatory response to stimulation of the ARCN. During the application, the excitatory response was still seen in seven of the eight cells. In the presence of NPY, these seven cells did not demonstrate statistically significant changes in the parameters of the response comparing to the initial control (before NPY). The latency of the responses was 7.43 ± 1.67 ms before the application, and the value was 8.00 ± 1.95 ms in the presence of NPY ($P = 0.569$: Student's paired t-test). There was also no change in the duration of the responses, with a median value of 20 ms before the application of NPY, and 16 ms in the presence of NPY in the perfusing chamber ($P = 0.844$: Mann–Whitney rank test). In two of these eight cells, qualitative changes of the responses were induced by NPY.

An initial simple excitatory response of the neurone in Figure 5A, transformed to a complex biphasic response in the presence of NPY. A first excitatory phase ($P = 0.009$: Mann–Whitney rank test) was followed by an additional inhibitory phase (P < 0.001: Mann–Whitney rank test) manifested during the application of NPY.

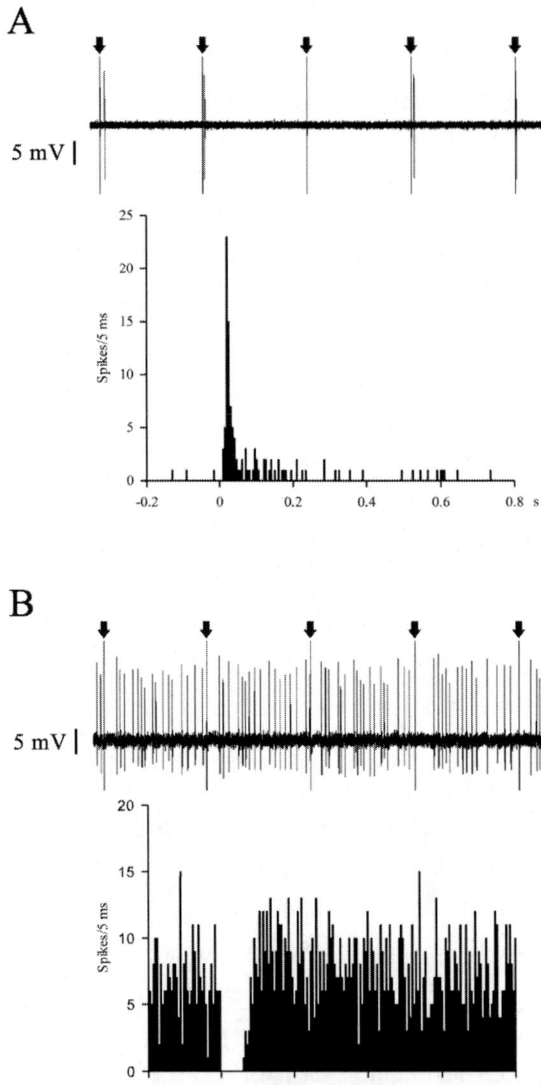

Figure 4. Examples of the responses of the two SCN neurones to stimulation of the ARCN. In the excerpts of raw traces, stimulation pulses are marked by arrows. Corresponding PSTHs are characteristic of a short-latency excitation (A), and a short-latency inhibition (B).

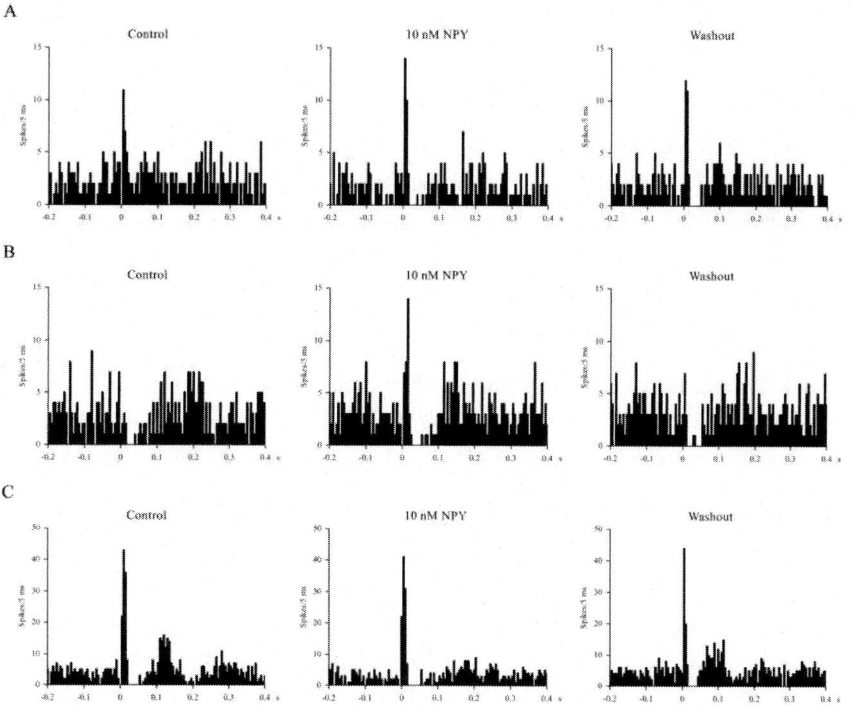

Figure 5. Representative PSTHs created from extracellular recordings from SCN neurones following stimulation of the ARCN in vitro. (A) The cell displayed transformation of a short-latency excitation (Control) to a complex biphasic response with an additional inhibitory phase in the presence of NPY. (B) The cell displayed transformation of a short-latency inhibition (Control) to a reversible complex biphasic response with an additional excitatory phase preceding the inhibition during NPY application. (C) The neurone initially displayed a response consisted of three consequential phases (a short-latency excitation, inhibition, and a long-latency excitation), showed transformation of the response to biphasic since the third phase of the response (long-latency excitation) was reversibly blocked in the presence of NPY.

The effect was not reversible since at the end of washout the response remained biphasic. In one additional neurone, application of NPY induced a complete disappearance of an initial simple orthodromic excitatory response (P = 0.316: Mann–Whitney rank test), but at the end of 15-min washout period, the cell displayed a recovery of the statistically significant excitatory response (P = 0.003: Mann–Whitney rank test) to stimulation of the ARCN.

Six (11.1%) of the SCN neurones tested, displayed a simple orthodromic inhibitory response to the ARCN stimulation. In the presence of NPY, inhibition persisted in all six neurons, though there were statistically significant changes in the parameters of the response comparing to the initial control. An increase in the latency of inhibition from 8.00 ± 2.13 ms to 12.00 ± 1.03 ms (P = 0.033: Student's paired t-test) was found along with an increase in the duration of inhibition from 123 ± 31.8 ms to 156 ± 41.8 ms (P = 0.046: Student's paired t-test). The effects were completely reversible since at the end of the washout period both, latency, and duration of the inhibition recovered to their initial values: 8.00 ± 1.86 ms (P = 1.000: Student's paired t-test) and 122 ± 36.5 ms (P = 0.840: Student's paired t-test), respectively. Two of the six neurones displayed qualitative changes in response to the ARCN stimulation in the presence of NPY. In both cases, an initial simple inhibitory response to stimulation transformed to biphasic during NPY application. A representative example of such transformation is shown in Figure 5B, where an initially displayed inhibitory phase of the response (P < 0.001: Mann–Whitney rank test) preceded by an additional excitatory phase (P < 0.001: Mann–Whitney rank test), manifested in the presence of NPY. At the end of 15-min washout period, a clear recovery of the initial monophasic inhibitory response was found in this cell, with no sign of excitation (Figure 5B). However, in the second case the effect was not reversible since additional excitatory phase of the response persisted even after 15-min washout of the slice with aCSF (P = 0.006: Mann–Whitney rank test).

Twelve (22.2%) of the neurones tested, displayed complex responses to stimulation of the ARCN. The most common complex response (n = 7) consisted of a short-latency orthodromic excitation followed by orthodromic inhibition. NPY application did not induce any qualitative changes in complex responses of this type. Most parameters of the PSTH did not alter by NPY either, except for the latency of orthodromic excitation. A significant reduction in value of this parameter from 6.29 ± 1.02 ms before the application to 3.71 ± 0.68 ms (P = 0.012: Student's paired t-test) in the presence of NPY was found. This effect of NPY was reversible since at the end of washout period the latency of excitation did not differ significantly from its initial control value ($P = 0.250$: Mann–Whitney rank test).

One of the SCN cells studied, displayed a biphasic response to stimulation consisted of a short-latency orthodromic inhibition ($P < 0.001$: Mann–Whitney rank test) followed by a long-latency excitation ($P < 0.001$: Mann–Whitney rank test). This response did not change significantly in the presence of NPY. The responses of three additional SCN neurones to the ARCN stimulation consisted of three consequential phases: a short-latency excitation, inhibition, and a long-latency excitation. Two of the three cells of this group displayed qualitative changes in the response during NPY application. The PSTHs, shown in Figure 5C, suggest that in this case, the third phase of the response to stimulation (long-latency excitation) was reversibly blocked by NPY. The complex triphasic response of another neurone was also transformed to biphasic, though in this case it was the second phase (orthodromic inhibition) rather than the third phase of the response, that was blocked by NPY. In this cell, there was no recovery since the response was still biphasic at the end of the 15-min washout.

Twenty-eight of the fifty-four (51.8%) SCN neurones tested, did not display any statistically significant response to stimulation of the ARCN before the application of NPY ($P > 0.02$: Mann–Whitney rank test). In twenty-seven of the twenty-eight cells of this group, there was no response during NPY application. In one of the twenty-eight neurones, a short-latency inhibition manifested in the presence of NPY ($P < 0.001$: Mann–Whitney rank test) and persisted at the end of washout period.

Conclusion

It is well established that in addition to light, nonphotic cues, such as meal timing, exercise, temperature, social contacts, can entrain the circadian clock (Mistlberger and Skene, 2005; Challet, 2007). The purpose of the present complex in vivo and in vitro study was to investigate circadian effects of NPY. The SCN receives afferent nonphotic NPYergic input from the IGL via geniculohypothalamic tract playing a prominent role in the nonphotic entrainment of the circadian clock (Biello et al. 1994; Marchant et al. 1997; Glass et al. 2010).

To study effects of NPY on circadian rhythms of behaviour, we compared changes in voluntary wheel running activity induced by intranasal administration of NPY at six projected daily time points (ZT 2, ZT 6, ZT 10, ZT 14, ZT 18 or ZT 22). Monitoring of wheel running activity is widely used behavioural assay, utilising the well-known fact that rats have a spontaneous tendency to run. In this study, animals had free access to running wheels for at least two weeks before the analysed DD periods, and were, therefore, admirably adapted to the wheels. Previous studies have demonstrated a clear phase relationship between the pattern of expression in the SCN of the clock genes belonging to the *Per* family and the locomotor activity rhythm (Fukuhara et al., 2000). A distinctive feature of the study was intranasal administration of NPY. The nasal route of administration has become an important and popular tool in human and rodent studies to reveal effects of neuropeptides on diverse behavioural aspects, emotionality, neuronal activity in health and disease (Hanson and Frey 2008; Meredith et al. 2015; Spetter and Hallschmid 2015; Wang et al. 2019). Intranasal delivery is used as a noninvasive and convenient method that rapidly targets therapeutics to the CNS, bypassing the blood-brain barrier and minimising systemic exposure (Dhuria et al., 2010). Previous studies show that peripheral levels of NPY are unaffected by intranasal administration (Serova et al. 2013), and that central and peripheral NPY levels may not be correlated (Baker et al. 2013). Thirty minutes after intranasal administration in rats, an increase in NPY level in cerebrospinal fluid and in several brain areas, including mediobasal hypothalamus (Serova et al. 2013; Laukova et al. 2014). Thus, it seems likely that NPY may directly affect the SCN circadian clock after intranasal administration.

At the beginning of the experiment, under LD schedule, wheel running activity predominated during the dark phase of the daily cycle, which coincides with that observed among nocturnal rodents (Jud et al. 2005; Majoy and Heideman 2000; Ocampo-Garces et al. 2005; Seroka et al. 2008; Verwey et al. 2013). The transition to free-running conditions under

DD schedule did not affect the rhythm of wheel running activity, so the voluntary locomotor activity was still predominated during the projected dark phase (Figure 1). This suggests a stability of the circadian clock function in rats even in the absence of the photic entraining mechanism. Intranasal administration of NPY induced a phase advance of wheel running activity of 3 h only in the experiments, where NPY was administered during the late dark period (ZT 22). Administration of NPY at any other projected daily time point did not alter the rhythm of wheel running activity. This is in agreement with the result of the in vitro study in rats (Medanic and Gillette 1993), where application of NPY to the SCN at CT 20 – 24 induced phase advances of neuronal electrical activity rhythm of up to 3.5 h. It should be noted that this period of sensitivity of the rat SCN to NPY preceded the transition from dark to light and occurred just before the SCN shows an increased sensitivity to pineal hormone, melatonin (Dubcovich et al. 2003). At dawn, melatonin induces a phase advance of the SCN neural firing rhythm via a protein kinase C-dependent signalling pathway acting downstream of the MT_2 receptor (McArthur et al. 1997; Hunt et al. 2001; Pfeffer et al. 2012). It was shown (Gamble et al. 2006) that microinjection of NPY to the SCN at CT 20 mimics the effects of light on pineal melatonin levels in hamsters and inhibits the light-induced phase advance in wheel running activity as well as increase in Per1 mRNA level. The phase shift of free-running locomotor activity rhythm induced by 15-min light pulse at CT 16 was reduced, and the period length of the rhythm was significantly greater in *npy–/–*mice relative to the wild type mice (Harrington et al. 2007). Since this sensitive period is driven by the SCN clock, the current study may suggest that the substrate underlying the increase in sensitivity to NPY is sequentially linked in the clock's mechanism, and the phase advance in wheel running activity induced by NPY at ZT 22 was due to the corresponding phase shift of neuronal activity in the SCN circadian clock. Phase shifts in response to NPY are mediated by the Y2 receptor, as indicated by the sensitivity to the Y2-selective agonists, insensitivity to other agonists, and by inability of NPY to reset the circadian clock in the Y2–/– mice (Golombek et al. 1996; Huhman et al. 1996; Gribkoff et al. 1998; Soscia and Harrington 2005).

Earlier in vitro study (Medanic and Gillette 1993) revealed an additional period of increased sensitivity of the SCN electrical activity rhythm to NPY at CT 5 – 9 with phase advance up to 4.8 h. This result was confirmed by finding that NPY injected to the SCN area of hamsters in vivo in the middle of the subjective day induces phase advances of about 3 h (Biello et al. 1994). Another study (Besing et al. 2012) demonstrated NPY-induced phase advances in PER2::LUC SCN cultures when NPY was given in the early part of the day (CT 0–6). However, we did not observe period of an increased sensitivity to NPY during the projected daytime. This inconsistency of findings may be explained by the difference between the circadian clock function in vivo and in vitro, between methods used to record circadian rhythms, and between species.

Previous studies indicated that NPY has significant effects on spike activity of the SCN neurones, though the changes in cell discharge have not been well characterised. In initial studies of the direct effects of NPY on the firing of single neurones in the SCN, NPY was reported to excite or to have no effect on the majority of cells tested in vitro (Mason et al. 1987; Shibata and Moore 1988). In latter studies with whole-cell recordings, NPY was found to induce significant inhibition of synaptic activity and Ca^{2+} levels in vitro (Chen and van den Pol 1996; van den Pol et al. 1996). In the present study, application of NPY to the hypothalamic slices, induced a decrease in the mean spike frequency of a population of the SCN neurones (43.2% of the cells tested). This effect was accompanied by an increase in both interspike interval irregularity and spike patterning in the presence of NPY. A decrease in mean spike frequency induced by NPY is consistent with previous studies in mice and hamsters that showed an inhibition of firing of the SCN cells during NPY application in vitro (Liou and Albers 1991; Besing et al. 2012). The proportion of cells displayed a decrease in spike frequency in the presence of NPY may reflect the proportion of the SCN neurones involved in expression of specific Y receptors. It was determined that at least two NPY receptor mRNA (Y1 and Y5) are expressed in the SCN (Larsen and Kristensen 1998; Yannielli et al. 2004; Kamiji and Inui 2007; Yi et al. 2018). Electrophysiological study (van den Pol et al. 1996) suggested a functional expression of Y2 and Y1 receptors at both presynaptic and postsynaptic sites of the SCN. Thus, the effects of NPY on wheel running activity and spike activity of the SCN neurones may be explained by a direct NPY action on the SCN cells through an interaction with Y receptors. In particular, NPY Y5 receptor subtype primarily contributes to direct NPY-mediated neuronal inhibition in the SCN (Gribkoff et al. 1998). The inhibitory effects of NPY are explained by a membrane hyperpolarisation, a substantial depression of cytosolic Ca^{2+} and a decrease in the frequency of glutamate-mediated mEPSCs (van den Pol et al. 1996). Induced by NPY inhibition of the SCN neuronal activity and phase shifts were accompanied by delayed suppression of PER2 protein, one of the key negative circadian clock regulators (Besing et al. 2012). One of the novel aspects of the present study not addressed in earlier works was an investigation of the effects of NPY on interval coding (Bhumbra et al. 2004, 2005; Bhumbra & Dyball 2005; Inyushkin et al. 2007, 2009), characterising spike firing more completely than frequency alone. In the present study, application of NPY to the hypothalamic slices induced an increase in both log interspike interval entropy and mutual information between the adjacent log interspike intervals. These parameters reflect a greater irregularity and patterning of spike activity in the presence of NPY. Because the increase in entropy and mutual information occurred in slice preparation, it is likely that the site of NPY action that was responsible for the changes in spike code was local to the SCN.

The possible cellular mechanisms of NPY action on the SCN neurones still need to be identified. Via specific Y receptors, NPY has been shown to affect several signal transduction pathways including inhibition of adenylate cyclase, mobilisation of intracellular Ca^{2+}, activation of protein kinase C, ERK1/2 (Schak et al. 2001; Holliday et al. 2004; Lecat et al. 2015). Modulation of Ca^{2+} and K^{+} channels by NPY has been demonstrated as well. In particular, activation of G protein-coupled inwardly rectifying K^{+} channels (GIRK) by NPY has been shown in the SCN (Hablitz et al. 2014). This study demonstrated, that NPY suppress spike firing in wild-type SCN neurons to a much greater extent than in the knockout, indicating that NPY exerts its action on the SCN spike activity via GIRK channels.

In another electrophysiological study (van den Pol et al. 1996), application of NPY to the SCN cells in vitro caused a long-term (more than 1.5 h) depression in both electrical activity and intracellular calcium levels. It was suggested that this long-term calcium depression is a result of the simultaneous release of glutamate and activation of Y receptors since the extended response to NPY was blocked by coapplication of glutamate receptor antagonists. The long-lasting effects of NPY in the SCN were also manifested in a depression of neuronal excitability in a phase-specific manner and suppression of PER2 levels accompanied by circadian phase advances in PER2::LUC SCN cultures (Besing et al. 2012). By contrast, postsynaptic actions of NPY, mediated by both Y1- and Y2-like receptors, were short term and recovered rapidly. These short-term effects of NPY were manifested in a suppression of spike activity seen in any phase of the circadian cycle (Gribkoff et al. 1998).

The results of the present study show that in addition to the direct postsynaptic actions on the SCN neurones, NPY might indirectly affect the excitability of the SCN cells by modulating their excitatory and/or inhibitory synaptic inputs from the ARCN. It has been shown previously that for the expression of physiological rhythms, the reciprocal interaction between the SCN and ARCN is essential (Buijs et al. 2017; Mendez-Hernandez et al. 2020). In the present study, projections from the ARCN to the SCN were characterised in vitro using electrophysiological technique. Forty-two per cent of SCN neurones responded to stimulation of the ARCN, indicating the existence of rich neural projections from the ARCN to the SCN. The data confirm earlier anatomical tracing studies (Krout et al. 2002; Yi et al. 2006). Stimulation of the ARCN elicited both excitatory and inhibitory responses in the SCN. Complex responses with several components (excitatory, inhibitory or both) were also observed in 22.2% of the SCN neurones in vitro. Such complexity may indicate that a substantial proportion of the SCN neurones receive two or more distinct inputs from cells in the ARCN. In earlier electrophysiological in vivo study (Saeb-Parsy et al. 2000), simple or complex responses to stimulation of the ARCN were found in 85.7% of the SCN neurones. The smaller proportion of responsive cells in the present in vitro study may reflect some damage to the projection during preparation of the slice. The in vitro experiments reported here demonstrate short-latency responses much more frequently than long-latency responses. The numbers of short-latency excitatory and inhibitory responses were essentially identical, suggesting the possibility of both stimulatory and inhibitory modulation of the SCN circadian clock from the ARCN. As most of the responses were characterised by short latency (< 20 ms), these afferent projections from the ARCN to the SCN were likely to be mono- or oligosynaptic, though in few cases complex responses included a long-latency excitation. This suggests that a small proportion of the SCN cells may receive excitatory afferent inputs from the ARCN *via* polysynaptic projections. These findings indicate the existence of afferent inputs to the SCN from the hypothalamic area, which is critical for the regulation of feeding and metabolism (Myers and Olson 2012). Afferent inputs from the ARCN may be involved in the mechanisms of nonphotic entrainment of the circadian clock accordingly the metabolic state, feeding schedule, and feeding motivation.

The SCN-ARCN interaction is essential to integrate metabolic information into the circadian system and thus adapt physiological, behavioural, and hormonal circadian rhythms (Buijs et al. 2017; Mendez-Hernandez et al. 2020). The close links between the feeding schedule, metabolism and circadian rhythms were demonstrated in several previous studies (Stephan 2002; Challet 2010; Mistlberger 2011; Bellet and Sassone-Corsi 2015). In the present in vitro study, application of exogenous NPY induced qualitative and/or quantitative changes in the responses to the ARCN stimulation in a proportion of the SCN neurones. Qualitative changes were found in 7 cells where the appearance of an additional phase or the disappearance of a preceding phase of the response occurred in the presence of NPY. Quantitative changes occurred more frequently (in 18 cells) and consisted of altered latency and/or duration of the response to the ARCN stimulation during NPY application. Thus, the modulatory effects of NPY on the afferent inputs to the SCN from the ARCN were revealed in the present study. These effects may be due to a modulation of the ARCN neurones by NPY via Y receptors. It was shown that a part of proopiomelanocortin (POMC) and GABAergic neurones in the ARCN express Y1 and Y5 receptors, via which NPY induces a membrane hyperpolarisation (Fetissov et al. 2004; Roseberry et al. 2004; Acuna-Goycolea et al. 2005; Kageyama et al. 2012). Another possibility for the effect of NPY would be a modulation of the synaptic transmission from the ARCN cells to the SCN neurones. Previous in vitro studies demonstrated both presynaptic and postsynaptic actions of NPY on the SCN cells (van den Pol et al. 1996).

Taken together, the present study confirms that NPY can play a modulatory role in the mammalian circadian system. In our in vivo experiments, we have shown that NPY has effects on circadian rhythm of wheel running activity in the rat being involved in time-specific entraining. The in vitro experiments have revealed the inhibitory effects of NPY on spike activity accompanied by an increase in both interspike interval irregularity and spike patterning of a significant proportion of the SCN neurons. The present results also provide evidence that NPY can modulate afferent inputs to the SCN from the ARCN essential to integrate metabolic information into the circadian system.

Acknowledgements

A.N. Inyushkin acknowledges RFBR for the research grant № 18-29-14073.

References

Acuna-Goycolea C, Tamamaki N, Yanagawa Y, Obata K, van den Pol AN (2005) Mechanisms of neuropeptide Y, peptide YY, and pancreatic polypeptide inhibition of identified green fluorescent protein-expressing GABA neurons in the hypothalamic neuroendocrine arcuate nucleus. *J Neurosci* 25: 7406 -7419.

Baker DG, Bertram TM, Patel PM, Barkauskas DA, Clopton P, Patel S, Geracioti TD, Haji U, O'Connor DT, Nievergelt CM, Hauger RL (2013) Characterization of cerebrospinal fluid (CSF) and plasma NPY levels in normal volunteers over a 24-h timeframe. *Psychoneuroendocrinology* 38: 2378-2382.

Bellet MM, Sassone-Corsi P (2010) Mammalian circadian clock and metabolism – the epigenetic link. *J Cell Sci* 123: 3837-3848.

Besing RC, Hablitz LM, Paul JR, Johnson RL, Prosser RA, Gamble KL (2012) NPY-induced phase shifts of PER2::LUC rhythms are mediated by long-term suppression of neuronal excitability in a phase-specific manner. *Chronobiol Int* 29: 91-102.

Bhumbra, GS, Dyball RE (2005) Spike coding from the perspective of a neurone. *Cogn Process* 6: 157-176.

Bhumbra GS, Inyushkin AN, Dyball RE (2004) Assessment of spike activity in the supraoptic nucleus. *J Neuroendocrinol* 16: 390-397.

Bhumbra GS, Inyushkin AN, Saeb-Parsy K, Hon A, Dyball RE (2005) Rhythmic changes in spike coding in the rat suprachiasmatic nucleus. *J Physiol* 563: 291-307.

Biello SM, Janik D, Mrosovsky N (1994) Neuropeptide Y and behaviorally induced phase shifts. *Neuroscience* 62: 213-219.

Borisov D, Blagov A, Inyushkin A (2020) Development of tools for processing and analysis of experimental data on the activity of laboratory animals. *CEUR Workshop Proc* 2667: 171-174.

Buijs FN, Guzmán-Ruiz M, León-Mercado L, Basualdo MC, Escobar C, Kalsbeek A, Buijs RM (2017) Suprachiasmatic nucleus interaction with the arcuate nucleus; essential for organizing physiological rhythms. *eNeuro* 4: e0028-17.

Challet E (2010) Interactions between light, mealtime and calorie restriction to control daily timing in mammals. *J Comp Physiol B* 180: 631-644.

Challet E (2007) Minireview: entrainment of the suprachiasmatic clockwork in diurnal and nocturnal mammals. *Endocrinology* 148: 5648-5655.

Chen G, van den Pol AN (1996) Multiple NPY receptors coexist in pre and postsynaptic sites: inhibition of GABA release in isolated selfinnervating SCN neurons. *J Neurosci* 16: 7711-7724.

Cui LN, Saeb-Parsy K, Dyball RE (1997) Neurones in the supraoptic nucleus of the rat are regulated by a projection from the suprachiasmatic nucleus. *J Physiol* 502: 149-159.

Dhuria SV, Hanson LR, Frey WH (2010) Intranasal delivery to the central nervous system: mechanisms and experimental considerations. *J Pharm Sci* 99: 1654-1673.

Dubocovich ML, Rivera-Bermudez MA, Gerdin MJ, Masana MI (2003) Molecular pharmacology, regulation and function of mammalian melatonin receptors. *Front Biosci* 8: 1093-1108.

Ellaway PH (1978) Cumulative sum technique and its application to the analysis of peri-stimulus time histograms. *Electroencephalogr Clin Neurophysiol* 45: 302-304.

Fetissov SO, Byrne LC, Hassani H, Ernfors P, Hökfelt T (2004) Characterization of neuropeptide Y Y2 and Y5 receptor expression in the mouse hypothalamus. *J Comp Neurol* 470: 256-265.

Fukuhara C, Dirden JC, Tosini G (2000) Circadian expression of Period1, Period 2 and AA-NAT in the rat pineal under different light conditions. *Neurosci Lett* 286: 167-170.

Gamble KL, Paul KN, Karom MC, Tosini G, Albers HE (2006) Paradoxical effects of NPY in the suprachiasmatic nucleus. *Eur J Neurosci* 23: 2488-2494.

Glass JD, Guinn J, Kaur G, Francl JM (2010) On the intrinsic regulation of neuropeptide Y release in the mammalian suprachiasmatic nucleus circadian clock. *Eur J Neurosci* 31: 1117-1126.

Golombek DA, Biello SM, Rendon, RA, Harrington ME (1996) Neuropeptide Y phase shifts the circadian clock in vitro via a Y2 receptor. *Neuroreport* 7: 1315-1319.

Golombek DA, Rosenstein RE (2010) Physiology of circadian entrainment. *Physiol Rev* 90: 1063-1102.

Gribkoff VK, Pieschl RL, Wisialowski TA, van den Pol AN, Yocca FD (1998) Phase shifting of circadian rhythms and depression of neuronal activity in the rat suprachiasmatic nucleus by neuropeptide Y: mediation by different receptor subtypes. *J Neurosci* 18: 3014-3022.

Hablitz LM, Molzof HE, Paul JR, Johnson RL, Gamble KL (2014) Suprachiasmatic nucleus function and circadian entrainment are modulated by G protein-coupled inwardly rectifying (GIRK) channels. *J Physiol* 592: 5079-5092.

Hannibal J (2002) Neurotransmitters of the retino-hypothalamic tract. *Cell Tissue Res* 309: 73-88.

Hanson L, Frey W (2008) Intranasal delivery bypasses the blood-brain barrier to target therapeutic agents to the central nervous system and treat neurodegenerative disease. *BMC Neurosci* 9: S5.

Harrington ME (1997) The ventral lateral geniculate nucleus and the intergeniculate leaflet: interrelated structures in the visual and circadian systems. *Neurosci Behav Rev* 21: 705-727.

Harrington M, Molyneux P, Soscia S, Prabakar C, McKinley-Brewer J, Lall G (2007) Behavioral and neurochemical sources of variability of circadian period and phase: studies of circadian rhythms of *npy–/–* mice mice. *Am J Physiol Regul Integr Comp Physiol* 292: R1306-R1314.

Harrington ME, Schak KM (2000) Neuropeptide Y phase advances the in vitro hamster circadian clock during the subjective day with no effect on phase during the subjective night. *Can J Physiol Pharmacol* 78: 87-92.

Hastings MH, Maywood ES, Brancaccio M (2018) Generation of circadian rhythms in the suprachiasmatic nucleus. *Nat Rev Neurosci* 19: 453-469.

Holliday ND, Michel MC, Cox HM (2004) NPY receptor subtypes and their signal transduction. In: *Neuropeptide Y and Related Peptides*: Michel MC, editor. Berlin, Heidelberg, Springer-Verlag, pp. 45-73.

Huhman KL, Albers HE (1994) Neuropeptide Y microinjected into the suprachiasmatic region phase shifts circadian rhythms in constant darkness. *Peptides* 15: 1475-1478.

Huhman KL, Gillespie CF, Marvel, CL, Albers HE (1996) Neuropeptide Y phase shifts circadian rhythms *in vivo* via a Y2 receptor. *Neuroreport* 7: 1249-1252.

Hunt AE, Al-Ghoul WM, Gillette MU, Dubocovich ML (2001) Activation of MT(2) melatonin receptors in rat suprachiasmatic nucleus phase advances the circadian clock. *Am J Physiol Cell Physiol* 280: C110-C118.

Inyushkin AN, Bhumbra GS, Dyball RE (2009) Leptin modulates spike coding in the rat suprachiasmatic nucleus. *J Neuroendocrinol* 21: 705-714.

Inyushkin AN, Bhumbra GS, Gonzalez JA, Dyball, RE (2007) Melatonin modulates spike coding in the rat suprachiasmatic nucleus. *J Neuroendocrinol* 19: 671-681.

Irwin RP, Allen, CN (2010) Neuropeptide mediated calcium signaling in the suprachiasmatic nucleus network. *Eu. J Neurosci* 32:1497-1506.

Janik D, Mrosovsky N (1994) Intergeniculate leaflet lesions and behaviorally-induced shifts of circadian rhythms. *Brain Res* 651: 174-182.

Jud C, Schmutz I, Hampp G, Oster H, Albrecht U (2005) A guideline for analyzing circadian wheel-running behavior in rodents under different lighting conditions. *Biol Proced Online* 7: 101-116.

Kageyama H, Takenoya F, Hirako S, Wada N, Kintaka Y, Inoue S, Ota E, Ogawa T, Shioda S (2012) Neuronal circuits involving neuropeptide Y in hypothalamic arcuate nucleus-mediated feeding regulation. *Neuropeptides* 46: 285-289.

Kamiji MM, Inui A (2007) Neuropeptide Y receptor selective ligands in the treatment of obesity. *Endocr Rev* 28: 664-684.

Krout KE, Kawano J, Mettenleiter TC, Loewy AD (2002) CNS inputs to the suprachiasmatic nucleus of the rat. *J Neurosci* 110: 73-92.

Larsen PJ, Kristensen P (1998) Distribution of neuropeptide Y receptor expression in the rat suprachiasmatic nucleus. *Mol Brain Res* 60: 69-76.

Laukova M, Alaluf LG, Serova LI, Arango V, Sabban EL (2014) Early intervention with intranasal NPY prevents single prolonged stress-triggered impairments in hypothalamus and ventral hippocampus in male rats. *Endocrinology* 155: 3920-3933.

Lecat S, Belemnaba L, Galzi JL, Bucher B (2015) Neuropeptide Y receptor mediates activation of ERK1/2 via transactivation of the IGF receptor. *Cell Signal* 27: 1297-1304.

Liou SY, Albers HE (1991) Single unit response of neurons within the hamster suprachiasmatic nucleus to neuropeptide Y. *Brain Res Bull* 27: 825-828.

Majoy SB, Heideman PD (2000) Tau differences between short-day responsive and short-day nonresponsive white-footed mice (*Peromyscus leucopus*) do not affect reproductive photoresponsiveness. *J Biol Rhythms* 15: 500-512.

Marchant EG, Watson NV, Mistlberger RE (1997) Both neuropeptide Y and serotonin are necessary for entrainment of circadian rhythms in mice by daily treadmill running schedules. *J Neurosci* 17: 7974-7987.

Mason R, Harrington ME, Rusak B (1987) Electrophysiological responses of hamster suprachiasmatic neurones to neuropeptide Y in the hypothalamic slice preparation. *Neurosci Lett* 80: 173-179.

McArthur, AJ, Hunt AE, Gillette MU (1997) Melatonin action and signal transduction in the rat suprachiasmatic circadian clock: activation of protein kinase C at dusk and dawn. *Endocrinology* 138: 627-634.

Medanic M, Gillette MU (1993) Suprachiasmatic circadian pacemaker of rat shows two windows of sensitivity to neuropeptide Y *in vitro*. *Brain Res* 620: 281-286.

Mendez-Hernandez R, Escobar C, Buijs RM (2020) Suprachiasmatic nucleus–arcuate nucleus axis: interaction between time and metabolism essential for health. *Obesity* 28: S10-S17.

Meredith ME, Salameh TS, Banks WA (2015) Intranasal delivery of proteins and peptides in the treatment of neurodegenerative diseases. *AAPS J* 17: 780-787.

Mistlberger RE (2011) Neurobiology of food anticipatory circadian rhythms. *Physio Behav* 104: 535-545.

Mistlberger RE, Antle MC, Glass JD, Miller JD (2000) Behavioral and serotonergic regulation of circadian rhythms. *Biol Rhythms Res* 31: 240-283.

Mistlberger RE, Skene DJ (2005) Nonphotic entrainment in humans? *J Biol Rhythms* 20: 339-352.

Moore RY, Speh JC, Leak RK (2002) Suprachiasmatic nucleus organization. *Cell Tissue Res* 309: 89-98.

Mrosovsky N (1996) Locomotor activity and non-photic influences on circadian clocks. *Biol Rev Camb Philos Soc* 71: 343-372.

Myers MG, Olson DP (2012) Central nervous system control of metabolism. *Nature* 491: 357-363.

Ocampo-Garces A, Hernandez F, Mena W, Palacios AG (2005) Wheel-running and rest activity pattern interaction in two octodontids (*Octodon degus*, *Octodon bridgesi*). *Biol Res* 38: 299-305.

Ono D, Honma KI, Honma S (2021) Roles of neuropeptides, VIP and AVP, in the mammalian central circadian clock. *Front Neurosci* 15: 650154.

Pfeffer M, Rauch A, Korf HW, von Gall C (2012) The endogenous melatonin (MT) signal facilitates reentrainment of the circadian system to light-induced phase advances by acting upon MT2 receptors. *Chronobiol Int* 29: 415-429.

Rieke F, Warland D, de Ruyter Van, Steveninck R, Bialek W (1999) In: *Spikes: exploring the neural code*. Cambridge, MA: Massachachusetts Institute of Technology Press.

Roseberry AG, Liu H, Jackson AC, Cai X, Friedman JM (2004) Neuropeptide Y-mediated inhibition of proopiomelanocortin neurons in the arcuate nucleus shows enhanced desensitization in ob/ob mice. *Neuron* 41: 711-722.

Saeb-Parsy K, Lombardelli S, Khan FZ, McDowall K, Au-Yong IT, Dyball RE (2000) Neural connections of hypothalamic neuroendocrine nuclei in the rat. *J Neuroendocrinol* 12: 635-648.

Schak, KM, Scordilis SP, Ferreyra G, Harrington ME (2001) Neuropeptide Y activates protein kinase C in hamster suprachiasmatic nuclei brain slices. *Biol Rhythm Res* 32: 201-206.

Seroka CD, Johnson CE, Heideman PD (2008) Variation in nocturnality and circadian activity rhythms between photoresponsive F344 and nonphotoresponsive Sprague Dawley rats. *J Circadian Rhythms* 6: 8.

Serova LI, Tillinger A, Alaluf LG, Laukova M, Keegan K, Sabban EL (2013) Single intranasal neuropeptide Y infusion attenuates development of PTSD-like symptoms to traumatic stress in rats. *Neuroscience* 236: 298-312.

Shibata S, Moore RY (1988) Neuropeptide Y and vasopressin effects on rat suprachiasmatic nucleus neurons *in vitro*. *J Biol Rhythms* 3: 265-276.

Soscia SJ, Harrington ME (2005) Neuropeptide Y does not reset the circadian clock in NPY Y2−/− mice. *Neurosci Lett* 373: 175-178.

Spetter, MS, Hallschmid M (2015) Intranasal neuropeptide administration to target the human brain in health and disease. *Mol Pharmaceutics* 12: 2767-2780.

Stephan FK (2002) The 'other' circadian system: food as a zeitgeber. *J Biol Rhythms* 17: 284-292.

Van Den Pol AN, Obrietan K, Chen G, Belousov AB (1996) Neuropeptide Y-mediated long-term depression of excitatory activity in suprachiasmatic nucleus neurons. *J Neurosci* 16: 5883-5895.

Verwey M, Robinson B, Amir S (2013) Recording and analysis of circadian rhythms in running-wheel activity in rodents. *JoVE* 71: e50186.

Wams EJ, Riede SJ, van der Laan I, ten Bulte T, Hut RA (2017) Mechanisms of non-photic entrainment. In: *Biological timekeeping: clocks, rhythms and behaviour*: Kumar V, editor. Springer, India pp. 395-404.

Wang Z, Xiong G, Tsang WC, Schatzlein AG, Uchegbu IF (2019) Nose-to-brain delivery. *J Pharm Exp Ther* 370: 593-601.

Wickland C, Turek FW (1994) Lesions of the thalamic intergeniculate leaflet block activity-induced phase shifts in the circadian activity rhythm of the golden hamster. *Brain Res* 660: 293-300.

Yannielli PC, Brewer JM, Harrington, ME (2004) Blockade of the NPY Y5 receptor potentiates circadian responses to light: complementary in vivo and in vitro studies. *Eur J Neurosci* 19: 891-897.

Yi CX, van der Vliet J, Dai J, Yin G, Ru L, Buijs RM (2006) Ventromedial arcuate nucleus communicates peripheral metabolic information to the suprachiasmatic nucleus. *Endocrinology* 147: 283-294.

Yi M, Li H, Wu Z, Yan J, Liu Q, Ou C, Chen M (2018) A promising therapeutic target for metabolic diseases: neuropeptide Y receptors in humans. *Cell Physiol Biochem* 45: 88-107.

In: Circadian Rhythms and Their Importance ISBN: 978-1-68507-547-7
Editor: Rajeshwar P. Sinha

Chapter 7

THE ROLE OF CIRCADIAN RHYTHMS DISTURBANCES IN THE PATHOPHYSIOLOGY AND CLINICAL EXPRESSION OF AFFECTIVE DISORDERS

Patrizia Moretti*[*], Giulia Menculini, Sara Bianchi, Lucia Gonfia, Luigi Maria Pandolfi, Filippo De Giorgi and Alfonso Tortorella

Department of Psychiatry, University of Perugia, Italy

ABSTRACT

Affective disorders, particularly Depressive and Bipolar Disorders, represent serious psychiatric conditions that may lie on a continuum and may thus be described under a dimensional aspect. Circadian rhythms, such as sleep-wake, eating, social, activity, and hormonal patterns, may be frequently disrupted in the context of affective spectrum disorders. Circadian rhythms disturbances were hypothesized to play a role in the pathophysiology of affective disorders, also underpinning alterations in

[*] Corresponding Author's E-mail: patrizia.moretti@unipg.it.

the hypothalamic-pituitary-adrenal axis. In this perspective, alterations in circadian rhythms may also be considered as mediators of allostasis and could therefore contribute to disease progression in affective spectrum disorders. In addition, disturbances in the above-mentioned systems present a relevant impact under a clinical point of view, contributing to the complex psychopathology of affective disorders and increasing their impact on global functioning. Indeed, circadian rhythms disturbances were associated with overall higher clinical severity, as well as with more prevalent subthreshold inter-episode symptomatology and worse treatment outcomes. In the present chapter, we will review evidence concerning circadian rhythms disruptions in the context of affective disorders, analyzing possible pathophysiological correlates and describing major clinical aspects that may be influenced by such disturbances.

Keywords: circadian rhythms, pathophysiology, affective disorders, environmental stimuli, HPA

INTRODUCTION

Rhythmicity plays a cardinal role in nature. Indeed, most physiological functions and behaviors of living beings – whether they are cyanobacteria, plants, animals or humans – follow periodic oscillations (Dibner et al. 2010; Kirlioglu and Balcioglu 2020). The presence of such processes among primitive organisms suggests that they may implicate a clear evolutionary advantage, allowing an adaptation to the surrounding environment (Mohawk et al. 2012; Garbazza and Benedetti 2018). The history of our comprehension of circadian clocks stems in the 18th century, when the French scientist Jean-Jacques d'Ortous de Mairan studied the movements of leaves, that remained the same regardless of illumination, thus indicating the existence of an endogenous mechanism regulating them. At the beginning of the 20th century, the idea of a *memory of time* emerged studying the behavior of honeybees that were able to remember the timing of food availability.

With the progress of research, it was understood that it wasn't just about memory, but there was an internal *production of time* mediated by biological clocks, whose molecular bases were mainly clarified through studies on *Drosophila melanogaster* and mice (Schultz 2007).

Based on their duration, biological rhythms can be divided into *circadian rhythms* (which are the most important and have an approximate 24 hours period, i.e., sleep-wake cycle), *ultradian rhythms* (whose duration is less than 24 hours, among others the rapid eye movement - REM and non-rapid eye movement - NREM phases of sleep), and *infradian rhythms* (whose duration is more than 24 hours, like the circa-monthly rhythm of the Hypothalamic-Pituitary-Gonadal axis) (Laje et al. 2018; Prendergast and Zucker 2020). We are primarily going to discuss circadian rhythms, due to their relevance. Circadian rhythms can be defined as endogenous (i.e., unprompted) daily variations in physical, mental and behavioral activity, that can nevertheless synchronize with geophysical time, thus showing entrainability (Zauderer and Gnazer 2015; Garbazza and Benedetti 2018). This is possible through the detection of external time givers, the so-called *zeitgebers*, which are mainly represented by photoperiod but also by other factors, such as meal-times and exercise. In this manner, once received light information from the environment, sleep-wake cycles can be precisely coordinated to light-dark cycles (Wulff et al. 2010; Pavlova 2017; Menculini et al. 2018).

Without this external adjustment, the circadian clock would actually have a periodicity of 25 hours on average, as was demonstrated by volunteers isolating themselves in bunkers or caves (Schultz 2007). In modern times, the "delay" of our internal clock can be indirectly experienced with jet lag: when flying across several time zones, our clock needs to reset according to rapidly changing external cues. Commonly, travelling eastwards determines a worse jet lag syndrome: this can be attributed to a greater difficulty for our "naturally delayed" clock to follow the phase advance of dark-light cycle, besides variations in daylength (Diekman and Bose 2018; Goldbeter and Leloup 2021).

Circadian Rhythm: An Overview

Physiologic and Genetic Bases

In mammals, the circadian system has a hierarchical organization, headed by a central clock, the suprachiasmatic nucleus (SCN), composed of nearly 20,000 neurons and situated in the anterior hypothalamus (Mohawk et al. 2012; Steardo et al. 2019). Within the SCN, two subnuclei can simplistically be distinguished, whose functions and cellular types appear to be different: the core and the shell. The shell, rich in arginine vasopressin (AVP)-positive neurons, is thought to be the actual pacemaker of self-sustaining circadian oscillations. On the other hand, the core, predominantly composed by vasoactive intestinal peptide (VIP)-positive and gastrin-releasing peptide (GRP)-positive cells, appears to gather afferent environmental signals, thus allowing entrainment (Rosenwasser and Turek 2015).

Important features of the SCN as a whole are robustness and plasticity. Each neuron of the SCN shows a cell-specific autonomous circadian oscillator, meaning that it is possible to find, among topographically organized regions of the SCN, cells showing different periods and phases: periods can range from 22 to 30 hours and phases tend to be, respectively, earlier or later. Therefore, what mainly determines the overall 24 hours periodicity of the clock is intercellular coupling, that confers robustness to the system: even if some neurons are lost, the network can still keep the pace. Moreover, this inter-cellular heterogeneity provides a better adaptation to environmental changes, i.e., the variation in photoperiod length according to different seasons (Mohawk et al. 2012; Sollars and Pickard 2015).

The SCN receives afferent direct inputs from approximately 35 different brain areas and indirect inputs (i.e., through multisynaptic pathways) from nearly 85 (Morin 2013). Among these, three can be considered to be the main: the *retinohypothalamic tract*, the *geniculohypothalamic tract* and the afferent system emerging from the *median raphe nucleus* (Morin 2013; Rosenwasser and Turek 2015), as schematized in Figure 1 and discussed later.

The *retinohypothalamic tract* (RHT) originates from intrinsically photoreceptive retinal ganglion cells (ipRGCs), whose axons converge in the optic nerve that reaches the SCN (Atkins et al. 2018; Logan et al. 2019). The ipRGCs are melanopsin-expressing, non-image-forming, short-wavelength-sensitive cells, diffusively placed in the retina. There are at least five different types of ipRGCs, the most represented being the M1 subtype, particularly rich in melanopsin and characterized by a strong response to light stimulation (Schnell et al. 2014; Sollars and Pickard 2015; Wirtz-Justice et al. 2020). When ipRGCs become activated by light, glutamate, substance P, and pituitary adenylate cyclase-activating polypeptide are released from RHT terminals into a subset of SCN neurons, thus evoking excitatory postsynaptic currents that spread to the rest of the SCN. This effect is mediated by electrical coupling, release of other neurotransmitters (such as gamma-aminobutyric acid - GABA), and probably by diffusible signaling by peptides like gastrin-releasing peptide - GRP and vasoactive-intestinal peptide - VIP (Rosenwasser and Tuker 2015; Sollars and Pickard 2015; Atkins et al. 2018). It is interesting to notice that both ipRGCs photosensitivity and SCN neurotransmitter receptors response can vary depending on external factors (such as time of the day or ambient light levels), thus adding complexity and efficiency to the circadian system (Sollars and Pickard 2015; Atkin et al. 2018).

The *geniculohypothalamic tract* (GHT) mostly arises from GABAergic and neuropeptide-Y expressing cells in the intergeniculate leaflet (IGL), part of the lateral geniculate complex, which is also component of the so-called "visual thalamus." The SCN indirectly receives light signals through this pathway, even if its role in circadian photoreception needs to be better investigated. What seems to be clearer is the contribution of the GHT in addressing non-photic stimuli to the SCN, particularly arousal-related ones (Rosenwasser and Turek 2015; Hanna et al. 2017). Moreover, the serotoninergic *median raphe nucleus* provides projections which are as well implied in the modulation of both photic and non-photic stimuli (Morin 2013; Rosenwasser and Turek 2015).

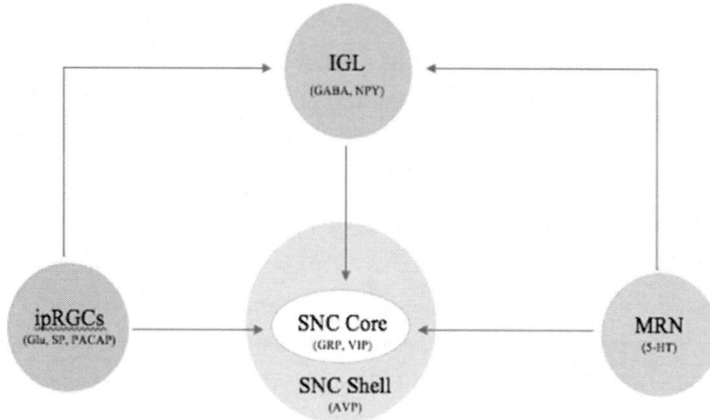

Figure 1. SCN organization and its major afferent systems. 5-HT, 5-hydroxy-tryptamine; AVP, arginine vasopressin; GABA, Gamma aminobutyric acid; Glu, glutamate; GRP, gastrin-releasing-peptide; IGL, intergeniculate leaflet; ipRGCs, intrinsically photoreceptive retinal ganglion cells; MRN, median raphe nucleus, NPY, neuropeptide Y; PACAP, pituitary adenyl cyclase–activating peptide; SP, substance P; VIP, vasoactive intestinal polypeptide.

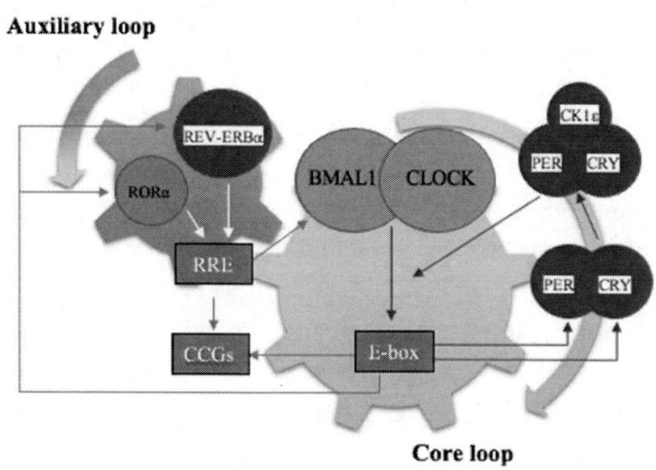

Figure 2. Major circadian genes and molecular loops.

At a molecular level, circadian rhythms are generated by a collection of interlocked transcriptional-translational feedback loops that regulate the expression of different genes. The core loop of the system is represented by the heterodimer composed by the proteins Circadian Locomotor Output

Cycles Kaput (CLOCK) and Brain and Muscle ARNT-Like protein 1 (BMAL1) activated during the daytime, that acts as transcriptional activator of *PERIOD* (*PER1, PER2* and *PER3*) and *CRYPTOCHROME* (*CRY1* and *CRY2*) genes. Their products (the proteins Per 1, 2 and 3 and Cry 1 and 2) accumulate during afternoon and evening and exert a negative feedback on their own production, inhibiting CLOCK and BMAL1 at night (Mohawk et al. 2012; Schnell et al. 2014; Takahashi 2017; Logan and McClung 2019; Steardo et al. 2019). Then, the suppressor complex is progressively degraded by the proteasome, thus stopping the negative feedback and allowing a new production cycle of CLOCK:BMAL1 to start in the morning (Takahashi 2017). Among the other loops coupled to this one, at least one needs to be mentioned: the heterodimer CLOCK:BMAL1 activates the transcription of RORα and REV-ERBα which, respectively, promote and inhibit BMAL1 transcription, additionally contributing to the rhythmic regulation of the circadian clock (Steardo et al. 2019; Xie et al. 2019). Core loop and main auxiliary loop are represented in Figure 2.

Besides transcriptional-translational feedback loops, also post-translational modifications play a relevant role in the circadian system. For instance, the interaction of Per and Cry proteins with Caseine Kinase 1δ (CK1δ) and ε (CK1ε) allows their nuclear translocation and, thus, the inhibition of CLOCK:BMAL1. Moreover, CK1 can also regulate the levels of per protein through its phosphorylation and proteasome-induced degradation (Schnell et al. 2014; Takahashi 2017; Kirlioglu and Balcioglu 2020). Not only do these proteins regulate their own transcription, but they also control that of many so-called clock-controlled genes, determining the systemic effects of the circadian system (Mills et al. 2018).

This same core molecular complex is present in the SCN, peripheral clocks – located both inside and outside the central nervous system – and, virtually, in every cell of the human body. In fact, it has been demonstrated that peripheral oscillators could keep an autonomous rhythm in absence of the SCN (although only for a few days) and that 5 to 10% of tissue cell transcriptome follows daily oscillations (Sollars and Pickard 2015). However, the SCN acts as a master pacemaker, synchronizing all peripheral clocks to keep the best adaptation to both external and internal settings, adjusting constantly thanks to the feedbacks received (Mohawk et al. 2012; Jones and Benca 2015; Rosenwaser and Turek 2015; Takahashi 2017; Xie et al. 2019). The communication between the SCN and the peripheral oscillators occurs through either humoral (especially glucocorticoids, potent transcriptional regulators) and non-humoral pathways, the latter mediated by autonomic nervous system, body temperature and even behavioral processes, such as feeding, that can influence peripheral clocks modulating local metabolism and signaling (Mohawk et al. 2012). Much about the nature of the link between central and peripheral clocks, however, is still poorly understood.

Turning to the master clock anatomical output pathways, the SCN projects directly to nearly 15 brain areas, most of which are located in the diencephalon and basal forebrain, releasing neurotransmitters (such as GABA and glutamate), peptides, and paracrine signals (like VIP, antidiuretic hormone - ADH, transforming growth factor α - TGF-α, and Prokineticin 2) (Dibner et al. 2010; Morin 2013; Rosenwasser and Turek 2015). Three of these pathways can be considered to be the major: a rostral one to the medial prefrontal cortex; a caudal route into the ventromedial nucleus; a dorsal and caudal pathway, the largest, directed to the paraventricular hypothalamic nucleus (PVN) and to the ventral subparaventricular zone, with some projections reaching also the ventrolateral preoptic nucleus. The third accounts for regulation of autonomic functions, body temperature, melatonin, and corticotropin releasing hormone secretion and facilitates sleep (Sollars and Pickard 2015). The SCN can thus regulate different processes, such as awareness, affectivity, memory, learning, cognition, motricity, sensitivity, neuroendocrine relay and autonomic responses (affecting both hormone secretion and hormonal target organs susceptibility), which explains why circadian clock desynchronization can be clinically relevant (Dibner et al. 2010; Krishnan and Lyons 2015; Rosenwasser and Turek 2015).

Main Types of Circadian Rhythms Disrupted in Psychiatric Disorders

Circadian misalignment embraces various conditions, ranging from inappropriate sleep or wake timing to desynchronization between either central and peripheral clocks (e.g., variations in food availability have no impact on the SCN, but can alter hepatic rhythm), or different types of circadian rhythms (e.g., the discrepancy between sleep-wake cycle and feeding cycle), as well as body rhythms and environmental cues (as happens in night-shift workers) (Baron and Reid 2014; Kirlioglu and Balcioglu 2020). During the last years, increasing attention has been paid to seasonality and circadian rhythm disruption in many psychiatric disorders, such as schizophrenia and eating disorders, but particularly affective disorders (Wulff et al. 2010; Bunney and Bunney 2013; Wirtz-Justice 2017; Menculini et al. 2019). The most frequently altered circadian rhythms in clinical subjects include sleep-wake, hormonal, feeding, activity, and social patterns (Mondin et al. 2017; Hong and Zhang 2019; Menculini et al. 2021). We are thus going to summarize their physiological functioning, prior to their pathological disruption in affective disorders.

Sleep-Wake Cycle

Sleep-wake cycle is currently explained as the final effect of two different processes: a homeostatic drive and a circadian mechanism. The first, called process S, represents sleep pressure that increases progressively with the duration of the waking state and is probably mediated by adenosine accumulation. The second, named process C, is the endogenous biological rhythm and balances the effect of the prior (Ono and Yamanaka 2017; Pavlova 2017).

The anatomical basis of the sleep-wake cycle mostly resides in brainstem and hypothalamus. In pons, mesencephalon, and posterior hypothalamus the ascending reticular activating system (ARAS or RAS) allows awareness through diffuse depolarizing projections to thalamus and neocortex (Luppi and Fort 2019). The RAS is mainly composed of four regions: locus coeruleus, noradrenergic; dorsal and median raphe nuclei, predominantly serotoninergic but also containing some dopaminergic neurons; pedunculopontine tegmentum, cholinergic; posterior tuberomammillary hypothalamus, histaminergic (Arguinchona and Tadi 2020). Additionally, also orexin/hypocretin neurons and melanin concentrating hormone-producing neurons in lateral hypothalamus, as well as other glutamatergic and GABAergic cells, play a role in maintaining the waking state (Ono and Yamanaka 2017). Instead, in the anterior hypothalamus there is what is referred to as the Hypothalamic Sleep System (HSS) that actively inhibits many RAS components at the beginning of sleeping. The SCN basically regulates sleep-wake cycle facilitating RAS and suppressing HSS activity during the daytime, then doing the opposite at nighttime (Conti 2010).

Sleep also shows an ultradian rhythm, which consists in the oscillation among four NREM or Slow-Wave Sleep (SWS) stages and a fifth REM stage, also called Paradoxical Sleep (PS) (Luppi and Fort 2019). This ultradian cycle - composed by the progression from the first to the fourth NREM stage, then back again and finally the REM phase - is repeated nearly 4 or 5 times each night.

NREM and REM phases are different mental experiences: while NREM sleep is characterized by reduced frequency and increased synchronization of neuronal activity, this resulting in progressively deeper loss of consciousness, REM sleep shows a brain activity very similar to the waking state, but no skeletal muscle tone and, in contrast, rapid eye movements and irregular breathing rate (Patel et al. 2020). NREM and REM stages are also controlled by different neuronal circuits that can be simplistically schematized as follows: hypothalamic ventrolateral and median preoptic nuclei contribute to NREM induction and stability, whereas neurons generating REM phases are in the pontine reticular formation (Chen et al. 2018; Luppi and Fort 2019). The transition from NREM to REM phases is possible through antagonistic interactions between the so-called REM-OFF cells (mainly hypothalamic histaminergic neurons and serotoninergic and noradrenergic cells in brainstem) and the REM-ON ones (mostly cholinergic, in brainstem and basal prosencephalon) (Chen et al. 2018). Bearing in mind that sleep is fundamental to maintain synaptic plasticity and superior brain functions (such as memory, cognition, mood and emotional regulation) as well as to regulate autonomous nervous system, hormones and metabolism, it is easily understandable why sleep-wake cycle disruptions can affect both mental and physical health (Tobaldini et al. 2016; de Vivo et al. 2017; Eidelman-Rothman et al. 2019; Klinzing et al. 2019).

Hormonal Circadian Rhythms

Several hormones show a circadian pulsatile secretion, this assuming particular relevance in light of the growing comprehension of circadian rhythm disruption consequences on hormonal secretion and core body temperature regulation (Resuehr et al. 2019). When taking into account the pathophysiology of hormonal rhythms in psychiatric disorders, the most important molecules to keep in mind are melatonin and glucocorticoids, also considering their role in synchronizing central and peripheral clocks; anyway, the others (such as growth hormone - GH, prolactin, insulin and gastro-intestinal peptide) must not be underestimated, especially when it comes to psychiatric disorders comorbidities and outcomes (Lopresti et al. 2013; Potter et al. 2016). Melatonin is a hormone secreted at night by the pineal gland under the regulation of the SCN and acts as an internal biological marker of darkness, promoting sleep and synchronizing central and peripheral clock through G-protein-coupled membrane receptors MT_1 and MT_2: for this reason, melatonin is also used as a therapeutic tool (Geoffroy et al. 2019; Agorastos et al. 2020). Melatonin thus has a systemic impact on energy metabolism, contributes to body temperature and blood pressure regulation, and shows neuroprotective, antioxidants, and immunoregulating properties (Gooley et al. 2011; Behn and de Gregorio 2020).

There is increasing evidence of a melatonin role in stress response: it can prevent adrenal response to adrenocorticotropic hormone - ACTH and modulate glucocorticoid receptors (Agorastos and Olff 2020). Glucocorticoids also show a rhythmic secretion, with cortisol blood peak happening in the early morning, which is under the control of the hypothalamic-pituitary-adrenal (HPA) axis, but also of the SCN, that has turned out to be necessary to maintain physiological glucocorticoids levels and pulsatility (Chung et al. 2011; Morris et al. 2012). Glucocorticoids have a part in a multitude of systemic effects through their impact on gene expression mediated by nuclear receptors, thus influencing stress response, metabolic and cardiovascular regulation, immunoregulation and superior functions, such as memory and cognition (Oster et al. 2017; Shimba and Ikuta 2020).

Alimentary Rhythms

As well as sleep, also food intake is regulated by both a homeostatic mechanism that keeps a balance between orexigenic and anorexigenic factors through hypothalamic pathways, and a circadian periodicity, with food intake and energy storing happening during the waking state, while fasting and energy store utilization occur during sleep (Challet 2019). This implies that sleep-wake and feeding-fasting cycles act in concert to modulate energy balance, mainly thanks to lateral hypothalamus (Schulka and Basheer 2016; Gazea et al. 2018). In fact, the central pacemaker of alimentary rhythms is once again the SCN, but it is assisted by other fundamental "food clocks": *hypothalamic arcuate nuclei* (ARC) and *paraventricular nuclei* (PVN) (Challet 2019). During fasting, gastric cells release a potent orexigenic signal, ghreline, that activates the ARC, this determining release of neuropeptide Y and Agouti-related peptide in lateral hypothalamus (which in turn secretes orexine/hypocretin and melanine-concentrating hormone) and in PVN (where oxytocin-expressing neurons are inhibited): this, as a whole, can be considered the main biochemical pathway that leads to food intake (Schulka and Basheer 2016; Challet 2019). On the opposite, hormones like insulin, glucagon-like peptide 1 and leptin reach the ARC after meals, where α-Melanocyte Stimulating Hormon is released towards PVN and other regions to promote satiation (Challet 2019).

Finally, also dopaminergic reward system and dorsal raphe nucleus play a relevant role in energy balance and alimentary rhythm regulation (Nectow et al. 2017; Volkow et al. 2017). In fact, increased central *dopaminergic signaling* tends to stimulate food intake, while dopamine action on different hypothalamic regions can both stimulate or reduce feeding (Folgueira et al. 2019). At the same time, the monoaminergic system is influenced by the circadian clock, particularly dopamine synthesis and degradation - through clock genes regulation on tyrosine hydroxylase and monoaminoxidase A, respectively (Albrecht 2017). The *dorsal raphe nucleus*, instead, regulates lateral hypothalamus activity, sending satiety-promoting serotoninergic inputs (Gazea et al. 2018).

Activity Rhythms

Even if every human being has the same circadian system, its activity peaks in terms of both physiological and superior brain functions can occur earlier or later during the day, thus determining a continuum of different activity patterns between two extremes, the so-called chronotypes (Fisher et al. 2017; Kivelä et al. 2018). Morning types or "larks" tend to wake up and go to bed earlier, having in the morning their cognitive performance peak, while the opposite is true for the evening types or "owls" (Randler et al. 2017). Nearly 40% of adults fall between one of the two extreme chronotypes, while the others show an intermediate chronotype (Kivelä et al. 2018). These activity patterns can be considered to be quite stable in each individual, even if there is a physiological evolution throughout the lifespan: the timing of sleep-wake cycle tends to be earlier during infancy, shifts during adolescence towards a later chronotype and then returns towards a morning chronotype with ageing (Randler et al. 2017). This can be explained by the fact that circadian rhythms, that emerge after birth, are initially entrained by light, then progressively by other physiological (for example gonadal hormones), environmental and social stimuli that are particularly relevant during adolescence. Finally, even the circadian system "grows old," loosing efficiency: this accounts for the poorer sleep quantity and quality often associated with older age (Logan and McClung 2019).

Social Rhythms

Social rhythms are defined as a habitual pattern by which each individual organizes its lifestyle and engages in social activities. Almost any routine interaction with the environment could be considered to be a social zeitgeber, even if performed alone, but five are the most variable behaviors in daily routine that should thus be taken into account: getting out from and going to bed, first contact with a human being, starting school or work, and eating dinner (Haynes et al. 2016). Social zeitgebers have been theorized to have a major role in mood disorders pathogenesis since the formulation of the social zeitgeber theory, that links life events to a disruption of the habitual social and then biological rhythmicity, potentially affecting mood as well as other circadian functions (Ehlers 1988; Sabet et al. 2021).

The term *social jet lag* has been coined to indicate the difference in sleep timing between working and free days: it gives an idea of how social rhythms can influence other circadian patterns, both directly and indirectly (i.e., modifying light exposure, for example in night-shift workers) (Roenneberg et al. 2012; Haynes 2016). The field of social patterns keeps being investigated. Indeed, the concept of *social synchrony* was recently introduced, indicating the alignment between the non-verbal communication of children and parents and its importance for children emotional and social development. A regular and rhythmic structure of social synchrony, in fact, is thought to enable the integration of social patterns and endogenous signals in the insula, in order to "predict" environmental influence on oneself and ensure allostasis (Yaniv et al. 2021). On the contrary, it has been demonstrated that not only may disturbances in social synchrony prevent this to happen, but they could also provoke a long-lasting disruption of circadian system related functions, due to the interconnections between insula and hypothalamus (Yaniv et al. 2021). The increasingly better shaping of social rhythms has prompted the development of social rhythm therapies that, as well as light therapies, aim to reset and stabilize circadian patterns and revert the effects of their disruption, especially on mood disorders (Haynes et al. 2016; Çalıyurt 2017).

Circadian Rhythm Disruptions in Affective Disorders

Affective disorders, namely Depressive Disorders (DD) and Bipolar Disorders (BD), represent heterogeneous clinical entities with relevant clinical impact on life quality and overall personal functioning. The etiopathogenesis of these conditions relies on the complex interaction among genetic factors and several environmental determinants (Lau and Eley 2010, Wittchen et al. 2011). Circadian rhythm disturbances were also hypothesized to be involved in the development of mood disorders. In this framework, biological rhythm abnormalities may be considered mediators at the interface between internal behavioral patterns and responses to environmental factors (Millar 2004). Indeed, as already stated, the circadian clock is involved in the modulation of different neurotransmitter and hormonal systems, such as HPA axis and the monoamine system, that may present major influence on mood states (Srinivasan et al. 2006; Schnell et al. 2014). Noteworthy, different parameters regulated by the circadian clock are altered in subjects affected by mood disorders, including special populations such as adolescents and the elderly (Bellivier et al. 2015; Dallaspezia and Benedetti 2015). The main circadian rhythm disruptions in the context of DD and BD will be analyzed in this section.

Circadian Rhythm Disruptions in Depressive Disorders

Circadian rhythm disruptions were demonstrated to be prevalent clinical features in the context of DD, leading to the hypothesis that alterations in the main biological rhythms may be core psychopathological characteristics of such disorders. Furthermore, circadian rhythm disturbances appeared to influence both the pathogenesis and the clinical presentation of DD, also affecting the disease course (Jones and Benca 2015). This relationship was hypothesized on the basis of common alterations in biological rhythms among subjects affected by Major Depressive Disorder (MDD), who frequently show sleep and eating disruptions, but also hormonal secretion abnormalities (Pinho et al. 2015). Moreover, circadian rhythm alterations appear to influence themselves mutually and present further effects on other circadian rhythms, such as body temperature and activity levels, thus shaping an intertwined association among different circadian patterns (Lam and Levitan 2000; Bunney and Bunney 2013). Furthermore, the relationship among DD and circadian rhythm disturbances is not univocal, since it was also postulated that depression itself could alter circadian rhythms.

Different hypotheses were formulated concerning the role of circadian rhythm dysfunction and the development of DD, but no consensus was reached. Specific biological alterations, particularly sleep disorders, may influence the maturation of neural systems, that results thus to be impaired in case of disruptions of the sleep-wake cycle, possibly facilitating the emergence of psychiatric disorders (Goldstone et al. 2018). Furthermore, circadian rhythm dysregulation influences the emergence of depression mainly due to its effects on the HPA axis (Bechtel 2015). Indeed, an hyperactivation of this hormonal system has extensively been described in the context of DD (Almeida et al. 2021) and awakening cortisol levels were detected to be significantly higher and phase-shifted in these subjects (Elder et al. 2014), which may also be associated with specific alterations in the immune system that were demonstrated in depressed populations. The interest concerning circadian rhythm disruptions and their role in the pathophysiology of DD grown also due to the presence of a specific nosographic entity mentioned in the previous editions of the Diagnostic and Statistical Manual of Mental Disorders (DSM), namely seasonal affective disorder (SAD), which is now defined as recurrent MDD with seasonal pattern (American Psychiatric Association 2013). Despite significant changes in its categorization, the seasonal affective model plays a key role in the understanding of the relationship between circadian rhythms and affective disorders. Subjects affected by SAD display depressive symptoms during winter, with a partial or total remission during spring/summer. The relationship between SAD and photoperiod was demonstrated by several studies, as confirmed by the higher prevalence of the disorder among subjects living in northern countries characterized by a short photoperiod (Albrecht 2010; Salgado-Delgado et al. 2011).

Circadian rhythm disturbances are also frequent in this population, with a major incidence of hypersomnia, carbohydrate craving, and weight gain (Bunney and Bunney 2000; Lewy et al. 2009). Such disruptions were hypothesized to underpin a phase delay in circadian rhythms (Lewy et al. 2006). The study of SAD also allowed clarifying specific associations with genes involved in the circadian clock regulation (Etain et al. 2011; Partonen 2012). Indeed, specific polymorphisms of genes such as *PER1, PER2, BMAL1, NPAS2 and CRY2* were associated with the emergence of SAD and, more in general, of DD (Schnell et al. 2014). These genes interact with several of the above-mentioned systems, in a complex interplay that involves monoaminergic transmission, HPA axis, and metabolic functions. This may explain the influence of clock genes also on DD course, since their disruption was associated not only with specific circadian rhythm abnormalities in this population, but also with worse treatment response (Ma et al. 2018) and illness relapses (Soria et al. 2010).

In the context of DD, different clinical features associated with circadian rhythm disruption may be identified. Sleep abnormalities are the most relevant among these disturbances and may be represented both by insomnia and hypersomnia, but also by frequent awakenings and disturbing dreams. Sleep-wake alteration in DD was causally linked to alterations in melatonin release (Lanfumey et al. 2013; Karatsoreos 2014). Particularly, melatonin levels appear to decrease at the beginning of a Major Depressive Episode (MDE) and increase again during the remission phase (Dmitrzak-Weglarz and Reszka 2017). Melatonin, which shows a circadian release pattern, appears to be desynchronized in the context of MDD. This is also confirmed by alterations in the release of the above-mentioned hormones, such as cortisol, that appears to be dissociated from melatonin secretion in these subjects (Germain and Kupfer 2008; Monteleone et al. 2011). Similarly, eating patterns are often disrupted and subjects affected by DD more often present reduced appetite, whilst hyperphagia is reported in a minority of cases. To note, in MDD subjects daily activities appear to be reduced, which is consistent with lack of motivation and fatigue that are frequent depressive symptoms. In the context of DD, daily activities do not seem to exert their physiological effects on cortisol release rhythm, thus contributing to the alteration of this hormonal pattern (Stetler et al. 2004). Furthermore, evening chronotype was potentially associated with the emergence of depressive symptoms (Hong and Zhang 2019). Circadian rhythm disruptions in DD are also associated to the phenomenology of depressive symptoms themselves, which may show certain rhythmicity in their presentation. Indeed, subjects affected by DD frequently report symptom exacerbation during the morning, although an opposite pattern is displayed in a minority of cases (Joyce et al. 2005; Wirz-Justice 2008). Patterns of activity in DD were measured by several studies using specific techniques such as actigraphy that allowed detecting a significant reduction of activity levels during the wake phase in this population when compared to healthy controls (Raoux et al. 1994).

Social rhythms were also significantly implicated in the emergence of depression. Indeed, according to the "social rhythm hypothesis," the disruption in social patterns may be involved in the occurrence of MDD (Stetler et al. 2004). As already introduced, this could be particularly due to specific life events that potentially alter social zeitgebers, i.e., work schedule, social demands, and interpersonal relationships, resulting in the disruption of individual social rhythms.

The emergence of circadian rhythm dysfunction in DD may modulate the clinical picture of such conditions. Particularly, DD where circadian rhythm disruptions are more prevalent appear to display higher symptom severity (Emens et al. 2009; Wirz-Justice 2009; Salgado-Delgado et al. 2011). Furthermore, abnormalities in biological rhythms may predict worse treatment outcomes, especially concerning response to antidepressants, as well as lower remission rates, higher suicide risk, and longer duration of untreated illness (Robillard et al. 2016; Rastad et al. 2017; Menculini et al. 2021). Furthermore, circadian rhythm disruptions represent a risk factor for higher comorbidity rates, particularly medical illnesses, such as endocrine and metabolic conditions, as demonstrated by higher rates of type II diabetes, hypertension, lipidic metabolism alterations, and obesity (Eckel et al. 2005; Pan et al. 2012; Valkanova and

Ebmeier 2013; Vancampfort et al. 2014; Barandas et al. 2015; Moreira et al. 2016).

Despite the overall impact of circadian rhythm disturbances in DD, these alterations in the clinical practice are often overlooked. Different tools for measuring biological rhythm abnormalities in DD may be identified, including serum hormone levels, body temperature monitoring, actigraphy, as well as psychometric tools, but none of these is routinely employed. Furthermore, specific interventions for such disruptions still represent an open issue. As for pharmacological agents, agomelatine may be particularly useful due to its agonist effect on melatonin receptors combined with its serotoninergic activity (Karatsoreos 2014). Agomelatine, together with its antidepressant effect, was indeed demonstrated to ameliorate circadian rhythm shift, mainly but not only in SAD (de Bodinat et al. 2010; Kasper et al. 2010; Menculini et al. 2018). Furthermore, non-pharmacological interventions were identified as potentially useful for treating both depressive symptoms and biological dysfunctions in subject with DD. This was particularly demonstrated for sleep deprivation that keeps subjects awake during most of the night-time and was associated with a rapid antidepressant effect in MDD subjects (Bunney and Bunney 2013). Similarly, bright light therapy (BLT) was proven to be efficacious on both depressive symptoms and circadian rhythms disturbances (Menculini et al. 2018; Menculini et al. 2019), possibly resulting in a phase-advance of the main disrupted biological rhythms (Therman et al. 2001). This is confirmed by the fact that BLT appears to be efficacious in the treatment of DD when administered in the morning, whilst evening administration did not display the same results (Burgess et al. 2008; Wrinkle et al. 2005). Psychotherapeutic approaches also demonstrated efficacy in ameliorating biological rhythms disruptions, together with depressive symptoms, in DD. Particularly, cognitive-behavioral therapy (CBT) showed promising effects on treating insomnia (Geoffroy and Palagini 2020), whilst the interpersonal and social rhythm therapy (ISRT) was efficacious in reducing the severity of depressive symptoms and regularizing circadian rhythms (Haynes et al. 2016).

Circadian Rhythm Disruptions in Bipolar Disorders

Circadian rhythm disturbances have been widely demonstrated in people affected by bipolar disorder (BD) (Takaesu 2018). In subjects suffering from BD, irregular circadian rhythms contribute to the pathophysiology of episodes of mania and depression (Gold and Kinrys 2019). Indeed, according to the already-mentioned social zeitgeber theory, in individuals vulnerable to BD life events that disrupt daily activity schedules or social rhythms, like bedtime, wake time, mealtimes, start of work, can disturb circadian rhythms and precipitate symptoms, thus culminating in bipolar affective episodes (Alloy et al. 2017). However, circadian abnormalities can arise not only from external perturbations, but also from endogenous dysfunction. According to genetic studies, several genes are implicated in BD chronobiological disturbances, like *CLOCK, GSK3-β*, *PER3*, *ARNTL*, *TIMELESS*, *NR1D1* and *ROR* (Charrier et al. 2017; Lai et al. 2015; Kripke et al. 2009; Shi et al. 2008), each one exerting a single modest effect with an additive fashion, supporting a polygenic heritability (Murray and Harvey 2010). Confirms arise from therapeutic findings: *GSK3-β* can indeed be targeted by lithium therapy, suggesting that some of lithium's clinical benefits are related to circadian clock mechanisms (Kaladchibachi et al. 2007; Li et al. 2012), and it is even suggested that chronotherapies, including sleep deprivation and sleep phase advance, modify clock gene machinery, which can represent a pathophysiological explanation to the improvement of mood symptoms (Bunney and Bunney 2013). Noteworthy, recent findings in literature suggest that lithium and sleep deprivation have similar effects at the molecular level (Andrabi et al. 2020).

Similarly to what has been described for DD, circadian disturbances identified in BD mainly concern sleep-wake rhythms, chronotype, melatonin secretion, and social time cues. Such dysfunctions may act as predictors of the disease course, from the first onset to the relapse of mood episodes (Takaesu 2018). As a consequence, reducing variability and consolidating social rhythms appears to constitute a protective factor against BD episodes recurrence (Levenson and Frank 2010).

Previous studies indicated that sleep pattern variations and sleep dysregulation may trigger manic/depressive symptoms, contributing to the onset of mood episodes (Bauer et al. 2006). Indeed, sleep disturbances are among the most prominent correlates of affective episodes and act as indicators of inadequate inter-episode recovery (Plante and Winkelman 2008). Within the latest edition of the DSM-5, sleep disturbance is listed as a diagnostic criterion of both manic and depressive episodes (American Psychiatric Association 2013). Manic episodes are frequently characterized by reduced need for sleep, while insomnia and hypersomnia occur frequently in depressive phases. Furthermore, sleep disturbances are present in inter-episode periods of BD and there was even coexistence in the types of experienced sleep disturbance, especially during affective episodes with mixed features (Soehnera 2015). According to a recent cohort study, 50% of BD subjects exhibit abnormal sleep, with 50% of them presenting further circadian rhythm disturbances. In fact, subjective reports of various sleep disturbances are described in all BD phases, including euthymia, but there is substantial between-study variability in sleep symptoms prevalence and it is currently unknown what proportion of sleep disorder relates to a circadian rhythm disorder or another disease like obstructive sleep apnea, although it is almost certain that abnormal sleep brings worse psychosocial functioning and quality of life (Bradley et al. 2017).

Interestingly, chronic rhythm desynchronization and sleep abnormalities can enhance inflammation, increasing the risk of cardiovascular and metabolic diseases: in BD, an association between lower sleep efficiency and sleep disturbances, higher body mass index (BMI) and metabolic syndrome has already been established (Boudebesse et al. 2015, Machado and Koike 2014). Abnormalities in genetics and secretion of hormones like leptin and ghrelin, that regulate appetite and hunger, may be the missing link between BD and metabolic syndrome, along with melatonin dysregulation and inflammation. Leptin and ghrelin secretion is regulated by the hypothalamus and by the effect of sleep (Brochard et al. 2016).

In BD subjects, there are alterations in daytime fluctuation of cortisol levels, one of the most widely used markers of the circadian cycles (Nováková et al. 2015). It seems that in BD there are larger cortisol fluctuations with flatter diurnal secretion slopes, and subjects with many previous episodes had higher overall cortisol levels and reduced cortisol reactivity to negative daily events, even in remission phases (Havermans et al. 2011). In BD depression, there is an anticipation of the normal daily pattern of body temperature, normalized during remission; changes have been detected even in other hormone patterns, including prolactin, thyroid-stimulating hormone - TSH, GH, and these variations apparently return to normal levels with improving clinical status (Benedetti and Dallaspezia 2009). When evaluating bipolar chronotypes findings in literature reveal a common pattern of evening preference in BD (Gold and Kinrys 2019) with a phase delay of circadian rhythms (Wood et al. 2009). Evening chronotype is linked to a delayed circadian phase, possibly caused by lower melatonin levels and delayed melatonin release (Robillard et al. 2013). Such circadian preference is already found in adolescent and drug-naïve BD subjects (Melo et al. 2017). Biological rhythm dysfunction, measured by the Biological Rhythms Interview for Assessment in Neuropsychiatry (BRIAN) scale, is more prominent in BD than in subjects suffering from MDD, even in euthymic phases, making circadian disturbances a potential trait marker to differentiate between the two conditions (Duarte Faria et al. 2015). Moreover, a study on the functional impact of circadian rhythms disturbances on subjects with BD concluded that most of the deterioration of the functioning during the inter-episode period was due to changes in circadian cycles. In agreement with previous research, a high prevalence of sleep disorders has also been demonstrated during remission phase (Giglio et al. 2010). Finally, another study concluded that during mania episodes the daily profiles of melatonin secretion in BD differed from depression and healthy controls, because it was elevated during daytime, thus accounting for a phase-shift advance (Nováková et al. 2015).

Circadian rhythms in subjects suffering from BD can be restored by both pharmacologic and non-pharmacologic interventions (Benedetti and Dallaspezia 2009). Controlled exposures to environmental stimuli that act on biological rhythms are labeled as chronotherapies. These include partial or total sleep deprivation, sleep phase advance, light therapy, and dark therapy (Wirz-Justice et al. 2009). The latter represents a widely discussed topic in the scientific literature, because of the hypersensitivity of BD subjects to light. Indeed, classic studies evidenced a greater suppression of melatonin synthesis, compared to healthy controls, after light source exposure during the night (Lewy et al. 1985). Such suppression is reduced by both lithium and valproate, drugs that have a chronobiological action (Hallam et al. 2005). It has been hypothesized that the therapeutic effects of light therapy in BD depression may work by shifting circadian rhythms phase. Phototherapy is considered to be well tolerated, although there are some reports concerning manic switches, suicidal ideation, and mood instability (Gonzalez 2014).

In parallel, dark therapy has been proposed for maniac episodes, leading to a faster reduction in symptoms compared to pharmacologic treatment alone (Barbini et al. 2005). In summary, chronotherapeutics offer mental health practitioners a set of nonpharmacologic, rapidly acting, and effective interventions for monotherapy or as adjuvants to conventional medication (Benedetti et al. 2007). However, some pharmacological agents, such as agomelatine and ramelteon, can enable resynchronization of disrupted circadian rhythms by MT1 and MT2 melatonin receptor agonism, with beneficial effect on sleep patterns and potential benefits on depressive symptoms in BD patients (Norman and Olver 2019; Kishi et al. 2021).

Finally, a specific psychosocial intervention for chronobiological disruptions has been developed: the Interpersonal and Social Rhythm Therapy (IPSRT), with the purpose of managing stressful life events, improving the disruptions of social and circadian rhythms, and increasing adherence to medications (Frank et al. 1997). These hypotheses have been confirmed by a recent real-world controlled trial, with a reduction in anxious, depressive and manic symptomatology, as well as in global functioning, along with a better response to mood stabilizers (Steardo et al. 2020).

CONCLUSION

In summary, our hierarchically organized circadian system headed by the SCN regulates many physiological functions, such as sleep-wake and feeding-fasting cycles, hormonal rhythmic secretion, as well as the more complex activity and social rhythms. A growing comprehension of the molecular basis underlying the circadian system, in addition to a progressively more accurate delineation of the clinical features related with affective disorders, is allowing outlining a two-sided relationship between circadian rhythms disturbances and disease physiopathology and course, especially in BD. Moreover, it is now possible to offer new effective therapeutical interventions, both pharmacological and non-pharmacological, targeting the circadian system and thus enhancing treatment opportunities for subjects suffering from affective disorders in a personalized perspective.

REFERENCES

Agorastos A, Nicolaides NC, Bozikas VP, Chrousos GP, Pervanidou P (2020) Multilevel interactions of stress and circadian system: Implications for traumatic stress. *Front Neurol* 10: 1003.

Agorastos A, Olff M (2020) Traumatic stress and the circadian system: neurobiology, timing and treatment of posttraumatic chronodisruption. *Eur J Psychotraumatol* 11(1): 1833644.

Albrecht U (2010) Circadian clocks in mood-related behaviors. *Ann Med* 42(4): 241-251.

Albrecht U (2017) Molecular mechanisms in mood regulation involving the circadian clock. *Front Neurol* 8: 30.

Alloy LB, Ng TH, Titone MK, Boland EM (2017) Circadian rhythm dysregulation in bipolar spectrum disorders. *Curr Psychiatry Rep* 19(4).

Almeida FB, Pinna G, Barros H (2021) The Role of HPA axis and allopregnanolone on the neurobiology of major depressive disorders and PTSD. *Int J Mol Sci* 22(11): 5495.

Andrabi M, Andrabi MM, Kunjunni R, Sriwastva MK, Bose S, Sagar R, Srivastava AK, Mathur R, Jain S, Subbiah V (2020) Lithium acts to modulate abnormalities at behavioral, cellular, and molecular levels in sleep deprivation-induced mania-like behavior. *Bipolar Disord* 22(3): 266-280.

Arguinchona JH, Tadi P (2020) Neuroanatomy, reticular activating system. In: *StatPearls.* Treasure Island (FL): StatPearls Publishing.

Atkins N Jr, Ren S, Hatcher N, Burgoon PW, Mitchell JW, Sweedler JV, Gilette MU (2018) Functional peptidomics: Stimulus- and time-of-day-specific peptide release in the mammalian circadian clock. *ACS Chem Neurosci* 9(8): 2001-2008.

Barandas R, Landgraf D, McCarthy MJ, Welsh DK (2015) Circadian clocks as modulators of metabolic comorbidity in psychiatric disorders. *Curr Psychiatry Rep* 17(12): 98.

Barbini B, Benedetti F, Colombo C, Dotoli D, Bernasconi A, Cigala-Fulgosi M, Florita M, Smeraldi E (2005) Dark therapy for mania: A pilot study. *Bipolar Disord* 7(1): 98-101.

Baron KL, Reid KJ (2014) Circadian misalignment and health. *Int Rev Psychiatry* 26(2): 139-154.

Bauer M, Grof P, Rasgon N, Bschor T, Glenn T, Whybrow PC (2006) Temporal relation between sleep and mood in patients with bipolar disorder. *Bipolar Disord* 8(2): 160-167.

Bechtel W (2015) Circadian rhythms and mood disorders: Are the phenomena and mechanisms causally related? *Front Psychiatry* 6: 118.

Behn C, De Gregorio N (2020) Melatonin relations with energy metabolism as possibly involved in fatal mountain road traffic accidents. *Int J Mol Sci* 21(6): 2184.

Bellivier F, Geoffroy PA, Etain B, Scott J (2015) Sleep- and circadian rhythm-associated pathways as therapeutic targets in bipolar disorder. *Expert Opin Ther Targets* 19(6): 747-763.

Benedetti F, Barbini B, Colombo C, Smeraldi E (2007) Chronotherapeutics in a psychiatric ward. *Sleep Med Rev* 11(6): 509-522.

Benedetti F, Dallaspezia S (2009) Melatonin, circadian rhythms, and the clock genes in bipolar disorder. *Curr Psychiatry Rep* 11(6): 488-493.

Boudebesse C, Geoffroy PA, Henry C, Germain A, Scott J, Lajnef M, Leboyer M, Bellivier F, Etain B (2015) Links between sleep and body mass index in bipolar disorders: An exploratory study. *Eur Psychiatry* 30(1): 89-93.

Bradley AJ, Webb-Mitchell R, Hazu A, Slater N, Middleton B, Gallagher P, McAllister-Williams H, Anderson KN (2017) Sleep and circadian rhythm disturbance in bipolar disorder. *Psychol Med* 47(9): 1678-1689.

Brochard H, Boudebesse C, Henry C, Godin O, Leboyer M, Étain B (2016) Metabolic syndrome and bipolar disorder: Is sleep the missing link?. *L'encephale* 42(6): 562-567.

Bunney BG, Bunney WE (2013) Mechanisms of rapid antidepressant effects of sleep deprivation therapy: clock genes and circadian rhythms. *Biol Psychiatry* 73(12): 1164-1171.

Burgess HJ, Fogg LF, Young MA, Eastman CI (2004) Bright light therapy for winter depression-is phase advancing beneficial? *Chronobiol Int* 21(4-5): 759-775.

Çalıyurt O (2017) Role of chronobiology as a transdisciplinary field of research: Its applications in treating mood disorders. *Balkan Med J* 34(6): 514-521.

Challet E (2019) The circadian regulation of food intake. *Nat Rev Endocrinol* 15(7): 393-405.

Charrier A, Olliac B, Roubertoux P, Tordjman S (2017) Clock genes and altered sleep–wake rhythms: Their role in the development of psychiatric disorders. *Int J Mol Sci* 18(5): 1-22.

Chen KS, Xu M, Zhang Z, Chang WC, Gaj T, Schaffer DV, Dan Y (2018) A hypothalamic switch for REM and non-REM sleep. *Neuron* 97(5): 1168-1176.

Chung S, Son GH, Kim K (2011) Circadian rhythm of adrenal glucocorticoid: its regulation and clinical implications. *Biochim Biophys Acta* 1812(5): 581-591.

Conti F (ed.) (2010) *Fisiologia Medica*, Volume 1. II Edizione, Edi. Ermes s.r.l. - Milano. [*Medical Physiology*]

Dallaspezia S, Benedetti F (2015) Chronobiology of bipolar disorder: therapeutic implication. *Curr Psychiatry Rep* 17(8): 606.

de Bodinat C, Guardiola-Lemaitre B, Mocaër E, Renard P, Muñoz C, Millan MJ (2010) Agomelatine, the first melatonergic antidepressant: discovery, characterization and development. *Nat Rev Drug Discov* 9(8): 628-642.

de Vivo L, Bellesi M, Marshall W, Bushong EA, Ellisman MH, Tononi G, Cirelli C (2017) Ultrastructural evidence for synaptic scaling across the wake/sleep cycle. *Science* 355(6324): 507-510.

Dibner C, Schibler U, Albrecht U (2010) The mammalian circadian timing system: organization and coordination of central and peripheral clocks. *Annu Rev Physiol* 72: 517-549.

Diekman CO, Bose A (2018) Reentrainment of the circadian pacemaker during jet lag: East-west asymmetry and the effects of north-south travel. *J Theor Biol* 437: 261-285.

Dmitrzak-Weglarz M, Reszka E (2017) Pathophysiology of depression: Molecular regulation of melatonin homeostasis - Current status. *Neuropsychobiology* 76(3): 117-129.

Dumay N, Gaskell G (2005) Do words go to sleep? Exploring consolidation of spoken forms through direct and indirect measures. *Behav Brain Sci* 28:69 - 70.

Faria AD, de Azevedo Cardoso T, Mondin TC, de Mattos Souza LD, da Silva Magalhaes PV, Zeni CP, da Silva RA, Kapczinski F, Jansen K (2015) Biological rhythms in bipolar and depressive disorders: A community study with drug-naïve young adults. *J Affect Disord* 186: 145-148.

Eckel RH, Grundy SM, Zimmet PZ (2005). The metabolic syndrome. *Lancet* 365(9468): 1415-1428.

Ehlers CL (1988) Social zeitgebers and biological rhythms: A unified approach to understanding the etiology of depression. *Arch Gen Psychiatry* 45(10): 948.

Eidelman-Rothman M, Ben-Simon E, Freche D, Keil A, Hendler T, Levit-Binnun N (2019) Sleepless and desynchronized: Impaired inter trial phase coherence of steady-state potentials following sleep deprivation. *NeuroImage* 202: 116055.

Emens J, Lewy A, Kinzie JM, Arntz D, Rough J (2009) Circadian misalignment in major depressive disorder. *Psychiatry Res* 168(3): 259-261.

Etain, B., Milhiet V, Bellivier F, Leboyer M (2011) Genetics of circadian rhythms and mood spectrum disorders. *Eur Neuropsychopharmacol* 21(4): S676-S682.

Folgueira C, Beiroa D, Porteiro B, Duquenne M, Puighermanal E, Fondevila MF, Barja-Fernández S, Gallego R, Hernández-Bautista R, Castelao C, Senra A (2019) Hypothalamic dopamine signaling regulates brown fat thermogenesis. *Nat Metab* 1(8): 811-829.

Fischer D, Lombardi DA, Marucci-Wellman H, Roenneberg T (2017) Chronotypes in the US - Influence of age and sex. *PloS ONE* 12(6): e0178782.

Frank E, Hlastala S, Ritenour A, Houck P, Tu XM, Monk, TH, Mallinger AG, Kupfer DJ (1997) Inducing lifestyle regularity in recovering bipolar disorder patients: Results from the maintenance therapies in bipolar disorder protocol. *Biol Psychiatry* 41(12): 1165-1173.

Garbazza C, Benedetti F (2018) Genetic factors affecting seasonality, mood, and the circadian clock. *Front Endocrinol* 9: 481.

Gazea M, Patchev AV, Anderzhanova E, Leidmaa E, Pissioti A, Flachskamm C, Almeida O, Kimura M (2018) Restoring serotonergic homeostasis in the lateral hypothalamus rescues sleep disturbances induced by early-life obesity. *J Neurosci* 38(2): 441-451.

Geoffroy PA, Franchi JA, Lopez R, Schroder CM (2019). The use of melatonin in adult psychiatric disorders: expert recommendations by the French institute of medical research on sleep (SFRMS). *L'Encephale* 45(5): 413-423.

Germain A, Kupfer DJ (2008) Circadian rhythm disturbances in depression. *Hum Psychopharmacol* 23(7): 571-585.

Giglio LM, Magalhães PV, Kapczinski NS, Walz JC, Kapczinski F (2010) Functional impact of biological rhythm disturbance in bipolar disorder. *J Psychiatry Res* 44(4): 220-223.

Gold AK, Kinrys G (2019) Treating circadian rhythm disruption in bipolar disorder. *Curr Psychiatry Rep* 21(3): 1-8.

Goldbeter A, Leloup JC (2021) From circadian clock mechanism to sleep disorders and jet lag: Insights from a computational approach. *Biochem Pharmacol* 114482.

Goldstone A, Willoughby AR, de Zambotti M, Franzen PL, Kwon D, Pohl KM, Pfefferbaum A, Sullivan EV, Müller-Oehring EM, Prouty DE, Hasler BP, Clark DB, Colrain IM, Baker FC (2018) The mediating role of cortical thickness and gray matter volume on sleep slow-wave activity during adolescence. *Brain Struct Funct* 223(2): 669-685.

Gonzalez R (2014) The relationship between bipolar disorder and biological rhythms. *J Clin Psychiatry* 75(4): 323-331.

Gooley JJ, Chamberlain K, Smith KA, Khalsa SB, Rajaratnam SM, Van Reen E, Zeitzer JM, Czeisler CA, Lockley SW (2011) Exposure to room light before bedtime suppresses melatonin onset and shortens melatonin duration in humans. *J Clin Endocrinol Metab* 96(3): E463–E472.

Hallam KT, Olver JS, Horgan JE, McGrath C, Norman TR (2005) Low doses of lithium carbonate reduce melatonin light sensitivity in healthy volunteers. *Int J Neuropsychopharmacol* 8(2): 255-259.

Hallam KT, Olver JS, Norman TR (2005) Effect of sodium valproate on nocturnal melatonin sensitivity to light in healthy volunteers. *Neuropsychopharmacol* 30(7): 1400-1404.

Hanna L, Walmsley L, Pienaar A, Howarth M, Brown TM (2017) Geniculohypothalamic GABAergic projections gate suprachiasmatic nucleus responses to retinal input. *J Physiol* 595(11): 3621-3649.

Haynes PL, Gengler D, Kelly M (2016) Social rhythm therapies for mood disorders: An update. *Curr Psychiatry Rep* 18(8): 75.

Havermans R, Nicolson NA, Berkhof J, deVries MW (2011) Patterns of salivary cortisol secretion and responses to daily events in patients with remitted bipolar disorder. *Psychoneuroendocrinology* 36(2): 258-265.

Hong W, Zhang Q (2019) Biological rhythms advance in depressive disorder. In: *Depressive Disorders: Mechanisms, Measurement and Managemen*: Yiru Fang, editor. Springer, pp. 117-133.

Jones SG, Benca RM (2015) Circadian disruption in psychiatric disorders. *Sleep Med Clin* 10(4): 481-493.

Joyce PR, Porter RJ, Mulder RT, Luty SE, McKenzie JM, Miller AL, Kennedy MA (2005) Reversed diurnal variation in depression: associations with a differential antidepressant response, tryptophan: large neutral amino acid ratio and serotonin transporter polymorphisms. *Psychol Med* 35(4): 511-517.

Kaladchibachi SA, Doble B, Anthopoulos N, Woodgett JR, Manoukian, AS (2007) Glycogen synthase kinase 3, circadian rhythms, and bipolar disorder: A molecular link in the therapeutic action of lithium. *J Circadian Rhythms* 5: 1-12.

Kanady JC, Soenhera AM, Harvey AG (2015) A retrospective examination of sleep disturbance across the course of bipolar disorder. *J Sleep Res* 4(2): 1000193.

Karatsoreos IN (2014) Links between circadian rhythms and psychiatric disease. *Front Behav Neurosci* 8: 162.

Kasper S, Hajak G, Wulff K, Hoogendijk WJ, Montejo AL, Smeraldi E, Rybakowski JK, Quera-Salva MA, Wirz-Justice AM, Picarel-Blanchot F (2010). Efficacy of the novel antidepressant agomelatine on the circadian rest-activity cycle and depressive and anxiety symptoms in patients with major depressive disorder: a randomized, double-blind comparison with sertraline. *J Clin Psychol* 71(2): 109-120.

Ketchesin KD, Becker-Krail D, McClung CA (2020) Mood-related central and peripheral clocks. *Eur J Neurosci* 51(1): 326-345.

Kirlioglu SS, Balcioglu YH (2020) Chronobiology revisited in psychiatric disorders: From a translational perspective. *Psychiatry Investig* 17(8): 725-743.

Kishi T, Nishiyama H, Kimura Y, Sadaka T, Terada A, Tatsumi Y, Wakahara M, Iwata N (2021) Melatonin receptor agonists for bipolar mania: A systematic review and meta-analyses of double-blind randomized placebo-controlled trials. *Bipol Disord* 1479-1486.

Kivelä L, Papadopoulos MR, Antypa N (2018) Chronotype and psychiatric disorders. *Curr Sleep Med* 4(2): 94-103.

Klinzing JG, Niethard N, Born J (2019) Mechanisms of systems memory consolidation during sleep. *Nat Neurosci* 22(10): 1598-1610.

Kripke DF, Nievergelt CM, Joo EJ, Shekhtman T, Kelsoe JR (2009) Circadian polymorphisms associated with affective disorders. *J Circadian Rhythms* 7: 1-10.

Krishnan HC, Lyons LC (2015) Synchrony and desynchrony in circadian clocks: impacts on learning and memory. *Learning & Memory (Cold Spring Harbor, NY)* 22(9): 426-437.

Lai YC, Kao CF, Lu ML, Chen HC, Chen PY, Chen CH, Shen WW, Wu JY, Lu RB, Kuo PH (2015) Investigation of associations between NR1D1, RORA and RORB genes and bipolar disorder. *PLoS ONE* 10(3): e0121245.

Laje R, Agostino PV, Golombek DA (2018) The times of our lives: Interaction among different biological periodicities. *Front Integr Neurosc* 12: 10.

Lam RW, Levitan RD (2000) Pathophysiology of seasonal affective disorder: a review. *J Psychiatry Neurosci* 25(5): 469-480.

Lanfumey L, Mongeau R, Hamon M (2013) Biological rhythms and melatonin in mood disorders and their treatments. *Pharmacol Ther* 138(2): 176-184.

Lau JY, Eley TC (2010) The genetics of mood disorders. *Annu Rev Clin Psychol* 6: 313-337.

Lewy AJ, Lefler BJ, Emens JS, Bauer VK (2006) The circadian basis of winter depression. *Proc Natl Acad Sci USA* 103(19): 7414-7419.

Lewy AJ, Emens JS, Songer JB, Sims N, Laurie AL, Fiala SC, Buti AL (2009) Winter Depression: Integrating mood, circadian rhythms, and the sleep/wake and light/dark cycles into a bio-psycho-social-environmental model. *Sleep Med Clin* 4(2): 285-299.

Li J, Lu WQ, Beesley S, Loudon A SI, Meng QJ (2012) Lithium impacts on the amplitude and period of the molecular circadian clockwork. *PLoS ONE* 7(3): 1-8.

Logan RW, McClung CA (2019) Rhythms of life: circadian disruption and brain disorders across the lifespan. *Nat Rev Neurosci* 20(1): 49-65.

Luppi PH, Fort P (2019) Sleep-wake physiology. *Handb Clin Neurol* 160: 359-370.

Lopresti AL, Hood SD, Drummond PD (2013) A review of lifestyle factors that contribute to important pathways associated with major depression: diet, sleep and exercise. *J Affect Disord* 148(1): 12-27.

Ma HY, Liu ZF, Xu YF, Hu XD, Sun N, Li XR, Zhang KR (2019) The association study of clock gene polymorphisms with antidepressant effect in Chinese with major depressive disorder. *J Pers Med* 16(2): 115-122.

Machado RM, Koike MK (2014) Circadian rhythm, sleep pattern, and metabolic consequences: An overview on cardiovascular risk factors. *Horm Mol Biol Clin Investig* 18(1): 47-52.

Melo MC, Abreu RL, Neto VB, de Bruin PF, de Bruin VM (2017) Chronotype and circadian rhythm in bipolar disorder: A systematic review. *Sleep Med Rev* 34: 46-58.

Menculini G, Verdolini N, Murru A, Pacchiarotti I, Volpe U, Cervino A, Steardo L, Moretti P, Vieta E, Tortorella A (2018) Depressive mood and circadian rhythms disturbances as outcomes of seasonal affective disorder treatment: A systematic review. *J Affect Disord* 241: 608-626.

Menculini G, Brufani F, Del Bello V, Moretti P, Tortorella A (2019) Circadian rhythms disruptions and eating disorders: clinical impact and possible psychopathological correlates. *Psychiatr Danub* 31(3): 497-502.

Menculini G, Verdolini N, Brufani F, Pierotti V, Cirimbilli F, Di Buò A, Spollon G, De Giorgi F, Sciarma T, Tortorella A, Moretti P (2021)

Comorbidities, depression severity, and circadian rhythms disturbances as clinical correlates of duration of untreated illness in affective disorders. *Medicina* 57(5): 459.

Millar AJ (2004) Input signals to the plant circadian clock. *J Exp Bot* 55(395): 277-283.

Mills SC, Enganti R, von Arnim AG (2018). What makes ribosomes tick? *RNA Biology* 15(1): 44-54.

Mohawk JA, Green CB, Takahashi JS (2012) Central and peripheral circadian clocks in mammals. *Annu Rev Neurosci* 35: 445-462.

Mondin TC, Cardoso TA, Souza L, Jansen K, da Silva Magalhães PV, Kapczinski F, da Silva RA (2017) Mood disorders and biological rhythms in young adults: A large population-based study. *J Psychiatr Res* 84: 98-104.

Monteleone P, Martiadis V, Maj M (2011) Circadian rhythms and treatment implications in depression. *Prog Neuropsychopharmacol Biol Psychiatry* 35(7): 1569-1574.

Moreira FP, Jansen K, Mondin TC, de Azevedo Cardoso T, da Silva Magalhães PV, Kapczinski F, Frey BN, Oses JP, de Mattos Souza LD, da Silva RA, Wiener CD (2016) Biological rhythms, metabolic syndrome and current depressive episode in a community sample. *Psychoneuroendocrinology* 72: 34-39.

Morin LP (2013) Neuroanatomy of the extended circadian rhythm system. *Exp Neurol* 243: 4-20.

Morris CJ, Yang JN, Scheer F (2012) The impact of the circadian timing system on cardiovascular and metabolic function. *Prog Brain Res* 199: 337-358.

Murray G, Harvey A (2010) Circadian rhythms and sleep in bipolar disorder. *Bipolar Disord* 12(5): 459-472.

Nectow AR, Schneeberger M, Zhang H, Field BC, Renier N, Azevedo E, Patel B, Liang Y, Mitra S, Tessier-Lavigne M, Han MH, Friedman JM (2017) Identification of a brainstem circuit controlling feeding. *Cell* 170(3): 429-442.

Norman TR, Olver JS (2019) Agomelatine for depression: expanding the horizons? *Expert Opin Pharmacother* 20(6): 647-656.

Nováková M, Praško J, Látalová K, Sládek M, Sumová A (2015) The circadian system of patients with bipolar disorder differs in episodes of mania and depression. *Bipolar Disord* 17(3): 303-314.

Ono D, Yamanaka A (2017) Hypotalamic regulation of sleep-wake cycle. *Neurosci Res* 118: 74-81.

Oster H, Challet E, Ott V, Arvat E, de Kloet ER, Dijk DJ, Lightman S, Vgontzas A, Van Cauter E (2017) The functional and clinical significance of the 24-hour rhythm of circulating glucocorticoids. *Endocr Rev* 38(1): 3-45.

Pan A, Keum N, Okereke OI, Sun Q, Kivimaki M, Rubin RR, Hu FB (2012) Bidirectional association between depression and metabolic syndrome: a systematic review and meta-analysis of epidemiological studies. *Diabetes Care* 35(5): 1171-1180.

Partonen T (2012) Clock gene variants in mood and anxiety disorders. *J Neural Transm* 119(10): 1133-1145.

Patel AK, Reddy V, Araujo JF (2020) Physiology, sleep stages. In: *StatPearls*: *Treasure Island* (FL), StatPearls Publishing.

Pavlova M (2017) Circadian rhythm sleep-wake disorders. *Continuum (Minneap Minn)* 23(4): 1051-1063.

Pinho M, Sehmbi M, Cudney LE, Kauer-Sant'anna M, Magalhães PV, Reinares M, Bonnín CM, Sassi RB, Kapczinski F, Colom F, Vieta E (2016) The association between biological rhythms, depression, and functioning in bipolar disorder: a large multi-center study. *Acta Psychiatr Scand* 133(2): 102-108.

Plante DT, Winkelman JW (2008) Sleep disturbance in bipolar disorder: Therapeutic implications. *Am J Psychiatry* 165(7): 830-843.

Prendergast BJ, Zucker I (2020) Ultradian rhythms in mammalian physiology and behavior. *Curr Curr Opin Neurobiol* 40: 150-154.

Potter GD, Skene DJ, Arendt J, Cade JE, Grant PJ, Hardie LJ (2016) Circadian rhythm and sleep disruption: Causes, metabolic consequences, and countermeasures. *Endocr Rev* 37(6): 584-608.

Randler C, Faßl C, Kalb N (2017) From Lark to Owl: developmental changes in morningness-eveningness from new-borns to early adulthood. *Sci Rep* 7: 45874.

Raoux N, Benoit O, Dantchev N, Denise P, Franc B, Allilaire, JF, Widlöcher D (1994) Circadian pattern of motor activity in major depressed patients undergoing antidepressant therapy: relationship between actigraphic measures and clinical course. *Psychiatry Res* 52(1): 85-98.

Rastad C, Wetterberg L, Martin C (2017) Patients' experience of winter depression and light room treatment. *Psychiatry J* 2017: 6867957.

Resuehr D, Wu G, Johnson RL Jr, Young ME, Hogenesch JB, Gamble KL (2019) Shift work disrupts circadian regulation of the transcriptome in hospital nurses. *J Biol Rhythms* 34(2): 167-177.

Robillard R, Naismith SL, Rogers NL, Scott, EM, Ip TKC, Hermens DF, Hickie IB (2013) Sleep-wake cycle and melatonin rhythms in adolescents and young adults with mood disorders: Comparison of unipolar and bipolar phenotypes. *EurPsychiatry* 28(7): 412-416.

Robillard R, Hermens DF, Lee RS, Jones A, Carpenter JS, White D, Naismith SL, Southan J, Whitwell B, Scott EM, Hickie IB (2016) Sleep-wake profiles predict longitudinal changes in manic symptoms and memory in young people with mood disorders. *J Sleep Res* 25(5): 549-555.

Roenneberg T, Allebrandt KV, Merrow M, Vetter C (2012) Social jetlag and obesity. *Curr Biol* 22(10): 939-943.

Rosenwasser AM, Turek FW (2015) Neurobiology of circadian rhythm regulation. *Sleep Med Clin* 10: 403-412.

Sabet SM, Dautovich ND, Dzierzewski JM (2021) The rhythm is gonna get you: Social rhythms, sleep, depressive, and anxiety symptoms. *J Affect Disord* 286: 197-203.

Salgado-Delgado R, Tapia Osorio A, Saderi N, Escobar C (2011) Disruption of circadian rhythms: a crucial factor in the etiology of depression. *Depress Res Treat* 2011: 839743.

Schnell A, Albrecht U, Sandrelli F (2014) Rhythm and mood: Relationship between the circadian clock and mood-related behavior. *Behav Neurosci* 128(3): 326-343.

Shi J, Wittke-Thompson JK, Badner JA, Hattori E, Potash JB, Willour VL, McMahon FJ, Gershon ES, Liu C (2008) Clock genes may influence bipolar disorder susceptibility and dysfunctional circadian rhythm. *Am J Med Genet B Neuropsychiatr Genet* 147(7): 1047-1055.

Shukla C, Basheer R (2016) Metabolic signals in sleep regulation: recent insights. *Nat Sci Sleep* 8: 9-20.

Schulz P (2007) Biological clocks and the practice of psychiatry. *Dialogues Clin Neurosci* 9(3): 237-255.

Shimba A, Ikuta K (2020) Glucocorticoids regulate circadian rhythm of innate and adaptive immunity. *Front Immunol* 11: 2143.

Sollars PJ, Pickard GE (2015) The neurobiology of circadian rhythms. *Psychiatr Clin North Am* 38(4): 645-655.

Soria V, Martinez-Amoros E, Escaramís G, Valero J, Pérez-Egea R, García C, Gutiérrez-Zotes A, Puigdemont D, Bayés M, Crespo JM, Martorell L (2010) Differential association of circadian genes with mood disorders: CRY1 and NPAS2 are associated with unipolar major depression and clock and VIP with bipolar disorder. *Neuropsychopharmacology* 35(6): 1279-1289.

Srinivasan V, Smits M, Spence W, Lowe AD, Kayumov L, Pandi-Perumal SR, Parry B, Cardinali DP (2006) Melatonin in mood disorders. *World J Biol Psychiatry* 7(3): 138-151.

Steardo Jr L, De Filippis R, Carbone EA, Segura-Garcia C, Verkhratsky A, De Fazio P (2019) Sleep disturbance in bipolar disorder: Neuroglia and circadian rhythms. *Front Psychiatry* 10(501).

Steardo L, Luciano M, Sampogna G, Zinno F, Saviano P, Staltari F, Garcia CS, de Fazio P, Fiorillo A (2020) Efficacy of the interpersonal and social rhythm therapy (IPSRT) in patients with bipolar disorder: Results from a real-world, controlled trial. *Ann Gen Psychiatry* 19(1): 1-7.

Stetler C, Dickerson SS, Miller GE (2004) Uncoupling of social zeitgebers and diurnal cortisol secretion in clinical depression. *Psychoneuroendocrinology* 29(10): 1250-1259.

Takaesu Y (2018) Circadian rhythm in bipolar disorder: A review of the literature. *Psychiatry Clin Neurosci* 72(9) 673-682.

Takahashi JS (2017) Transcriptional architecture of the mammalian circadian clock. *Nat Rev Genet* 18(3): 164-179.

Terman JS, Terman M, Lo ES, Cooper TB (2001) Circadian time of morning light administration and therapeutic response in winter depression. *Arch Gen Psychiatry* 58(1): 69-75.

Tobaldini E, Costantino G, Solbiati M, Cogliati C, Kara T, Nobili L, Montano N (2017) Sleep, sleep deprivation, autonomic nervous system and cardiovascular diseases. *Neurosci Biobehav Rev* 74: 321-329.

Yaniv AU, Salomon R, Waidergoren S, Shimon-Raz O, Djalovski A, Feldman R. Synchronous caregiving from birth to adulthood tunes humans' social brain. *Proc Natl Acad Sci USA* 118(14): e2012900118.

Valkanova V, Ebmeier KP (2013) Vascular risk factors and depression in later life: a systematic review and meta-analysis. *Biol Psychiatry* 73(5): 406-413.

Vancampfort D, Correll CU, Wampers M, Sienaert P, Mitchell AJ, De Herdt A, Probst M, Scheewe TW, De Hert M (2014) Metabolic syndrome and metabolic abnormalities in patients with major depressive disorder: a meta-analysis of prevalences and moderating variables. *Psychol Med* 44(10): 2017-2028.

Volkow ND, Wise RA, Baler R (2017) The dopamine motive system: implications for drug and food addiction. *Nat Rev Neurosci* 18(12): 741-752.

Winkler D, Pjrek E, Praschak-Rieder N, Willeit M, Pezawas L, Konstantinidis A, Stastny J, Kasper S (2005) Actigraphy in patients with seasonal affective disorder and healthy control subjects treated with light therapy. *Biol Psychiatry* 58(4): 331-336.

Wirz-Justice A (2008) Diurnal variation of depressive symptoms. *Dialogues Clin Neurosci* 10(3): 337-343.

Wirz-Justice A (2009) From the basic neuroscience of circadian clock function to light therapy for depression: on the emergence of chronotherapeutics. *J Affect Disord* 116(3): 159-160.

Wirz-Justice A, Skene DJ, Münch M (2020) The relevance of daylight for humans. *Biochem Pharmacol* 114304.

Wirz-Justice A (2018) Seasonality in affective disorders. *General and Comparative Endocrinology* 258: 244-249.

Wittchen HU, Jacobi F, Rehm J, Gustavsson A, Svensson M, Jönsson B, Olesen J, Allgulander C, Alonso J, Faravelli C, Fratiglioni L (2011) The size and burden of mental disorders and other disorders of the brain in Europe 2010. *Eur Neuropsychopharmacol* 21(9): 655-679.

Wood J, Birmaher B, Axelson D, Ehmann M, Kalas C, Monk K, Kupfer DJ, Brent D, Monk TH, Nimgainkar VL (2009) Bipolar disorder patients and control individuals. *Psychiatry* 166: 201-209.

Wulff K, Gatti S, Wettstein JG, Foster RG (2010) Sleep and circadian rhythm disruption in psychiatric disorders and neurodegenerative disease. *Nat Rev Neurosci* 11: 589-599.

Xie Y, Tang Q, Chen G, Xie M, Yu S, Zhao J, Chen L (2019) New insight into the circadian rhythm and its related diseases. *Front Physiol* 10: 682.

Zauderer C, Ganzer CA (2015) Seasonal affective disorder: an overview. *Ment Health Pract* 18 (9): 21-24.

In: Circadian Rhythms and Their Importance ISBN: 978-1-68507-547-7
Editor: Rajeshwar P. Sinha

Chapter 8

INSIGHTS INTO CIRCADIAN MODULATION OF INNATE IMMUNITY: THE ROLE OF MELATONIN AND PROENKEPHALIN-DERIVED OPIOID PEPTIDES

Héctor Solís-Chagoyán[1], Edgar Flores-Soto[2], Luis M. Montaño[2], Bettina Sommer[3], Germán O. López-Riquelme[4], Zuly A. Sánchez-Florentino[1], Jesús Argueta[1], Ramón Alvarado-Álvarez[5] and Marcela Valdés-Tovar[6,*]

[1]Laboratorio de Neurofarmacología, Instituto Nacional de Psiquiatría Ramón de la Fuente Muñiz, CDMX, México

[2]Departamento de Farmacología, Facultad de Medicina, Universidad Nacional Autónoma de México, CDMX, México

[3]Departamento de Investigación en Hiperreactividad Bronquial, Instituto Nacional de Enfermedades Respiratorias, CDMX, México

---

[*] Corresponding Author's E-mail: mvaldes@imp.edu.mx.

[4]Laboratorio de Socioneurobiología, Centro de Investigación en Ciencias Cognitivas, Universidad Autónoma del Estado de Morelos, Morelos, México.
[5]Departamento de Fisiología, Facultad de Medicina, Universidad Nacional Autónoma de México, CDMX, México
[6]Departamento de Farmacogenética, Instituto Nacional de Psiquiatría Ramón de la Fuente Muñiz, CDMX, México

ABSTRACT

The environmental day-night cycle is a condition imposed to all living beings on Earth. Through evolution both anatomy and physiology have adapted to this cycle by development of circadian rhythms. Additionally, as a process of vital importance, innate immunity should be optimum and thus, it is subject to tight regulation, including circadian modulation. Although there is evidence of circadian oscillations in diverse immune parameters, most of the underlying mechanisms and the role of the biochemical mediators involved in them have not been completely elucidated. Among biomolecules that could influence immune physiology, melatonin and proenkephalin-derived opioid peptides have been described as modulators of innate immunity. Both mediators can be synthesized in the central nervous system, as well as in peripheral organs following a nocturnal circadian rhythm. Melatonin and Met-Enkephalin (MEnk) rhythms are blunted by exposure to experimental constant light (LL). Apparently, melatonin induces the MEnk-containing peptides synthesis and release, and these opioid peptides are key factors for the action of macrophages. Interestingly, LL disrupts the circadian systems and impairs the innate immune response. Moreover, LL ablated the synthesis of MEnk-containing peptides in splenic macrophages and exogenous melatonin ameliorated the peptides release. Since macrophage functions are crucial to innate-to-adaptive transition of immune responses, the impaired enkephalins production in LL could reveal a novel role of these bioactive peptides as key circadian immunomodulatory effectors. Also, the consequences of the presence of light at night on the immune and circadian systems can be further studied to design strategies to counteract the impact of these environmental alterations on health.

Keywords: circadian rhythms, melatonin, enkephalin, macrophages, innate immunity, adaptive immunity, cytokines

INTRODUCTION

Organisms are immersed in an environment that is composed by abiotic and biotic factors. Among the former, the rotational movement of the Earth causes the cycle of day and night with a 24h-period. Interestingly, all unicellular and multicellular organisms have deeply adapted to this factor, and a core of genes (called clock genes) have evolved to generate and coordinate the rhythmic functions synchronizing them with the photoperiod (Buijs et al. 2021; Xiao et al. 2021). Regarding the biotic factors, different long-term symbiotic inter-specific interactions have been established within the ecological communities that can be positive, neutral, or negative. These ecological interactions can benefit both species (mutualism), can benefit one and be neutral regarding the other (commensalism), or can harm one and benefit the other species (parasitism) (Zeigler 2014). Parasitic interactions between the human as a host and pathogens can be counteracted by activation of the immune system mechanisms to avoid or diminish the damages caused by the infection (Abbas et al. 2018). Evidence has indicated that the photoperiod can influence different components of the immune system, suggesting that some features of this system are regulated by the clock genes.

Additionally, multicellular organisms are made up of a variety of individual organ systems, such as the cardiovascular, respiratory and gastrointestinal systems. Just as the different organs and/or tissues of the same system are closely communicated, the components of different systems interact at different levels by the action of modulating mechanisms such as biochemical messengers that maintain homeostasis and the proper functioning of organisms. Two systems that are evolutionally intended to optimize physiology and protect the organism from environmental challenges or threats are the circadian and the immune systems. The aim of this chapter is to analyze the evidence that suggests the interaction between these systems by two broadly released modulators, melatonin and endogenous opioid peptides, in particular, the MEnk-containing bioactive peptides. First, we present an overview of these systems and both modulators to contextualize the specific actions of melatonin and MEnk-containing peptides on macrophages that are a key cellular component of the immune system. The action of both modulators has been proved by housing animals under constant light; this condition blocks their synthesis and allows studying their function. This environmental condition also mimics the nighttime light pollution which has become pervasive in industrialized cities and highlights the importance of the rhythmic release and action of both circadian messengers to maintain human health.

Overview of Circadian Rhythms: Environmental Factors Influence on Physiology

Cyclical environmental changes generated by Earth's rotation and revolution have imposed adaptive pressure on living organisms through evolution. This has led to the development of an internal system of temporal organization that allows physiology and behavior to be synchronized with the environment to optimize energy and favour vital functions. In this way, most biological and physiological processes have periodic variations known as rhythms. At the molecular level, an extremely precise clock machinery has evolved, consisting of a set of proteins that act as transcription factors that positively and negatively self-regulate through feedback loops and give rise to rhythms in gene expression (Amaral et al. 2014). Thus, biological rhythms, and particularly the circadian variations, manifest at all organization levels: from the molecular to the whole organism (Aguilar et al. 2008). The circadian system of a complex organism is composed of various oscillators (cells or cellular processes) that repeat themselves with a period close to 24 h, synchronization pathways (of phase and period with respect to the light-dark cycle), transmission or communication pathways towards rhythm effector systems, and oscillator-coupling mechanisms (Aguilar et al. 2008).

In this context, the period is defined as the time interval in which a rhythmic variation repeats itself. Endogenous rhythms have an intrinsic period that can be synchronized to environmental cycles. Therefore, an endogenous rhythm with a period of about 24 h is called a circadian rhythm (from the Latin phrase *circa diem*, meaning "about a day"). Cyclical variations in environmental factors provide the body with information regarding the external background, i.e., environmental cues. In fact, the most relevant, constant and predictable of the environmental cycles and one of the main synchronizers (*Zeitgebers*) of the endogenous temporal system of organisms is the day-night cycle (light-dark, under experimental conditions) (Pittendrigh 1993). Thus, the photoperiod variation, i.e., the day and night duration differences along a year, and the intensity and quality of the light provide the organism with information indicating the time of day and the season of the year.

In this regard, in the mammalian eye, light is detected by retinal specialized photoreceptors. In particular, the retinal ganglion cells that contain the photopigment melanopsin that absorbs the blue light (Berson et al. 2002); these photosensible cells transmit photic information from the retino-hypothalamic tract to the suprachiasmatic nucleus (SCN), considered the main circadian clock in vertebrates. The SCN functions as the central pacemaker, that is, it can impose its period and phase on other peripheral oscillators (Aguilar et al. 2008). To achieve this, the circadian rhythmic signal must be communicated to the rest of the organism either through the neural pathway including different efferences of the SCN and/or through the secretion of melatonin by the pineal gland as will be described further on. A circadian rhythm is said to be in synchrony with the external light-dark cycle when a stable phase-relationship is established between the cyclical function and the ambient lighting cycle (Aguilar et al. 2008). If the organisms are subjected to constant dim light or to constant darkness conditions, the endogenous rhythms persist but adopt their own period, close to but not equal to 24 h, and this is called free running (Aschoff 1981). Evidence indicates that circadian rhythmicity can be altered or lost in some environmental conditions such as continuous exposure to bright light generating physiological and behavioral disorders.

In this sense, the use of electric light has provided humanity with countless benefits. Nonetheless, modern society and its 24/7 lifestyle, have been consequently exposing human population and animals to unnatural light conditions inducing some alterations in their circadian systems. The internal desynchronization originated in the aforementioned organisms has been noticed by the scientific community and is known as chronodisruption; it has been defined as "a relevant disturbance of the circadian organization of physiology, endocrinology, metabolism and behavior" (Erren and Reiter 2009). The presence of light at night (LAN) is one of the main drivers of such chronodisruption and has been associated with a higher prevalence of illnesses such as affective and metabolic disorders (Bedrosian and Nelson 2013; Bedrosian et al. 2013b; Escobar et al. 2011; Reiter et al. 2012), and cancer (Jasser et al. 2006; Reiter et al. 2007). It is not a minor issue, contrary to that, twenty years ago Cinzano et al. estimated that in industrialized cities, more than 90% of the population is repeatedly exposed to LAN-pollution (Cinzano et al. 2001).

Exposure to constant light (LL) has been used as an experimental strategy to study disruption of circadian rhythms due to LAN, and the pathophysiologic effects it may produce at different levels. For instance, evidence indicates that exposure to a chronic LL regime increases spontaneous tumor formation and shortens life expectancy in young rats (Vinogradova et al. 2010). Chronic LL exposure in rats also increased carcinogenesis induced by chemical substances and development of metabolic syndrome (Anisimov et al. 2012). Both tumorigenesis and development of metabolic disorders imply an imbalance of the immune system functions. In agreement, at least two studies observed and described an internal desynchronization that impairs the immune responsiveness and resistance to a challenge with toxic chemicals at the molecular level, after a two-week exposure to LL in mice (Trufakin et al. 2007; Shurlygina et al. 2010). Additionally, LAN pollution has been reported to alter the immune system functioning (Bedrosian et al. 2011) and the response to stress in hamsters. Specifically, modulation of delayed-type hypersensitivity (DTH) response was differential among animals that were exposed to acute restraint stress and to artificial LAN regarding animals housed under a light-dark cycle (LD) (Bedrosian et al. 2013a). In non-stressed LAN-exposed animals, the DTH response lasted longer than in LD-exposed animals, meanwhile in stressed animals LAN exposure seemed to prevent an intense DTH reaction. This evidence indicates that immunologic mechanisms involved in DTH response are deeply influenced by environmental factors such as stress and lightning conditions (Bedrosian et al. 2011; Bedrosian et al. 2013a). This evidence together highlights the influence of environmental factors such as the photoperiod regime on physiology, particularly altering the homeostasis and the immune response by disruption of the circadian system.

Overview of the Innate Immune Response and the Transition to Adaptive Immunity

The immune response is a series of processes carried out in a delicately concerted manner, basically including from the antigen recognition to its elimination. The innate immune response encompasses the epithelial barriers that prevent the entry of microorganisms or exogenous particles into the body, as well as the effector mechanisms of soluble molecules and cells that specifically recognize these antigens if they manage to enter the sub-epithelial tissues or the bloodstream. Antigen recognition is centered on general molecular patterns associated with microorganisms or tissue damage that are known respectively as PAMPs (Pathogen-Associated Molecular Patterns) or DAMPs (Danger-Associated Molecular Patterns). The receptors that specifically recognize PAMPs and/or DAMPs are called PRRs (Pattern Recognition Receptors) and are present on plasma and endosomal membranes as well as in the cytoplasm, therefore being capable of recognizing antigens in different cell compartments (Abbas et al. 2018). Neutrophils, macrophages, and dendritic cells (DCs) express the broadest repertoire of PRRs in the body and consequently play a key role in innate immune response, as will be discussed later.

Antigen recognition by the soluble molecules present in the blood or in the extracellular fluids triggers its elimination by direct lysis after activation of the complement system, or by opsonization which targets and facilitates its elimination by phagocytosis. Additionally, chemoattractant and pro-inflammatory substances generated during this process induce the recruitment of leukocytes that recognize and eliminate the antigen through phagocytosis or cytotoxicity; neutrophils, macrophages and DCs are able to develop phagocytosis, while natural killer (NK) and mast cells have cytotoxic activity. Both resident leukocytes and those recruited to the lesion site, as well as epithelial and endothelial cells release cytokines, chemokines, and vasoactive substances that promote inflammation. Epithelial cells also secrete peptides with microbicidal activity (Abbas et al. 2018).

Succinctly, the innate immune response prevents and controls all type of infections, eliminates damaged, transformed or dead cells and stimulates the adaptive immune response. Actually, innate immunity is essential for a proper establishment of adaptive immunity, which is much more versatile in terms of specificity and generates memory cells which are capable of developing a more rapid and stronger response to subsequent encounters with the same antigen (Abbas et al. 2018). Induction of the adaptive immune response is carried out by antigen presenting cells (APCs) that have the ability to recognize and internalize antigens, process them biochemically and couple them to molecules of the Major Histocompatibility Complex (MHC), to externalize and present them to T lymphocytes. Besides, APCs express costimulatory molecules on their membranes and secrete cytokines that induce the activation of T lymphocytes. Both DCs and macrophages have this capacity and therefore have a key role in the transition from the innate to the adaptive immune response.

Adaptive immunity can be cellular or humoral depending on the type of effectors that are produced. In overall, the cellular immune response engenders specific clones of effector T lymphocytes that eliminate the antigen through cell-cell interactions. The effector T lymphocytes of the $CD4^{(+)}$ lineage, also called "helper" (T_H: T helper), can recruit and activate leukocytes specialized in pathogen removal. For instance, T_H1 lymphocytes induce activation of macrophages for enhancing destruction of intracellular microorganisms; T_H17 lymphocytes recruit neutrophils and monocytes and stimulate their activity for the elimination of extracellular pathogens such as bacteria and fungi, while T_H2 lymphocytes recruit and activate eosinophils and mast cells for the elimination of helminths (Abbas et al. 2018). On the other hand, the $CD8^{(+)}$ or cytotoxic lineage of effector T lymphocytes kill infected cells directly through the controlled release of perforin and granzymes, or by activating the Fas receptor through its ligand (FasL). Both mechanisms induce apoptosis of infected cells; this type of response is prototypical for viral-type infections or for microorganisms able to escape from phagosomes into the cell cytoplasm (Abbas et al. 2018).

On the other hand, the humoral immune response stimulates the generation of specific clones of B lymphocytes that differentiate into antibody-producing plasma cells and memory B lymphocytes. The antibodies produced are the response effectors; therefore, this process is called "humoral". This response is mainly directed against extracellular microorganisms and toxins that can be neutralized by antibodies for their further elimination. In case of proteins acting as antigens, the humoral immune response requires the cooperation of T_H lymphocytes (Abbas et al. 2018). A T-cell dependent response is then carried out and it involves APCs that exhibit the antigen in association with class II MHC molecules. This complex is recognized by the T cell receptor (TCR) and this recognition and the binding of costimulatory molecules, trigger the clonal proliferation of specific T_H lymphocytes against that specific antigen (Abbas et al. 2018). This process happens in the T-cell zones from secondary lymphoid organs. Likewise, in the follicles of lymphoid organs, antigen-specific B lymphocytes internalize and process the foreign protein to further present it, associated with class II MHC molecules. Interestingly, B lymphocytes are able to change their chemokine receptor pattern, which allows them to migrate towards the T lymphocyte zone. Finally, B lymphocytes present the antigen to the T_H lymphocytes, which cooperate with B lymphocytes to carry out an efficient humoral immune response.

T_H cells activate B cells by secreting cytokines and by ligand-receptor interactions through membrane molecules such as CD40L and CD40. T_H lymphocytes optimize the humoral immune response facilitating processes such as immunoglobulin isotype change, affinity maturation, and generation/differentiation of memory B lymphocytes. These processes develop in the germinal centers inside the B-cell follicles in secondary lymphoid organs such as lymphatic nodes and the spleen (Abbas et al. 2018).

The adaptive immune response is known as "primary" the first time that the organism comes into contact with an antigen, and consequently, it will be called "secondary" at any other subsequent time when memory cells capable of recognizing the antigen will respond quickly and efficiently to the threat.

As already mentioned, macrophages play a key role in modulating the immune response as they have a wide repertoire of receptors that allow them to identify various damage signals. Macrophages are distributed throughout the body and contribute to the homeostasis of the tissue in which they are located. When activated, they can differentiate into various functional phenotypes because of their plasticity and thus influence the type of immune response necessary in accordance to the nature of the threat. Clearly, macrophage physiology must be precisely regulated to avoid unwanted tissue damage.

Macrophages differentiate from myeloid precursors from the yolk sac or the fetal liver at early stages of development. Later in life, or at inflammatory conditions, they may turnover from monocytes found in blood circulation, which, when incorporated into the tissues, differentiate into macrophages. They specialize according to the tissue to which they belong and even receive different names depending on their location, e.g., microglia at the central nervous system (CNS), osteoclasts at the bone, and Kupffer cells at the liver (Davies et al. 2013). Macrophages participate in the immune response and modulate it at different levels. As was mentioned before, they can act directly as effectors of innate immunity by engulfing pathogenic microorganisms, or infected, aged or malignant cells. Macrophages also secrete pro-inflammatory cytokines and have the ability to transfer or present antigens to lymphocytes (Martínez-Pomares and Gordon 2007), therefore they participate actively in the transition from innate to adaptive immune responses. They could be also effectors of adaptive cell-mediated immunity, when their phagocytic activity becomes enhanced by stimulation of T_H1-derived cytokines. Interestingly, macrophages can also contribute to the resolution of the immune response by secreting anti-inflammatory cytokines. On the other hand, macrophages participate in functions aimed at preserving homeostasis such as wound healing, tissue remodeling, immune surveillance, and maintenance of self-tolerance (Murray and Wynn 2011).

To fulfill their wide repertoire of functions, macrophages possess plasticity to differentiate into diverse functional phenotypes, according to the signals they receive from the extracellular milieu and the interactions they establish with other cells. Macrophages are sensitive to and respond to endogenous damage signals through activation and differentiation. The most studied activation phenotypes are those found at opposite physiological extremes, that is, the M1 and M2 phenotypes, with pro-inflammatory and anti-inflammatory characteristics, respectively (Mosser and Edwards 2008). The M1 phenotype develops once macrophages are activated through the "classical pathway" that comprises the presence of IFN-γ and TNF-α and/or Toll-like receptor (TLR) ligands. M1 macrophages possess increased microbicidal and tumoricidal capabilities and secrete various pro-inflammatory cytokines and other biochemical mediators involved in the cellular-type immune response (Murray and Wynn 2011). Consequently, these macrophages play a key role in the defense of the organism, especially against intracellular microorganisms and malignant cells. Understandably, their activation must be finely tuned, since the cytokines they produce (i.e., IL-1, IL-6, IL-23) can favor the development of autoimmune inflammatory pathologies.

In contrast, the so-called "regulatory" phenotype (Mreg) develops in the presence of glucocorticoids, prostaglandins, immune complexes, TLRs ligands and apoptotic cells, among other factors (Lichtnekert et al. 2013; Mosser and Edwards 2008). The presence of two of these signals is required to induce macrophages´ differentiation towards this phenotype characterized by the secretion of anti-inflammatory cytokines such as IL-10 and TGF-β; Mreg also contribute to the decrease in IL-12 production. Development of this phenotype is favored during the late stages of the adaptive immune response, and its function contributes to the resolution phase of the response to limit inflammation and return to homeostasis. However, some pathogenic microorganisms can emulate the aforementioned activation signals and induce the development of Mreg with deficient phagocytic and microbicidal capacities, thus facilitating the microorganism´s proliferation and spread. In a similar manner, tumor-associated macrophages (TAMs) adopt a regulatory-like phenotype and may promote tumor growth and metastasis (Cassetta et al. 2011).

The M2 phenotype is also known as "wound-healing" because it plays a key role in that regenerative process (Mosser and Edwards 2008; Murray and Wynn 2011). M2 macrophages respond to early signs of tissue damage, including IL-4, produced by basophils and mast cells. In turn, this cytokine induces the production of extracellular matrix components and, along with IL-13, it is secreted in the context of the T_H2-type immune response. Macrophages exposed to these cytokines develop the wound-healing phenotype characterized by diminished phagocytic, microbicidal, and antigen-presenting capabilities. Conceivably, impaired or dysregulated M2 activity is detrimental to the organism resulting in tissue fibrosis and a health hazard since these macrophages are more susceptible to be infected by intracellular microorganisms (Cassetta et al. 2011; Mosser and Edwards 2008). It is important to highlight that macrophages retain the plasticity so they can continuously differentiate into one phenotype or another including intermediate or hybrid phenotypes that share characteristics of each one of them. This plasticity allows an immediate and efficient response to micro-environmental stimuli.

Among biochemical mediators that may influence macrophage activation and differentiation towards particular phenotypes are melatonin and opioid peptides. Interestingly, these modulators are synthesized and released following a nocturnal circadian rhythm as will be described further on, suggesting that modulation exerted by these chemical messengers could be related to the control of macrophage function (as well as other cell types of the immune system) by the circadian system (Okuliarova et al. 2021); hence these molecules can be a pathway to coordinate the function of key cells of the immune system.

CIRCADIAN REGULATION OF INNATE IMMUNITY: BIOCHEMICAL MODULATORS RHYTHMICALLY RELEASED THAT AFFECT IMMUNE CELLS

Melatonin

The indoleamine N-acetyl-5-methoxytryptamine, better known as melatonin (Figure 1), is the main biological product of the pineal gland in vertebrates. It was first isolated by Aaron B. Lerner from bovine pineal glands (Lerner et al. 1958), and, as mentioned earlier, one of the efferent systems of SCN is responsible of melatonin synthesis regulation by the pineal gland. The SCN is located in the anterior hypothalamus and has several nerve fibers projected into the spinal cord. The neurons of this component of the CNS innervate the superior cervical ganglion from where derive the sympathetic post-ganglionic fibers that innervate the pineal gland. Thus, the SCN transmits the stimulus to the pineal gland for melatonin production when the dark phase begins (Reiter 1993).

The sympathetic post-ganglionic terminals innervating the pineal gland release norepinephrine (NE). This neurotransmitter binds to both α- (α-AR) and β-adrenergic receptors (β-AR) on the pinealocyte membrane. The adrenergic signal induces melatonin synthesis by stimulating the expression of N-acetyltransferase (AANAT) and increasing its enzymatic activity (Gupta et al. 2005). AANAT transforms serotonin into N-acetylserotonin. This molecule is converted to melatonin by the hydroxyindole-O-methyltransferase (HIOMT). NE binding to α-AR results in an increase in intracellular calcium concentration that has been related to an upgraded response to stimulation of β-AR (Klein and Berg 1970). The enzymatic activity of pineal AANAT in mammals, and consequently the production and secretion of melatonin, has a circadian rhythm, i.e., the maximum plasma levels of the hormone are reached during the dark phase, regardless of the activity pattern of the species (Reiter 1993).

Recently, the crystallized structure of the melatonin membrane receptors MT1 (Stauch et al. 2019) and MT2 (Johansson et al. 2019) were determined. These proteins are heptahelical class A G-protein coupled receptors (GPCRs) and both have a very compact hydrophobic binding pocket (Stauch et al. 2020). The pharmacological characterization of MT1 and MT2 with synthetic ligands has been considered a useful tool for determining melatonin membrane receptor-mediated effects (Cecon et al. 2018; Jockers et al. 2016); a usual strategy consists on the initial application of a competitive non-selective antagonist such as luzindole or a non-selective agonist such as 6-Chloromelatonin, followed by the discrimination between MT1- and MT2-mediated effects by the use of a MT2-selective antagonist such as 4P-PDOT (Cecon et al. 2018) or DH97, as well as an MT2-selective agonist such as 8-M-PDOT (Jockers et al. 2016).

Melatonin membrane receptors can be coupled to the pertussis toxin-sensitive Gi-type proteins (Figure 3); hence, melatonin triggers the inhibition of the adenylyl cyclase activity inducing the reduction of cAMP concentration. Evidence suggests that MT1 receptors are coupled preferentially to Gi2 and Gi3 (Cecon et al. 2018; Liu et al. 2018; Dubocovich and Markowska 2005). However, in some tissues or cell lines, melatonin membrane receptors can be coupled to the Gq/11 subunit; under this situation, melatonin binding activates the enzyme phospholipase C (PLC) leading to inositol 1,4,5-trisphosphate (IP3) and diacylglycerol (DAG) synthesis and the release of Ca^{2+} from the intracellular stores as well as the activation of the protein kinase C (PKC) (Cecon et al. 2018; Liu et al. 2018; Dubocovich and Markowska 2005).

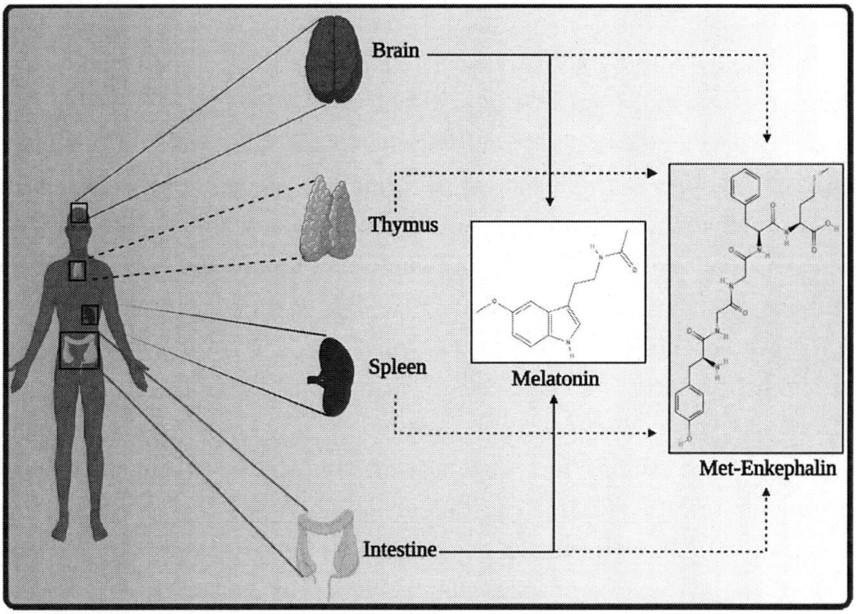

Figure 1. Main sites of Melatonin and Met-Enkephalin production. Although both biochemical messengers and their specific receptors are widely distributed along the diverse organs and tissues of the organism, the main sources of melatonin are the pineal gland and the gut, meanwhile Met-Enkephalin is mainly produced at the CNS and the gut, as well as in lymphoid organs such as the thymus and the spleen.

Exposure to visible light (specially blue light) is an environmental factor that immediately inhibits the enzymatic mechanisms for the production of melatonin. The intensity and wavelength of the light required for this inhibition vary according to the species, and it is recognized that nocturnal animals are more sensitive to these changes. For further detail see Reiter 1993. As widely known, modern society produces intense light pollution that has already been associated with the increase in the prevalence of some diseases, including cancer. In this sense, the group of Blask et al. has shown that the suppression of nocturnal melatonin due to LAN exposure, even with a very dim light intensity (0.2 lux), is associated with an increase in tumor growth and metabolism (Blask et al. 2009), as well as metastasis (Mao et al. 2012).

In this sense, it has also been shown that disorders such as obesity and metabolic syndrome may be associated with LAN seemingly because it generates chronodisruption (Fonken et al. 2013a; Reiter et al. 2012).

The development and progression of cancer as well as metabolic diseases involves an imbalance of the immune system. In this sense, it has been shown that melatonin has various immunomodulatory functions and can have endocrine, paracrine and autocrine actions on the cells of the immune system (Carrillo-Vico et al. 2013). It modulates the production of cytokines and pro-inflammatory enzymes, as well as other mediators of inflammation such as leukotrienes. It also contributes to the proliferation of lymphocytes and/or modulates their apoptosis. Furthermore, due to its capacity to scavenge free radicals, it contributes to redox balance restoration in inflamed tissues (Radogna et al. 2010). Melatonin has pro- or anti-inflammatory effects depending on the stage of the immune response, that is, it enhances its activation in the early phases but contributes to attenuate the response during the resolution phase, to prevent chronic inflammation (Carrillo-Vico et al. 2013).

Proenkephalin-Derived Opioid Peptides

Endogenous opioids are peptides able to bind and activate receptors for opium and its derivatives (opiates). Morphine receptors were discovered in 1973 (Pert and Snyder 1973; Simon et al. 1973; Terenius 1973) and triggered the study of the endogenous opioid neurotrans-mitters. The search for the endogenous molecules with chemical structure similar to morphine was unsuccessful, until in 1975 when the studies of Terenius et al. (1975) suggested the peptidic nature of the endogenous ligands. Enkephalins were the first two opioid peptides identified and characterized. The aminoacid sequence of these pentapeptides was published by Hughes et al. (1975).

The sequence of the first four amino acids is identical in both peptides and is often referred to as the "opioid core". The only difference between enkephalins' sequence is in the fifth, carboxyl terminal amino acid, i.e., methionine for Met-Enkephalin (MEnk) (Table 1), and leucine for Leu-Enkephalin.

Table 1. Sequence of methionine for Met-Enkephalin (MEnk) and leucine for Leu-Enkephalin

Met-Enkephalin	NH_2- Tyr^1-Gly^2-Gly^3-Phe^4-**Met^5**-COOH
Leu-Enkephalin	NH_2- Tyr^1-Gly^2-Gly^3-Phe^4-**Leu^5**-COOH

Currently, three main families of opioid peptides are known, all derived from large precursor proteins (Figure 2) enzymatically cleavaged to generate biologically active peptides. β-endorphin derives from proopiomelanocortin, while dynorphins A and B are generated from prodynorphin, and finally, the aforementioned enkephalins are generated from proenkephalin A (PENK), as well as the octapeptide Met^5-Enkephalin-Arg^6-Gly^7-Leu^8 (MERGL) and the heptapeptide Met^5-Enkephalin-Arg^6-Phe^7 (MERF). The amino acid sequence of PENK has a total of six copies of the MEnk sequence, two of which are part of MERGL and MERF (Udenfriend and Kilpatrick 1983).

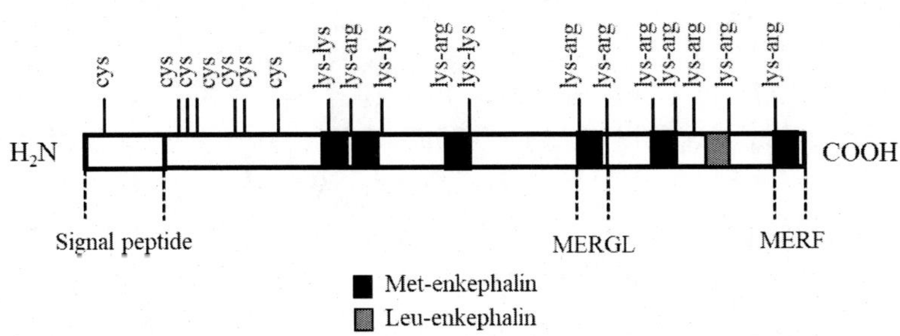

Figure 2. Representation of the preproenkephalin A protein and the sites of proteolytic cleavage into bioactive opioid peptides. PENK has one copy of Leu-Enkephalin and a total of six copies of the MEnk sequence, two of them are part of the octapeptide MERGL and the heptapeptide MERF. Several MEnk-containing peptides (MECPs) of different molecular weights can be generated from cleavage of the PENK precursor.

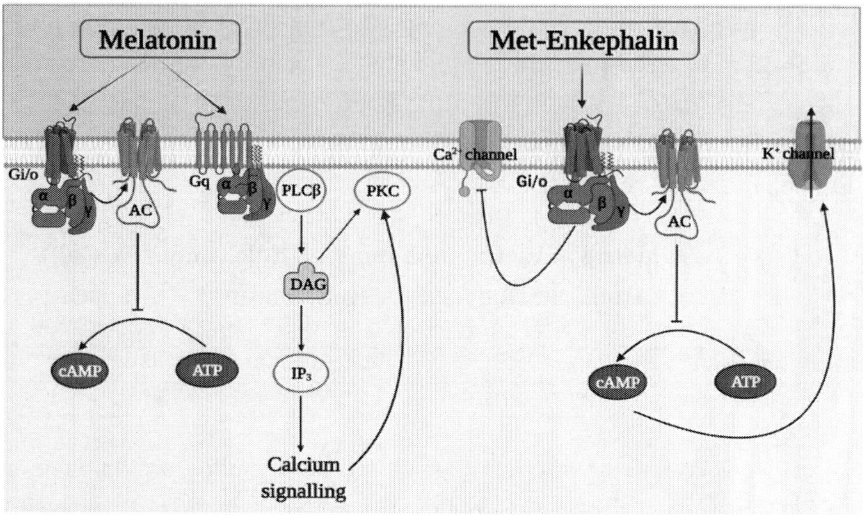

Figure 3. Representation of the main signaling pathways triggered by melatonin and Met-Enkephalin. Both melatonin and MEnk bind to their corresponding class A GPCRs. Melatonin receptors could be coupled to either G_i or G_q proteins meanwhile the three main opioid receptors (δ, μ, and κ, in order of MEnk binding affinity) are coupled to G_i, so they inhibit adenylate-cyclase activity and thus reduce cAMP levels.

Three main types of opioid receptors have been recognized: μ, δ, and κ. These are also class A GPCRs that inhibit the adenylate-cyclase activity reducing the level of cAMP (Figure 3). PENK-derived opioid peptides bind with higher affinity to the δ-opioid receptor, although they can also bind to the μ-opioid receptor. MEnk is also known as the opioid growth factor (OGF) for its functions after binding to the OGF receptor (OGF-R; formerly called ζ-opioid receptor). The study of the OGF/OGF-R system has gained relevance in the context of cancer (McLaughlin and Zagon 2012). Both opioid peptides and their receptors are widely distributed in the body (Figure 1) and are involved in various functions, therefore, they are considered homeostasis modulators (Akil et al. 1998). Among the functions described for the endogenous opioid system are: drug tolerance and/or dependence, food and liquid intake, learning, memory, gastrointestinal, renal and cardiovascular function, thermoregulation, respiration, pregnancy, development, aging, movement, and pain perception (Bodnar 2012). Likewise, it has been associated with pathologies such as depression, epilepsy and cancer. Besides, the endogenous opioid system is also involved in immunomodulation processes and in neuroendocrine-immunological communication.

Many cell lines of the immune system are capable of producing and releasing opioid peptides derived from PENK. The expression of this precursor and the presence of biologically active peptides have been reported in rat lymphocytes (Behar et al. 1991; Rosen et al. 1989) and human thymocytes (Kavelaars and Heijnen 2000). As well, mitogenic activation of human lymphocytes induced the release of enkephalins (Padros et al. 1989). Likewise, Zurawski et al. (1986) demonstrated that antigenic or mitogenic stimulation of a mouse T_H lymphocyte line induces the expression of preproenkephalin A mRNA and the release of opioid material to the culture medium. Polymorphonuclear cells (granulocytes) also produce and release PENK-derived peptides (Vindrola et al. 1990). In monocytes of human origin, these peptides are produced and released in response to stimulation with the phorbol 12-myristate 13-acetate (PMA), and their post-translational processing is cell-type specific (Kuis et al. 1991). Monocytes release MEnk and peptides containing its sequence such as peptide F, while higher molecular weight material is generated in T lymphocytes. This material is not immunoreactive to MEnk until undergoing proteolytic treatment. Similarly, Kamphuis et al. (1998) reported that stimulation of human blood T lymphocytes with anti-CD2/CD28 monoclonal antibodies, as well as stimulation of monocytes with lipopolysaccharide (LPS), induce the production and secretion of MEnk and other intermediate peptides of higher molecular weight containing the MEnk sequence (MECPs: MEnk-containing peptides). The expression of MEnk, as well as of two of the proteolytic enzymes involved in PENK cleavage was demonstrated in human monocyte-derived macrophages (LaMendola et al. 1997). The expression of preproenkephalin A mRNA has also been demonstrated in lymphoid and myeloid cell lines, as well as in macrophages and mast cells of murine origin (Martin et al. 1987). The μ, δ and κ opioid receptors are also expressed in the different cells of the immune system (Sharp 2003, 2006), indicating that the immunomodulatory actions of opioid peptides can be autocrine or paracrine.

Modulation of Spleen Macrophage Physiology and Immune Response by Melatonin and Enkephalins

Spleen Macrophages

The spleen is a secondary lymphoid organ anatomically organized into sections known as red and white pulp, respectively. The region that surrounds and delimits the above is called the marginal zone. In each of these regions, distinct subpopulations of macrophages with specific functions related to their location and the cell types they relate with are present (den Haan and Kraal 2012). Red pulp macrophages specialize in the removal of aged erythrocytes and in iron recycling. Macrophages in the white pulp are located both in the T-cell zone (the periarteriolar lymphatic sheath) and in the B-cell follicles. In these follicles, the macrophages eliminate apoptotic B cells generated in the germinal centers during somatic hypermutation and immunoglobulin isotype change.

In the splenic marginal zone, two populations of macrophages exist: the so-called "marginal zone macrophages" (MZM) and the marginal metalophilic macrophages (MMM). The anatomy of this marginal zone allows a slower blood flow and macrophages strategically located there recognize and eliminate pathogenic microorganisms present in the blood. To achieve this, both MZM and MMM express a wide repertoire of receptors and surface molecules specialized in antigen recognition. Furthermore, both subtypes of macrophages produce cytokines that contribute to the response to these pathogens, such as type I interferon (IFN), interleukin (IL)-1α and tumoral necrosis factor (TNF)-α (den Haan and Kraal 2012).

Also, at the splenic marginal zone resides a specific population of B lymphocytes and DCs that can be stimulated by antigens and migrate to the white pulp. These cells as well as lymphocytes and granulocytes that continuously transit through the marginal zone, can interact with resident macrophages (den Haan and Kraal 2012). Meanwhile, recent studies indicate that macrophages located in this zone are essential for the development of germinal centers, since they recruit and organize antigen-activated B lymphocytes in the follicular perimeter to generate the germinal center reaction (Nikbakht et al. 2013).

In the rat, both MZM and MMM express the surface marker CD169 (O'Neill et al. 2013), also known as sialoadhesin. As its name implies, this marker binds to sialic acid molecules present on various particulate antigens. Splenic MMMs may act in a similar way to macrophages in the subcapsular sinuses of lymph nodes that are capable of binding particulate antigens and transport them from the membrane surface to the interior of the B-cell follicles, where they can recognize the antigen and activate the adaptive immune response. This mechanism has been considered a specific mechanism of antigen presentation of this macrophage subpopulation (Martínez-Pomares and Gordon 2007). Likewise, MMMs can transport antigens from the marginal zone to the white pulp and transfer them to DCs so that they present the antigen to the T cells. In this way, CD169$^{(+)}$-macrophages contribute to induce the cytotoxic immune response (Backer et al. 2010).

In general, CD169$^{(+)}$-macrophages carry out the function of recognition and elimination of pathogens and secretion of pro-inflammatory cytokines, classic of the innate immune response. However, through sialoadhesin, these macrophages also capture and concentrate antigens and establish interactions with B and T lymphocytes to induce the adaptive immune response (den Haan and Kraal 2012). Besides, these macrophages also play a key role in maintaining self-tolerance by not eliciting an inflammatory response after phagocytosis of apoptotic cells or cellular debris (McGaha et al. 2011; O'Neill et al. 2013).

Since their functions are mainly pro-inflammatory, CD169$^{(+)}$-macrophages are involved in the anti-tumor immune response and, according to various studies they might be effective in the development of vaccines containing tumor antigens. In contrast, in autoimmune diseases CD169$^{(+)}$-macrophages might produce detrimental consequences since their absence induces the proliferation of Treg lymphocytes correlating with a lower severity of the condition. However, it has not been established whether the expression of CD169 in macrophages has a causal effect on the initiation or progression of autoimmune diseases or is simply a consequence of the inflammatory process (O'Neill et al. 2013).

Melatonin and Enkephalins Modulation of Macrophage Physiology and Immune Response

Different immunomodulatory functions of peptides derived from PENK, in particular MEnk, have been described. The latter participates in both innate and adaptive immunity, modulating functions of various immunocompetent cells in the early stages of the response. This modulation has been proposed as a survival strategy for the organism (Salzet and Tasiemski 2001). It has been documented that, in human blood mononuclear cells, MEnk stimulates chemotaxis to the site of tissue injury (van Epps and Saland 1984). In mouse peritoneal macrophages, MEnk stimulates the secretion of pro-inflammatory cytokines such as IL-1β (Yang and Li 1989) and IL-6 (Kowalski et al. 2000). Similar results have been found from microglia in the CNS (Das et al. 1995; Kowalski et al. 2002). Likewise, both MEnk and selective μ, δ and κ opioid receptor agonists modulate the metabolic activity of macrophages by stimulating the production of nitric oxide (Kowalski 1998a; Marotti et al. 1998; Vujic et al. 2004) and hydrogen peroxide (Radulovic et al. 1995; Vujic-Redzic et al. 2000). These oxidizing agents are necessary for the efficient destruction of phagocytosed microorganisms.

On the other hand, in rat peritoneal macrophages, the group of Foris et al. (1986) reported a stimulating effect of MEnk (in the range of 10^{-9} to 10^{-7}M) on IgG2a antibody-dependent cytotoxicity and simultaneously an inhibitory effect of Fcγ receptor-mediated phagocytosis. This effect was bidirectional and concentration-dependent since the opposite effect was found with 10^{-6} to 10^{-5}M of MEnk. The increase in phagocytosis induced by MEnk was also demonstrated in a murine malaria model, in which a combined in vivo treatment with granulocyte-macrophage colony stimulating factor (GM-CSF) and MEnk almost eliminated parasitemia. In treated animals, the phagocytic capacity of peritoneal macrophages increased 100 times, while the percentage of macrophages in the spleen and their phagocytic capacity increased approximately 7 times compared to animals that did not receive the treatment (Singh and Singh 2001). MEnk also increases the tumoricidal activity of murine peritoneal macrophages in the presence of IFN-γ and LPS (Hagi et al. 1994). Recent studies have shown that the increase in macrophage tumoricidal activity may be due to the fact that MEnk is capable of inducing functional polarization of TAMs towards the M1 phenotype in vivo. Consistent with this evidence, *in vitro* assays with bone marrow-derived macrophages showed that treatment with MEnk increased their tumoricidal activity, the release of reactive oxygen species (ROS), and the production of IL-12p40 and TNF-α, while the production of the anti-inflammatory cytokine IL-10 decreased (Chen et al. 2012).

Regarding the functions of MEnk in the adaptive immune response, this enkephalin is capable of inducing the maturation of DCs towards a phenotype specialized in the presentation of antigens. By acting on the function of APCs such as DCs and macrophages, MEnk contributes to induce the $CD4^{(+)}$-T cell response (Shan et al. 2011). MEnk can also directly stimulate T lymphocytes since murine (Kowalski 1998b) or human lymphocytes (Hucklebridge et al. 1990; Singh et al. 1999) have been shown to express δ opioid receptors, through which MEnk and other agonists trigger proliferation, intracellular calcium mobilization, and cytokine production. In T cells of the Jurkat line transfected with the δ opioid receptor, an agonist induced IL-2 expression by increasing the NF-AT/AP-1 transcription factor activity (Hedin et al. 1997; Sharp et al. 1998). Likewise, Yang and Li (1992) found that treatment with MEnk increases the production of IL-2 in vivo and in vitro in parallel with the augmented proliferation of T lymphocytes. This effect was not associated with an upregulated expression of the IL-2 receptor.

Consistent with evidence mentioned above, activation of T lymphocytes in vivo or in vitro induced the expression of δ opioid receptors (Sharp et al. 2000; Sharp et al. 1998). The induction of these receptors is strictly mediated by the costimulatory signal produced by CD28-binding at the T lymphocyte membrane during its activation (Nguyen and Miller 2002). Likewise, in other cell lines such as macrophages and NK cells, MEnk (Marotti et al. 1994) or other agonists of the δ opioid receptor (House et al. 1996) induce cytokine production and cytotoxic activity, respectively.

Rat exposure to dim LL for 15 days is a methodological strategy that allows studying the effects of chronodisruption. Although it is a relatively short time exposure, it is enough to completely disrupt the general activity rhythm of the rats (Valdés-Tovar et al. 2015). Other groups have also reported the loss of this circadian rhythm after several days in LL in rats (Cambras et al. 2011, 2012; Novakova et al. 2011; Tapia-Osorio et al. 2013). The locomotor activity rhythm has been considered a reference of the circadian system functioning, therefore, a condition that abolishes the circadian activity pattern of the animals may also alter other circadian rhythms and general functions. In this context, it has been found that LL decreases the amplitude or disorganizes various physiological rhythms such as the expression of *mPER2* in the SCN (Sudo et al. 2003), the neuronal activity of the SCN, the metabolic output and insulin resistance (Coomans et al. 2013), glucocorticoid and melatonin production (Claustrat et al. 2008), intestinal hormone secretion (Gil-Lozano et al. 2014), and the phagocytic capacity of neutrophils (Hriscu et al. 2002), among others. Moreover, this specific dim LL regime drastically reduced nocturnal circulating melatonin levels, reaching similar levels to those found at daytime (Asai et al. 2007; Valdés-Tovar et al. 2015). By using this methodology, Valdés-Tovar et al. (2003) found that the total tissue content of MEnk and other bioactive PENK-derived opioid peptides in lymphoid organs such as the thymus and the spleen were significantly reduced regarding animals exposed to a regular 12:12 LD-cycle. MEnk determination in extracts of isolated mononuclear splenocytes showed a similar significant reduction in LL-exposed rat samples regarding the LD-exposed rats (unpublished results). This single time-point determination was further validated when total splenic MEnk content was determined during a 24h-period. In this case, LD-exposed rats showed a biphasic pattern of splenic MEnk, while in LL no oscillations were observed and splenic MEnk content was significantly lower at all the analyzed hours (Valdés-Tovar et al. 2015). This is consistent with what was previously observed in CNS structures (Asai et al. 2007) and supports the hypothesis that enkephalins' synthesis is affected in this constant lighting condition.

Regarding the effect of dim LL-exposure on MEnk production in immune cells, our group found that *de novo* synthesis of MEnk-containing peptides (MECPs), their intracellular content and the presence of granular structures immunoreactive to MEnk were significantly decreased in splenic CD169$^{(+)}$-macrophages derived from animals exposed to LL (Valdés-Tovar et al. 2015). Evaluation was performed by immunodetection with a specific antiserum against MEnk that showed less than 0.01% cross-reactivity with respect to structure-related peptides (Asai et al. 1998). Given this antiserum's high specificity, peptides derived from enzymatic proteolysis of PENK that have one or more amino acids after the MEnk sequence (e.g., MERF or MERGL) are not recognized. Thus, the peptides immunoreactive to the antiserum can be the MEnk pentapeptide itself, as well as fragments of higher molecular weight with the MEnk sequence at the carboxyl terminal end. As already mentioned, MEnk is generated by the sequential enzymatic cleavage of PENK, its protein precursor. Several enzymes can carry out this process, as for instance pro-hormone convertases 1/3 and 2 (PC1/3 and PC2) (Hook et al. 1996; Rouille et al. 1995), and cathepsin L (Yasothornsrikul et al. 2003). These enzymes generate different types of MECPs (Hook et al. 2004; Hwang et al. 2007). To date, the proteolytic pathway of PENK in immune cells and its regulation have not been fully characterized; notwithstanding, the presence of PENK, MEnk, PC1 and PC2 convertases, and carboxypeptidase E has been detected in human alveolar macrophages (Krajnik et al. 2010).

Furthermore, macrophages also express cathepsin L (Beers et al. 2003), so it is possible that splenic macrophages have both pathways of proteolytic processing of PENK.

The MEnk immunoreactivity pattern distributed into granule-like structures that were also immunoreactive to the vesicle-associated membrane protein VAMP1/2, suggested that the MECPs could be stored in secretory granules in splenic macrophages. In neuroendocrine cells, endogenous opioid peptides are stored within individual granules (Fischer-Colbrie and Frischenschlager 1985; Flemingeret al. 1983; Loh et al. 2002), whereas in cells of the immune system, β-endorphin was shown to be contained into membranous vesicular structures (Mousa et al. 2004). VAMP-related proteins are expressed in murine macrophages and their participation in exocytosis have been documented before (Pitzurra et al. 1996; Shukla et al. 2000). However, as far as we know, the specific mechanisms and regulation of MEnk storage and secretion in macrophages have not been elucidated. The significantly reduced co-localization of MEnk and VAMP1/2 in macrophages derived from animals in LL suggest that these cells have a lower amount of stored MECPs, in contrast to what was observed in macrophages obtained from animals in LD (Valdés-Tovar et al. 2015).

As previously mentioned, MEnk modulates various aspects of macrophage physiology, as well as the physiology of the cells with which macrophages interact in the context of the immune response. In the rat spleen, both MZM and MMM express the CD169 surface marker (O'Neill et al. 2013). Due to their strategic location on the border between the white pulp and the red pulp, these macrophages carry out the identification and elimination of pathogens found in the blood (den Haan and Kraal 2012) and interact with both DCs (Backer et al. 2010) and antigen-activated B lymphocytes (Nikbakht et al. 2013). To induce the transition from the innate to the adaptive immune response, macrophages may act directly as APCs or they could transfer the antigens to DCs so that they in turn perform this function (Backer et al. 2010; Martínez-Pomares and Gordon 2007).

Macrophages also enhance the activation of lymphocytes through the secretion of several cytokines. In particular, CD169$^{(+)}$-macrophages play a key role in inducing adaptive immunity in the spleen.

Early critical events in the development of the adaptive immune response such as activation of T lymphocytes, stimulation of their clonal expansion and their functional polarization, must be properly carried out to get an efficient outcome. In this regard, we found that the primary humoral immune response of rats exposed to constant light is compromised (Valdés-Tovar et al. 2015). After an antigenic challenge with SRBC (sheep red blood cells), there was a lower number of splenic IgM antibody-forming cells, as well as a reduced number of cells secreting IL-2 and IFN-γ cytokines (Valdés-Tovar et al. 2015). These two cytokines are essential in the clonal expansion of T lymphocytes and their functional polarization (Gagro and Gordon, 1999; Kasprowicz et al., 2003; Nishimura et al., 2002). Since MEnk is a direct modulator of lymphocyte activation that also participates in the interaction between DCs and T lymphocytes (Li et al. 2014; Li et al. 2012; Liu et al. 2012; Shan et al. 2011), it is possible that the decrease in the synthesis of MECPs in splenic macrophages of LL-exposed rats contributes to the impairment of the primary immune response.

IL-2 is the main lymphocyte mitogen that allows for clonal expansion after an antigen has been properly presented and recognized. The synthesis of IL-2 and its receptor IL-2R are induced shortly after lymphocyte activation. Once secreted, IL-2 exerts its effect in an autocrine and paracrine fashion. In this context, results from *in vitro* antigenic stimulation of splenocytes suggested that the amount of secreted IL-2 could be modulated by immune-derived opioid peptides such as MEnk, acting through opioid receptors (Valdés-Tovar et al. 2015). Further research is definitely needed for a better understanding of the precise opioidergic modulatory mechanisms over cytokine secretion.

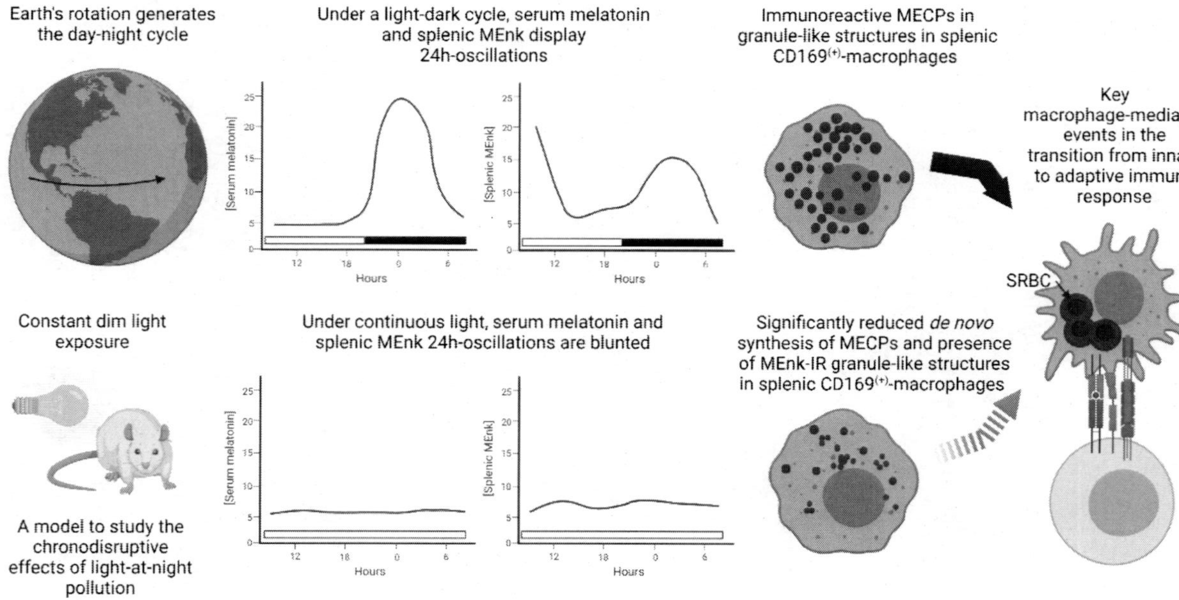

Figure 4. Schematic representation of our research on the immunomodulatory role of melatonin and MEnk as circadian messengers. Constant light significantly blunted melatonin circulating levels and reduces the *de novo* synthesis of MEnk-containing peptides (MECPs) in splenic macrophages, its intracellular content, the presence of granules that contain it, and, in general, lowers its splenic 24h-content. Since MEnk is a modulator of macrophage physiology and lymphocyte activation, its decrease may contribute to the deficiency of the primary humoral immune response seen in this environmental condition. Properly timed exogenous melatonin administration or reestablishment of the light-dark cycle ameliorate the immune-derived MEnk and the impaired immune response (Valdés-Tovar et al. 2015).

The mechanisms underlying the regulation of enkephalin synthesis in macrophages in relation to ambient lighting have not been characterized yet. However, melatonin might be participating in them since its nocturnal secretion constitutes an efferent pathway of the circadian system. Exposure to constant light has been reported to suppress melatonin production by the pineal gland causing a clear decrease in the amplitude of its circadian rhythm (Blask et al. 2009; Claustrat et al. 2008; Mendez et al. 2012). As mentioned before, the experimental 15-day LL regime employed by our group drastically reduced the melatonin serum concentration at the subjective night (Asai et al. 2007; Valdés-Tovar et al. 2015). Moreover, previous results indicate that the decrease in enkephalins tissue content in both CNS structures and lymphoid organs of animals exposed to LL is reversed by the administration of a single dose of exogenous melatonin after 4 h (Asai et al. 2007; Asai et al. 2010; Valdés-Tovar et al. 2003). The increase in enkephalins tissue content was partially blocked by the administration of luzindole, an antagonist of melatonin membranal receptors (Asai et al. 2007; Asai et al. 2010). Predictably, exposure to 6 h of darkness, and the consequent increase in the melatonin serum concentration, restored the enkephalins tissue content in the hypothalamus and hippocampus of animals exposed to LL (Asai et al. 2007). Taken together, evidence suggest that melatonin endogenously modulates enkephalin synthesis at the central and peripheral levels in response to changes in environmental light (Figure 4).

The functional relationship between endogenous opioids and melatonin is associated with the immune system physiology. Melatonin is an important immunomodulator (Carrillo-Vico et al. 2013). Moreover, experimental or pathophysiological diminution of its synthesis has been associated with dysregulation of the immune system (Cardinali et al. 2008; Maestroni et al. 1986). In this context, the studies by Maestroni et al. at the late 80's strongly suggest that melatonin's immunostimulatory effects are mediated by an opioidergic mechanism, as they were blocked by the opioid antagonist naltrexone (Maestroni et al. 1987, 1988). In the first study, authors showed that the evening administration of melatonin stimulated the humoral immune response of mice to an antigenic challenge with

SRBC, while the administration of naltrexone blocked it (Maestroni et al. 1987). Subsequently, they demonstrated that acute restriction stress in mice decreases the thymus weight and the humoral immune response against SRBC; furthermore, the administration of a sub-lethal dose of encephalomyocarditis virus to these animals increased their mortality by 80% in comparison to animals without stress. Contrastingly, evening melatonin administration reversed the immune-suppressive effects of acute stress, while naltrexone administration abolished the melatonin effect (Maestroni et al. 1988). Based on this evidence, Maestroni and Conti (1992) proposed the existence of a pineal-immune-opioid network. It would optimize the responsivity of the immune system and counteract the immunosuppressive effects of stress and other adverse environmental conditions. In agreement, our results indicate that evening melatonin administration for 4 days after an antigen challenge with SRBC counteracts the reduction in the number of cytokine-secreting cells (IL-2 and IFN-γ) in LL-exposed rats. The effect of melatonin was partially blocked by luzindole, an antagonist of its membrane receptors, as well as by the opioid receptor antagonist naloxone (unpublished data).

In this kind of studies, the effects observed in animals exposed to a constant illumination regime can be associated with the alteration in the function of the central and peripheral oscillators, and also to the lack of the light-dark cycle *per se*. In this sense, a diurnal variation in LPS-induced mortality and survival time only occurred in animals that were exposed to a LL, while in constant darkness conditions, the animals showed significantly increased mortality independently of the time of LPS administration (Marpegan et al. 2009). Based on this evidence, these authors propose that being under a light-dark cycle represents a benefit when facing an immunological challenge. In agreement, our results indicate that 5 LD cycles after a 30-day-LL regime are enough to restore MEnk content in macrophages and the secretion of IL-2 from splenocytes stimulated with SRBC, to levels similar to those found in animals exposed to LD (unpublished data). Other studies have shown that the light-dark cycle reestablishment not only influences the immune response capacity but also counteracts the metabolic alterations induced by the presence of dim LAN (dLAN) (Fonken et al. 2013b). Reversibility of pathophysiologic changes induced by exposure to chronic non-natural lightning conditions has also been proved in the context of a depressive-like phenotype in hamsters exposed to dLAN for four weeks. In these animals, the significantly dampened general locomotor activity, reduced BDNF expression and increased TNF-α expression found at the hippocampus of dLAN-exposed animals returned to levels similar to the LD-exposed hamsters within one week of LD-reestablishment. Reversal of the significant reduction in spine density on apical dendrites of hippocampal CA1 pyramidal neurons, and of depressive-like behavioral traits such as increased immobility in the forced swim test and decreased sucrose preference, took two weeks of LD-exposure to reach similar levels than animals that were not exposed to dLAN (Bedrosian et al. 2013b).

Notably, the close relationship between metabolic diseases and a state of chronic inflammation, plus the current high prevalence of these conditions and their behavioral consequences (Tapia-Osorio et al. 2013), become a matter of public health, granting the need for measures to counteract light pollution at night (light hygiene). Understandably, more basic research studies are required to elucidate the precise mechanisms underlying the effects of exposure to altered light regimes on enkephalin-mediated immunomodulation.

Conclusion

Both melatonin and PENK-derived opioid peptides, specifically MEnk, are important biochemical messengers with important immunomodulatory functions. Both messengers are widespread in the organism as well as their receptors, and their functional relationship has been proposed before in the context of immunomodulation as the pineal-immuno-opioid axis. Both messengers are released following a circadian rhythm; however, evidence indicates that exposure of animals to constant light induces suppression of nocturnal melatonin synthesis as well as disruption of the circadian rhythm of general activity and lowers daily oscillations and magnitude of splenic MEnk content and inhibits its de novo synthesis in splenic CD169(+)-macrophages. Furthermore, this environmental condition induces a deficiency in the primary immune response that could be due to the suboptimal participation of macrophages in early events of the adaptive response, which are modulated by enkephalins. The data indicate that the lack of the light-dark cycle affects the synthesis of immunomodulators such as the macrophage-derived MECPs resulting in significant detriment of the immune response. Finally, we suggest that to counteract the potential health consequences that environmental lighting changes might bring to general population, light hygiene habits ought to become a priority in modern public health policies. Further research is needed to unveil the molecular mechanisms underlying the chronodisruptive effects of LAN-pollution which has become pervasive for our modern 24/7 society, not without the concurrent detrimental effects on health.

ACKNOWLEDGEMENTS

This work was supported by the Fondo Sectorial de Investigación para la Educación FSIE-SEP-CONACyT, grant 287115 to MV-T.

REFERENCES

Abbas AK, Lichtman AH, Pillai S (2018) In: *Cellular and Molecular Immunology*, Ninth Edition: Elsevier, Inc, Philadelphia, PA, USA.

Aguilar R, Carmona V, Fanjul ML (2008) Circadian rhythms. In: *Functional Biology of Animals. Comparative Neurophysiology*: Fanjul ML, Hiriart M, editors. Siglo XXI, Mexico, DF, pp. 327-352.

Akil H, Owens C, Gutstein H, Taylor L, Curran E, Watson S (1998) Endogenous opioids: overview and current issues. *Drug Alcohol Depend* 51: 127-140.

Amaral FG, Castrucci AM, Cipolla-Neto J, Poletini MO, Mendez N, Richter HG, Sellix MT (2014) Environmental control of biological rhythms: effects on development, fertility and metabolism. *J Neuroendocrinol* 26: 603-612.

Anisimov VN, Vinogradova IA, Panchenko AV, Popovich IG, Zabezhinski MA (2012) Light-at-night induced circadian disruption, cancer and aging. *Curr Aging Sci* 5: 170-177.

Asai M, Mayagoitia L, Garcia D, Matamoros-Trejo G, Valdes-Tovar M, Leff P (2007) Rat brain opioid peptides-circadian rhythm is under control of melatonin. *Neuropeptides* 41: 389-397.

Asai M, Valdés-Tovar M, Matamoros-Trejo G, García D (2010) Melatonin induces enkephalins synthesis and release in the rat brain. *Salud Mental* 33: 123-131.

Asai M, Zubieta M, Matamoros-Trejo G, Linares G, Agustin P (1998) Diurnal variations of opioid peptides and synenkephalin in vitro release in the amygdala of kindled rats. *Neuropeptides* 32: 293-299.

Aschoff J (1981) Free running and entrained circadian rhythms. In: *Handbook of Behavioral Neurology*, Aschoff J, editor. Plenum Press, New York, USA, pp. 81-93.

Backer R, Schwandt T, Greuter M, Oosting M, Jungerkes F, Tuting T, Boon L, O'Toole T, Kraal G, Limmer A, den Haan JM (2010) Effective collaboration between marginal metallophilic macrophages and CD8+ dendritic cells in the generation of cytotoxic T cells. *Proc Natl Acad Sci USA* 107: 216-221.

Bedrosian TA, Aubrecht TG, Kaugars KE, Weil ZM, Nelson RJ (2013a) Artificial light at night altersdelayed-type hypersensitivity reaction in response to acute stress in Siberian hamsters. *Brain Behav Immun* 34: 39-42.

Bedrosian TA, Fonken LK, Walton JC, Nelson RJ (2011) Chronic exposure to dim light at night suppresses immune responses in Siberian hamsters. *Biol Lett* 7: 468-471.

Bedrosian TA, Nelson RJ (2013) Influence of the modern light environment on mood. *Mol Psychiatry* 18: 751-757.

Bedrosian TA, Weil ZM, Nelson RJ (2013b) Chronic dim light at night provokes reversible depression-like phenotype: possible role for TNF. *Mol Psychiatry* 18: 930-936.

Beers C, Honey K, Fink S, Forbush K, Rudensky A (2003) Differential regulation of cathepsin S and cathepsin L in interferon gamma-treated macrophages. *J Exp Med* 197: 169-179.

Behar OZ, Ovadia H, Polakiewicz RD, Abramsky O, Rosen H (1991) Regulation of proenkephalin A messenger ribonucleic acid levels in normal B lymphocytes: specific inhibition by glucocorticoid hormones and superinduction by cycloheximide. *Endocrinology* 129: 649-655.

Berson DM, Dunn FA, Takao M (2002) Phototransduction by retinal ganglion cells that set the circadian clock. *Science* 295: 1070-1073.

Blask DE, Dauchy RT, Brainard GC, Hanifin JP (2009) Circadian stage-dependent inhibition of human breast cancer metabolism and growth by the nocturnal melatonin signal: consequences of its disruption by light at night in rats and women. *Integr Cancer Ther* 8: 347-353.

Bodnar RJ (2012) Endogenous opiates and behavior: 2011 *Peptides* 38: 463-522.

Buijs RM, Soto Tinoco EC, Hurtado Alvarado G, Escobar C (2021). The circadian system: From clocks to physiology. *Handb Clin Neurol* 179: 233-247.

Cambras T, Castejon L, Diez-Noguera A (2011) Social interaction and sex differences influence rat temperature circadian rhythm under LD cycles and constant light. *Physiol Behav* 103: 365-371.

Cambras T, Castejon L, Diez-Noguera A (2012) Social interaction with a rhythmic rat enhances the circadian pattern of the motor activity and temperature of LL-induced arrhythmic rats. *Physiol Behav* 105: 835-840.

Cardinali DP, Esquifino AI, Srinivasan V, Pandi-Perumal SR (2008) Melatonin and the immune system in aging. *Neuroimmunomodulation* 15: 272-278.

Carrillo-Vico A, Lardone PJ, Alvarez-Sanchez N, Rodriguez-Rodriguez A, Guerrero JM (2013) Melatonin: buffering the immune system. *Int J Mol Sci* 14: 8638-8683.

Cassetta L, Cassol E, Poli G (2011) Macrophage polarization in health and disease. *Scientific World Journal* 11: 2391-2402.

Cecon E, Oishi A, Jockers R (2018) Melatonin receptors: molecular pharmacology and signalling in the context of system bias. *British J Pharmacol* 175: 3263-3280.

Chen W, Liu J, Meng J, Lu C, Li X, Wang E, Shan F (2012) Macrophage polarization induced by neuropeptide methionine enkephalin (MENK) promotes tumoricidal responses. *Cancer Immunol Immunother* 61: 1755-1768.

Cinzano P, Falchi F, Elvidge CD (2001) The first World Atlas of the artificial night sky brightness. *Monthly Notices of the Royal Astronomical Society* 328: 689-707.

Claustrat B, Valatx JL, Harthe C, Brun J (2008) Effect of constant light on prolactin and corticosterone rhythms evaluated using a noninvasive urine sampling protocol in the rat. *Horm Metab Res* 40: 398-403.

Coomans CP, van den Berg SA, Houben T, van Klinken JB, van den Berg R, Pronk AC, Havekes LM, Romijn JA, van Dijk KW, Biermasz NR, Meijer JH (2013) Detrimental effects of constant light exposure and high-fat diet on circadian energy metabolism and insulin sensitivity. *FASEB J* 27: 1721-1732.

Das KP, McMillian MK, Bing G, Hong JS (1995) Modulatory effects of [Met5]-enkephalin on interleukin-1 beta secretion from microglia in mixed brain cell cultures. *J Neuroimmunol* 62: 9-17.

Davies LC, Jenkins SJ, Allen JE, Taylor PR (2013) Tissue-resident macrophages. *Nat Immunol* 14: 986-995.

den Haan JM, Kraal G (2012) Innate immune functions of macrophage subpopulations in the spleen. *J Innate Immun* 4: 437-445.

Erren TC, Reiter RJ (2009) Defining chronodisruption. *J Pineal Res* 46: 245-247.

Escobar C, Salgado-Delgado R, Gonzalez-Guerra E, Tapia Osorio A, Angeles-Castellanos M, Buijs RM (2011) Circadian disruption leads to loss of homeostasis and disease. *Sleep Disord* 2011: 964510.

Fischer-Colbrie R, Frischenschlager I (1985) Immunological characterization of secretory proteins of chromaffin granules:

chromogranins A, chromogranins B, and enkephalin-containing peptides. *J Neurochem* 44: 1854-1861.

Fleminger G, Ezra E, Kilpatrick DL, Udenfriend S (1983) Processing of enkephalin-containing peptides in isolated bovine adrenal chromaffin granules. *Proc Natl Acad Sci USA* 80: 6418-6421.

Fonken LK, Lieberman RA, Weil ZM, Nelson RJ (2013a) Dim light at night exaggerates weight gain and inflammation associated with a high-fat diet in male mice. *Endocrinology* 154: 3817-3825.

Fonken LK, Weil ZM, Nelson RJ (2013b) Dark nights reverse metabolic disruption caused by dim light at night. *Obesity (Silver Spring)* 21: 1159-1164.

Foris G, Medgyesi GA, Hauck M (1986) Bidirectional effect of met-enkephalin on macrophage effector functions. *Mol Cell Biochem* 69: 127-137.

Gagro A, Gordon J (1999) The interplay between T helper subset cytokines and IL-12 in directing human B lymphocyte differenti-ation. *Eur J Immunol* 29: 3369-3379.

Gil-Lozano M, Mingomataj EL, Wu WK, Ridout SA, Brubaker PL (2014) Circadian secretion of the intestinal hormone, glucagon-like peptide-1, by the rodent L-cell. *Diabetes* 63: 3674-3685

Gupta BB, Spessert R, Vollrath L (2005) Molecular components and mechanism of adrenergic signal transduction in mammalian pineal gland: regulation of melatonin synthesis. *Indian J Exp Biol* 43: 115-149.

Hagi K, Uno K, Inaba K, Muramatsu S (1994) Augmenting effect of opioid peptides on murine macrophage activation. *J Neuroimmunol* 50: 71-76.

Hedin KE, Bell MP, Kalli KR, Huntoon CJ, Sharp BM, McKean DJ (1997) Delta-opioid receptors expressed by Jurkat T cells enhance IL-2 secretion by increasing AP-1 complexes and activity of the NF-AT/AP-1-binding promoter element. *J Immunol* 159: 5431-5440.

Hook V, Yasothornsrikul S, Greenbaum D, Medzihradszky KF, Troutner K, Toneff T, Bundey R, Logrinova A, Reinheckel T, Peters C, Bogyo M (2004) Cathepsin L and Arg/Lys aminopeptidase: a distinct prohormone processing pathway for the biosynthesis of peptide neurotransmitters and hormones. *Biol Chem* 385: 473-480.

Hook VY, Schiller MR, Azaryan AV, Tezapsidis N (1996) Proenkephalin-processing enzymes in chromaffin granules: model for neuropeptide biosynthesis. *Ann N Y Acad Sci* 780: 121-133.

House RV, Thomas PT, Bhargava HN (1996) A comparative study of immunomodulation produced by in vitro exposure to delta opioid receptor agonist peptides. *Peptides* 17: 75-81.

Hriscu M, Saulea G, Ostriceanu S, Baciu I (2002) Circadian phagocytic activity in rats under lightdark and constant light regimens. *Rom J Physiol* 39: 17-26.

Hucklebridge FH, Hudspith BN, Lydyard PM, Brostoff J (1990) Stimulation of human peripheral lymphocytes by methionine enkephalin and delta-selective opioid analogues. *Immunopharmacology* 19: 87-91.

Hughes J, Smith TW, Kosterlitz HW, Fothergill LA, Morgan BA, Morris HR (1975) Identification of two related pentapeptides from the brain with potent opiate agonist activity. *Nature* 258: 577-580.

Hwang SR, O'Neill A, Bark S, Foulon T, Hook V (2007) Secretory vesicle aminopeptidase B related to neuropeptide processing: molecular identification and subcellular localization to enkephalin- and NPY-containing chromaffin granules. *J Neurochem* 100: 1340-1350.

Jasser SA, Blask DE, Brainard GC (2006) Light during darkness and cancer: relationships in circadian photoreception and tumor biology. *Cancer Causes Control* 17: 515-523.

Jockers R, Delagrange P, Dubocovich ML, Markus RP, Renault N, Tosini G, Cecon E, Zlotos DP (2016) Update on melatonin receptors: IUPHAR Review 20. *British J Pharmacol* 173: 2702-2725.

Kavelaars A, Heijnen CJ (2000) Expression of preproenkephalin mRNA and production and secretion of enkephalins by human thymocytes. *Ann N Y Acad Sci* 917: 778-783.

Kamphuis S, Eriksson F, Kavelaars A, Zijlstra J, van de Pol M, Kuis W, Heijnen CJ (1998) Role of endogenous pro-enkephalin A-derived peptides in human T cell proliferation and monocyte IL-6 production. *J Neuroimmunol* 84: 53-60.

Kasprowicz DJ, Smallwood PS, Tyznik AJ, Ziegler SF (2003) Scurfin (FoxP3) controls T-dependent immune responses in vivo through regulation of CD4+ T cell effector function. *J Immunol* 171: 1216-1223.

Klein DC, Berg GR (1970) Pineal gland: stimulation of melatonin production by norepinephrine involves cyclic AMP-mediated stimulation of N-acetyltransferase. *Adv Biochem Psychopharmacol* 3: 241-263.

Kowalski J (1998a) Augmenting effect of opioids on nitrite production by stimulated murine macrophages. *Neuropeptides* 32: 287-291.

Kowalski J (1998b) Immunomodulatory action of class mu-, delta- and kappa-opioid receptor agonists in mice. *Neuropeptides* 32: 301-306.

Kowalski J, Gabryel B, Labuzek K, Herman ZS (2002) Methionine-enkephalin and leucine-enkephalin increase interleukin-1 beta release in mixed glia cultures. *Neuropeptides* 36: 401-406.

Kowalski J, Makowiecka K, Belowski D, Herman ZS (2000) Augmenting effect of methionine enkephalin on interleukin-6 production by cytokine-stimulated murine macrophages. *Neuropeptides* 34: 187-192.

Krajnik M, Schafer M, Sobanski P, Kowalewski J, Bloch-Boguslawska E, Zylicz Z, Mousa SA (2010) Enkephalin, its precursor, processing enzymes, and receptor as part of a local opioid network throughout the respiratory system of lung cancer patients. *Hum Pathol* 41: 632-642.

Kuis W, Villiger PM, Leser HG, Lotz M (1991) Differential processing of proenkephalin-A by human peripheral blood monocytes and T lymphocytes. *J Clin Invest* 88:817-824.

LaMendola J, Martin SK, Steiner DF (1997) Expression of PC3, carboxypeptidase E and enkephalin in human monocyte-derived macrophages as a tool for genetic studies. *FEBS Lett* 404: 19-22.

Lax P, Zamora S, Madrid JA (1998) Coupling effect of locomotor activity on the rat's circadian system. *Am J Physiol* 275: R580-R587.

Lax P, Zamora S, Madrid JA (1999) Food-entrained feeding and locomotor circadian rhythms in rats under different lighting conditions. *Chronobiol Int* 16: 281-291.

Lerner AB, Case JD, Takahashi Y, Lee TH, Mori W (1958) Isolation of melatonin, the pineal gland factor that lightens melanocytes. *J Am Chem Soc* 80: 2587-2587.

Li W, Chen W, Herberman RB, Plotnikoff NP, Youkilis G, Griffin N, Wang E, Lu C, Shan F (2014) Immunotherapy of cancer via mediation of cytotoxic T lymphocytes by methionine enkephalin (MENK). *Cancer Lett* 344: 212-222.

Li W, Meng J, Li X, Hua H, Yiming M, Wang Q, Wang E, Shan F (2012) Methionine enkephalin (MENK) improved the functions of bone marrow-derived dendritic cells (BMDCs) loaded with antigen. *Hum Vaccin Immunother* 8: 1236-1242.

Lichtnekert J, Kawakami T, Parks WC, Duffield JS (2013) Changes in macrophage phenotype as the immune response evolves. *Curr Opin Pharmacol* 13: 555-564.

Liu J, Chen W, Meng J, Lu C, Wang E, Shan F (2012) Induction on differentiation and modulation of bone marrow progenitor of dendritic cell by methionine enkephalin (MENK). *Cancer Immunol Immunother* 61: 1699-1711.

Liu J, Clough SJ, Hutchinson AJ, Adamah-Biassi EB, Popovska-Gorevski M, Dubocovich ML (2016) MT1 and MT2 Melatonin Receptors: A Therapeutic Perspective. *Ann Rev Pharmacol Toxicol* 56: 361-383.

Loh YP, Maldonado A, Zhang C, Tam WH, Cawley N (2002) Mechanism of sorting proopiomelanocortin and proenkephalin to the regulated secretory pathway of neuroendocrine cells. *Ann N Y Acad Sci* 971: 416-425.

Maestroni GJ, Conti A (1992) The pineal-immuno-opioid network. Mechanisms and significance. *Ann N Y Acad Sci* 650: 56-59.

Maestroni GJ, Conti A, Pierpaoli W (1986) Role of the pineal gland in immunity. Circadian synthesis and release of melatonin modulates the antibody response and antagonizes the immunosuppressive effect of corticosterone. *J Neuroimmunol* 13: 19-30.

Maestroni GJ, Conti A, Pierpaoli W (1987) Role of the pineal gland in immunity: II. Melatonin enhances the antibody response via an opiatergic mechanism. *Clin Exp Immunol* 68: 384-391.

Maestroni GJ, Conti A, Pierpaoli W (1988) Role of the pineal gland in immunity. III. Melatonin antagonizes the immunosuppressive effect of acute stress via an opiatergic mechanism. *Immunology* 63: 465-469.

Mao L, Dauchy RT, Blask DE, Slakey LM, Xiang S, Yuan L, Dauchy EM, Shan B, Brainard GC, Hanifin JP, Frasch T, Duplessis TT, Hill SM (2012) Circadian gating of epithelial-to-mesenchymal transition in breast cancer cells via melatonin-regulation of GSK3beta. *Mol Endocrinol* 26: 1808-1820.

Marotti T, Balog T, Mazuran R, Rocic B (1998) The role of cytokines in MET-enkephalin modulated nitric oxide release. *Neuropeptides* 32: 57-62.

Marotti T, Burek B, Rabatic S, Balog T, Hrsak I (1994) Modulation of lipopolysaccharide-induced production of cytokines by methionine-enkephalin. *Immunol Lett* 40: 43-47.

Marpegan L, Leone MJ, Katz ME, Sobrero PM, Bekinstein TA, Golombek DA (2009) Diurnal variation in endotoxin-induced mortality in mice: correlation with proinflammatory factors. *Chronobiol Int* 26: 1430-1442.

Martin J, Prystowsky MB, Angeletti RH (1987) Preproenkephalin mRNA in T-cells, macrophages, and mast cells. *J Neurosci Res* 18: 82-87.

Martínez-Pomares L, Gordon S (2007) Antigen presentation the macrophage way. *Cell* 131: 641-643.

McGaha TL, Chen Y, Ravishankar B, van Rooijen N, Karlsson MC (2011) Marginal zone macrophages suppress innate and adaptive immunity to apoptotic cells in the spleen. *Blood* 117: 5403-5412.

McLaughlin PJ, Zagon IS (2012) The opioid growth factor-opioid growth factor receptor axis: homeostatic regulator of cell proliferation and its implications for health and disease. *Biochem Pharmacol* 84: 746-755.

Mendez N, Abarzua-Catalan L, Vilches N, Galdames HA, Spichiger C, Richter HG, Valenzuela GJ, Seron-Ferre M, Torres-Farfan C (2012) Timed maternal melatonin treatment reverses circadian disruption of the fetal adrenal clock imposed by exposure to constant light. *PLoS One* 7: e42713.

Mosser DM, Edwards JP (2008) Exploring the full spectrum of macrophage activation. *Nat Rev Immunol* 8: 958-969.

Mousa SA, Shakibaei M, Sitte N, Schafer M, Stein C (2004) Subcellular pathways of betaendorphin synthesis, processing, and release from immunocytes in inflammatory pain. *Endocrinology* 145: 1331-1341.

Murray PJ, Wynn TA (2011) Protective and pathogenic functions of macrophage subsets. *Nat Rev Immunol* 11: 723-737.

Nguyen K, Miller BC (2002) CD28 costimulation induces delta opioid receptor expression during anti-CD3 activation of T cells. *J Immunol* 168: 4440-4445.

Nikbakht N, Shen S, Manser T (2013) Cutting edge: Macrophages are required for localization of antigen-activated B cells to the follicular perimeter and the subsequent germinal center response. *J Immunol* 190: 4923-4927.

Nishimura Y, Hosokawa T, Hosono M, Baba M, Hosokawa M (2002). Insufficient interleukin-2 production from splenic CD4+ T cells causes impaired cell proliferation and early apoptosis in SAMP1, a strain of senescence-accelerated mouse. *Immunology* 107: 190-198.

Novakova M, Polidarova L, Sladek M, Sumova A (2011) Restricted feeding regime affects clock gene expression profiles in the suprachiasmatic nucleus of rats exposed to constant light. *Neuroscience* 197: 65-71.

Okuliarova M, Mazgutova N, Majzunova M, Rumanova VS, Zeman M (2021). Dim Light at Night Impairs Daily Variation of Circulating Immune Cells and Renal Immune Homeostasis. *Front Immunol* 11: 614960.

O'Neill AS, van den Berg TK, Mullen GE (2013) Sialoadhesin - a macrophage-restricted marker of immunoregulation and inflammation. *Immunology* 138: 198-207.

Padros MR, Vindrola O, Zunszain P, Fainboin L, Finkielman S, Nahmod VE (1989) Mitogenic activation of the human lymphocytes induce the release of proenkephalin derived peptides. *Life Sci* 45: 1805-1811.

Pert CB, Snyder SH (1973) Opiate receptor: demonstration in nervous tissue. *Science* 179: 1011-1014.

Pittendrigh CS (1981) Circadian systems: General perspective. In: *Handbook of Behavioral Neurology*: Aschoff J, editor. Plenum Press, New York, USA, pp. 57-80.

Pittendrigh CS (1993) Temporal organization: reflections of a Darwinian clock-watcher. *Annu Rev Physiol* 55:16-54.

Pitzurra L, Rossetto O, Chimienti AR, Blasi E, Bistoni F (1996) Tetanus toxin-sensitive VAMP related proteins are present in murine macrophages. *Cell Immunol* 169: 113-116.

Radogna F, Diederich M, Ghibelli L (2010) Melatonin: a pleiotropic molecule regulating inflammation. In: *Biochem Pharmacol*: Elsevier Inc, England, pp. 1844-1852.

Radulovic J, Dimitrijevic M, Laban O, Stanojevic S, Vasiljevic T, Kovacevic-Jovanovic V, Markovic BM (1995) Effect of Met-enkephalin and opioid antagonists on rat macrophages. *Peptides* 16: 1209-1213.

Reiter RJ (1993) The melatonin rhythm: both a clock and a calendar. *Experientia* 49: 654-664.

Reiter RJ, Tan DX, Korkmaz A, Erren TC, Piekarski C, Tamura H, Manchester LC (2007) Light at night, chronodisruption, melatonin suppression, and cancer risk: a review. *Crit Rev Oncog* 13: 303-328.

Reiter RJ, Tan DX, Korkmaz A, Ma S (2012) Obesity and metabolic syndrome: association with chronodisruption, sleep deprivation, and melatonin suppression. *Ann Med* 44: 564-577.

Rosen H, Behar O, Abramsky O, Ovadia H (1989) Regulated expression of proenkephalin A in normal lymphocytes. *J Immunol* 143: 3703-3707.

Rouille Y, Duguay SJ, Lund K, Furuta M, Gong Q, Lipkind G, Oliva AA, Jr., Chan SJ, Steiner DF (1995) Proteolytic processing mechanisms in the biosynthesis of neuroendocrine peptides: the subtilisin-like proprotein convertases. *Front Neuroendocrinol* 16: 322-361.

Salzet M, Tasiemski A (2001) Involvement of pro-enkephalin-derived peptides in immunity. *Dev Comp Immunol* 25: 177-185.

Shan F, Xia Y, Wang N, Meng J, Lu C, Meng Y, Plotnikoff NP (2011) Functional modulation of the pathway between dendritic cells (DCs) and CD4+T cells by the neuropeptide: methionine enkephalin (MENK). *Peptides* 32: 929-937.

Sharp BM (2003) Opioid receptor expression and intracellular signaling by cells involved in host defense and immunity. *Adv Exp Med Biol* 521: 98-105.

Sharp BM (2006) Multiple opioid receptors on immune cells modulate intracellular signaling. *Brain Behav Immun* 20: 9-14.

Sharp BM, Li MD, Matta SG, McAllen K, Shahabi NA (2000) Expression of delta opioid receptors and transcripts by splenic T cells. *Ann N Y Acad Sci* 917: 764-770.

Sharp BM, McKean DJ, McAllen K, Shahabi NA (1998) Signaling through delta opioid receptors on murine splenic T cells and stably transfected Jurkat cells. *Ann N Y Acad Sci* 840: 420-424.

Shukla A, Berglund L, Nielsen LP, Nielsen S, Hoffmann HJ, Dahl R (2000) Regulated exocytosis in immune function: are SNARE-proteins involved? *Respir Med* 94: 10-17.

Shurlygina AV, Mitshurina SV, Verbitskaja LV, Belkin AD, Shanturova TV, Bitchaeva MV, Kovshik IG, Trufakin VA (2010) Structure of nuclear chromatin in cells of lymphoid organs and blood from mice maintained under abnormal light-dark cycle and treated with benz(a)pyrene. *Bull Exp Biol Med* 149: 29-32.

Simon EJ, Hiller JM, Edelman I (1973) Stereospecific binding of the potent narcotic analgesic (3H) Etorphine to rat-brain homogenate. *Proc Natl Acad Sci USA* 70: 1947-1949.

Singh PP, Singh S (2001) Protection of mice from malaria after co-administration of recombinant mouse granulocyte-macrophage colony- stimulating factor and methionine-enkephalin. *Eur Cytokine Netw* 12: 528-536.

Singh VK, Bajpai K, Narayan P, Yadav VS, Dhawan VC, Haq W, Mathur KB, Agarwal SS (1999) Delta-opioid receptor antagonist inhibits immunomodulation by Met-enkephalin analogs. *Neuroimmunomodulation* 6: 355-360.

Stauch B, Johansson LC, Cherezov V (2020) Structural insights into melatonin receptors. *FEBS J* 287: 1496-1510.

Stauch B, Johansson LC, McCorvy JD, Patel N, Han GW, Huang XP, Gati C, Batyuk A, Slocum ST, Ishchenko A, Brehm W, White TA, Michaelian N, Madsen C, Zhu L, Grant TD, Grandner JM, Shiriaeva A, Olsen RHJ, Tribo AR, Yous S, Stevens RC, Weierstall U, Katritch V, Roth BL, Liu W, Cherezov V (2019) Structural basis of ligand recognition at the human MT(1) melatonin receptor. *Nature* 569: 284-288.

Stephan FK (2002) The "other" circadian system: food as a Zeitgeber. *J Biol Rhythms* 17: 284-292.

Sudo M, Sasahara K, Moriya T, Akiyama M, Hamada T, Shibata S (2003) Constant light housing attenuates circadian rhythms of mPer2 mRNA and mPER2 protein expression in the suprachiasmatic nucleus of mice. *Neuroscience* 121: 493-499.

Tapia-Osorio A, Salgado-Delgado R, Angeles-Castellanos M, Escobar C (2013) Disruption of circadian rhythms due to chronic constant light leads to depressive and anxiety-like behaviors in the rat. *Behav Brain Res* 252: 1-9.

Terenius L (1973) Characteristics of the "receptor" for narcotic analgesics in synaptic plasma membrane fraction from rat brain. *Acta Pharmacol Toxicol (Copenh)* 33: 377-384.

Terenius L, Wahlstrom A (1975) Search for an endogenous ligand for the opiate receptor. *Acta Physiol Scand* 94: 74-81.

Trufakin VA, Shurlygina AV, Michurina SV, Verbitskaja LV, Litvinenko GI, Kovshik IG, Panteleeva NG, Melnikova EV, Bitchaeva MV (2007) The influence of experimental desynchronosis on the morphofunctional characteristics of mouse immune system. *Alaska Med* 49: 169-176.

Udenfriend S, Kilpatrick DL (1983) Biochemistry of the enkephalins and enkephalin-containing peptides. *Arch Biochem Biophys* 221: 309-323.

Valdés-Tovar M, Asai M, Matamoros-Trejo G (2003) Melatonin effect over the opioid peptides tissue content in the rat immune system. *Salud Mental* 26: 46-56.

Valdés-Tovar M, Escobar C, Solís-Chagoyán H, Asai M, Benítez-King G (2015) Constant light suppresses production of Met-enkephalin-containing peptides in cultured splenic macrophages and impairs primary immune response in rats. *Chronobiol Int* 32: 164-177.

van Epps DE, Saland L (1984) Beta-endorphin and met-enkephalin stimulate human peripheral blood mononuclear cell chemotaxis. *J Immunol* 132: 3046-3053.

Vindrola O, Padros MR, Sterin-Prync A, Ase A, Finkielman S, Nahmod V (1990) Proenkephalin system in human polymorphonuclear cells. Production and release of a novel 1.0-kD peptide derived from synenkephalin. *J Clin Invest* 86: 531-537.

Vinogradova IA, Anisimov VN, Bukalev AV, Ilyukha VA, Khizhkin EA, Lotosh TA, Semenchenko AV, Zabezhinski MA (2010) Circadian disruption induced by light-at-night accelerates aging and promotes tumorigenesis in young but not in old rats. *Aging (Albany NY)* 2: 82-92.

Vujic V, Stanojevic S, Dimitrijevic M (2004) Methionine-enkephalin stimulates hydrogen peroxide and nitric oxide production in rat peritoneal macrophages: interaction of mu, delta and kappa opioid receptors. *Neuroimmunomodulation* 11: 392-403.

Vujic-Redzic V, Dimitrijevic M, Stanojevic S, Kovacevic-Jovanovic V, Miletic T, Radulovi J (2000) Peripheral effects of methionine-enkephalin on inflammatory reactions and behavior in the rat. *Neuroimmunomodulation* 8: 70-77.

Xiao Y, Yuan Y, Jimenez M, Soni N, Yadlapalli S (2021). Clock proteins regulate spatiotemporal organization of clock genes to control circadian rhythms. *Proc Natl Acad Sci USA* 118: e2019756118.

Yamanaka Y, Honma S, Honma K (2013) Daily exposure to a running wheel entrains circadian rhythms in mice in parallel with development of an increase in spontaneous movement prior to running-wheel access. *Am J Physiol Regul Integr Comp Physiol* 305: R1367-R1375.

Yang SX, Li XY (1989) Enhancement of interleukin-1 production in mouse peritoneal macrophages by methionine-enkephalin. *Zhongguo Yao Li Xue Bao* 10: 266-270.

Yang SX, Li XY (1992) Influence of methionine-enkephalin on interleukin-2 production and interleukin-2 receptor expression. *Zhongguo Yao Li Xue Bao* 13: 131-136.

Yasothornsrikul S, Greenbaum D, Medzihradszky KF, Toneff T, Bundey R, Miller R, Schilling B, Petermann I, Dehnert J, Logvinova A, Goldsmith P, Neveu JM, Lane WS, Gibson B, Reinheckel T, Peters C, Bogyo M, Hook V (2003) Cathepsin L in secretory vesicles functions as a prohormone processing enzyme for production of the enkephalin peptide neurotransmitter. *Proc Natl Acad Sci USA* 100: 9590-9595.

Zeigler D (2014) Symbiosis. In: *Evolution:* Zeigler D editor. Elsevier Inc, Philadelphia, PA, USA, pp. 101-110.

Zhang B, Sannajust F (2000) Diurnal rhythms of blood pressure, heart rate, and locomotor activity in adult and old male Wistar rats. *Physiol Behav* 70: 375-380.

Zurawski G, Benedik M, Kamb BJ, Abrams JS, Zurawski SM, Lee FD (1986) Activation of mouse T helper cells induces abundant preproenkephalin mRNA synthesis. *Science* 232: 772-775.

In: Circadian Rhythms and Their Importance ISBN: 978-1-68507-547-7
Editor: Rajeshwar P. Sinha

Chapter 9

FEATURES OF CIRCADIAN RHYTHMS OF HEART RATE VARIABILITY, ARTERIAL STIFFNESS AND OUTPATIENT MONITORING OF BLOOD PRESSURE IN DIABETES MELLITUS: DATA, MECHANISMS AND CONSEQUENCES

Victoria A. Serhiyenko**, Ludmila M. Serhiyenko and Alexandr A. Serhiyenko
Department of Endocrinology, Danylo Halytsky Lviv National Medical University, Lviv, Ukraine

ABSTRACT

Abnormalities in blood pressure (BP) variability and heart rate (HR) variability (HRV) may increase the risk of cardiovascular disease over the next six years from 4% to 100%. It is reported that the probability of

* Corresponding Author's E-mail: serhiyenkoa@gmail.com.

stroke or cardiac death can be immediately reduced by chronobiologically assessing BP and HRV and optimizing timed treatment efficacy. Physiological cardiovascular activities are under the control of the cardiac autonomic nervous system (ANS). Damage to the autonomic nerves results in dysfunction in HR control and vascular dynamics, particularly in cardiac autonomic neuropathy (CAN). Autonomic imbalance between the sympathetic (SNS) and parasympathetic nervous systems (PSNS) regulation of cardiovascular function contributes to metabolic abnormalities and significant morbidity and mortality for individuals with diabetes mellitus (DM). Misalignment of circadian rhythms has been evidenced in patients with DM, and there is a close relationship between alterations in neuroendocrine sleep architecture, circadian clock oscillations, glucose metabolism, autonomic function, and diurnal profiles of BP and HR.

Metabolic syndrome, arterial hypertension, myocardial infarction, and DM are characterized by increased SNS activity and decreased PSNS activity. However, type 2 DM (T2DM) patients had a decrease in both PSNS and SNS activity. The latter can be explained by T2DM which is a metabolic disease responsible for a CAN that affects both sympathetic and parasympathetic fibers. The pathophysiological relationship between arterial stiffness and autonomic dysfunction in DM and whether impaired cardiac autonomic function induces arterial stiffness or elevated arterial stiffness leads to the impairment of the autonomic function remains unclear. Both arterial stiffness and cardiac autonomic dysfunction share common pathogenetic pathways, including chronic hyperglycemia and hyperinsulinemia, the formation of advanced glycation end products and protein kinase C activation, low grade-inflammation, and endothelial dysfunction.

Patients with T2DM are noted having a relationship between changes in HR, "nondipping" during 24-h ambulatory BP monitoring (ABPM) and increased mortality. This may represent various pathophysiologic processes: preferential parasympathetic denervation in the CAN of long-standing DM, heart failure; the increased sympathetic activity in chronic kidney disease; and obstructive sleep apnea, etc. Several cross-sectional studies revealed that in patients with T2DM and T1DM there is a connection between arterial stiffness and CAN. One of the possible reasons for the detected changes is dysfunction of baroreceptors with subsequent attachment and/or progression of the arterial wall "stiffness." On the other hand, the ANS dysfunction can affect vascular tone and, consequently, the elasticity of large arteries' walls. Another probable mechanism of attachment and/or development of "arterial stiffness" is an increase in HR and autonomic dysfunction's corresponding progression.

The purpose of the charter is to analyze the current state of the problem and the results of our own research in the relationship between

DM and circadian rhythm disorders, HRV, arterial stiffness and ABPM. Particular attention is paid to the analysis, interpretation, and application of the results obtained in patients with T2DM with diabetic CAN.

Keywords: diabetes mellitus, diabetic cardiac autonomic neuropathy, 24-h heart rate variability monitoring, 24-h arterial stiffness monitoring, 24-h ambulatory blood pressure monitoring

INTRODUCTION

A circadian rhythm is an oscillation of a physiological process over a 24-hour period. Many cardiovascular (CV) variables, including heart rate (HR), heart rate variability (HRV), electrocardiogram (ECG) waveforms, and blood pressure (BP), demonstrate a robust circadian rhythm. Strictly speaking, an oscillation is considered circadian only if it persists over 24 h of darkness. In many cases, this has not been tested for cardiovascular variables, but nevertheless the oscillations are here described as circadian because the term is in common usage. Many cardiovascular diseases (CVD's) vary in prevalence according to the time of day, including myocardial infarction (MI), supraventricular/ventricular arrhythmias, and sudden cardiac death (SCD) (Portaluppi and Hermida 2007; Black et al. 2019).

Circadian rhythms are controlled by circadian clocks, which drive day-night oscillations with a free-running period of ~24 hrs. The molecular machinery of the circadian clock has been extensively characterized and reviewed elsewhere (Takahashi 2017). At its core is a negative feedback loop involving four basic helix-loop-helix or per-arnt-sim domain transcription factors: CLOCK, BMAL1, PER, and CRY3. CLOCK and BMAL1 form a heterodimer within the nucleus and bind to canonical enhancer or E-box regions in the promoters of the *Per*, *Cry* and *Rev-erb* genes, thus activating their transcription. As a consequence, PER and CRY proteins gradually accumulate in the cytoplasm, and together they reenter the nucleus and suppress CLOCK/BMAL1-mediated transcription of the

Per and *Cry* genes. At the same time, the accumulation of REV-ERB suppresses the transcription of *Clock* and *Bmal1*. Together, this establishes a negative feedback loop. This full cycle takes ~24 hrs (Halberg et al. 2006; Halberg et al. 2009; Durgan et al. 2005; Durgan and Young 2010; Takahashi 2017).

Circadian clocks have been identified in the majority of mammalian cells. The suprachiasmatic nucleus (SCN) in the hypothalamus contains the central circadian clock and is synchronized with the environment by external cues such as light. In turn, the central clock entrains clocks within peripheral tissues *via* neurohumoral factors (autonomic tone, body temperature, and glucocorticoid signaling) (Mohawk et al. 2012). There is known to be a peripheral (or local) clock in the heart: oscillations in the expression of core circadian clock genes have been observed in the intact heart and are known to persist in isolated cultured myocardial tissue and cardiomyocytes (Davidson et al. 2005; Durgan and Young 2010; Takahashi 2017). Up to 10% of the cardiac transcriptome is controlled by the heart's local clock (Martino et al. 2004). As a result, key processes in the heart (including electrical excitability, signal transduction, and metabolism) vary in a circadian manner (Durgan and Young 2010; Hansen et al. 2011; Black et al. 2019).

Continuous 24-h ECG recordings from healthy volunteers have shown a circadian rhythm in the ECG. The RR interval increases at night, corresponding to a slowing of the HR. This nocturnal bradycardia seems to be independent of the nocturnal fall in BP. At night, there is also a lengthening of the PQ interval, QRS duration, and both uncorrected and corrected QT intervals. This indicates slower atrioventricular (AV) node conduction, His-Purkinje conduction, and ventricular repolarization, respectively. A similar circadian rhythm in the ECG is seen in rodents. Hence, the normal electrical properties of the sinus node, AV node, His-Purkinje system, and ventricular muscle change over a 24-h period (Sheward et al. 2010; Black et al. 2019).

It is commonly believed that the SCN clock controls the circadian rhythm in HR through a circadian variation in autonomic tone to the sinus node, in particular an increase in vagal tone at night. The SCN controls the circadian release of other neurohumoral factors, but the effect on the circadian rhythm in HR is not well understood. Apparent support for the role of autonomic tone comes from studies of day-night variations in HRV, which is widely used as an indirect measure of cardiac autonomic tone (Sztajzel 2004). However, biophysical analysis of HRV has demonstrated an exponential-like relationship between HRV and HR, and changes in HRV observed in humans and rodents are mainly attributable to the accompanying changes in HR (Monfredi et al. 2014).

Interestingly, direct recordings of stellate ganglion nerve activity and vagal nerve activity in an ambulatory dog model of heart failure show a circadian rhythm in sympathetic but not vagal tone to the heart (Ogawa et al. 2007). These studies suggest that circadian sympathovagal balance may be largely due to diurnal fluctuations in sympathetic tone. Catecholamine secretion from the adrenal medulla also exhibits a prominent circadian rhythm, and this may contribute to the circadian rhythm in HR (Scheer et al. 2010).

An alternative way to test the involvement of the autonomic nervous system (ANS) is to block autonomic control of the heart. The autonomic tone is the dominant mechanism by which the SCN drives circadian changes in HR (Tong et al. 2013; Black et al. 2019). Black et al. (2019) hypothesize that the ANS influences the circadian rhythm in HR by synchronizing the local cardiac clock to drive circadian oscillations in gene expression.

Diabetes Mellitus and Circadian Rhythms of Heart Rate Variability

Adaptation to stress is characterized by an increase in sympathetic activity and a decrease in parasympathetic activity, inducing a state of alertness (Hufnagel et al. 2017). Common diseases such as obesity and metabolic syndrome (Thorp and Schlaich 2015), MI etc, are associated with a decrease in parasympathetic activity and an activation of sympathetic activity.

Tables 1 and 2 show the parameters of the time and frequency domain measures of HRV described by the Task Force of the European Society of Cardiology and the North American Society of Pacing and Electrophysiology (1996).

Table 1. Time domain measures of HRV. The four measures marked in grey were recommended for time domain HRV assessment (Task Force of the European Society of Cardiology and the North American Society of Pacing and Electrophysiology 1996)

Variable	Units	Description
Statistical Measures		
SDNN	ms	Standard deviation (SD) of all normal-to-normal (NN) intervals
SDANN	ms	Standard deviation (SD) of the averages of normal-to-normal (NN) intervals in all 5-min segments of the entire recording
RMSSD	ms	Root-mean square of the differences of successive normal-to-normal (NN) intervals
SDNN index	ms	Mean of the standard deviations (SDs) of all normal-to-normal (NN) intervals for all 5-min segments of the entire recording
SDSD	ms	SD of differences between adjacent normal-to-normal (NN) intervals
NN50 count	ms	Number of pairs of adjacent normal-to-normal (NN) intervals differing by more than 50 ms in the entire recording (counting all such NN intervals pairs or only pairs in which the first or the second interval is longer)
pNN50	%	NN50 count divided by the total number of all normal-to-normal (NN) intervals
Geometric Measures		
Heart rate variability triangular index		Total number of all normal-to-normal intervals divided by the height of the histogram of all normal-to-normal (NN) intervals measured on a discrete scale with bins of 7.8125 ms (1/128 sec)
Triangular interpolation (TINN)	ms	Baseline width of the minimum square difference triangular interpolation of the highest peak of the histogram of all normal-to-normal (NN) intervals
Differential index	ms	Difference between the widths of the histogram of differences between adjacent normal to-normal intervals measured at selected heights
Logarithmic index	ms	Coefficient φ of the negative exponential curve k x eφt, which is the best approximation of the histogram of absolute differences between adjacent NN intervals

Table 2. Frequency domain measures of HRV (Task Force of the European Society of Cardiology and the North American Society of Pacing and Electrophysiology 1996)

Variable	Units	Description
Analysis of Short-Term Recordings		
5-min total power	ms^2	The variance of normal-to-normal (NN) intervals over the temporal segment
VLF	ms^2	Power in very low frequency (VLF) range
LF	ms^2	Power in low frequency (LF) range
LF norm	nu	LF power in normalized units LF/(total power-VLF)x100
HF	ms^2	Power in high frequency (HF) range
HF norm	nu	HF power in normalized units HF/(total power-VLF)x100
LF/HF	ms^2	Ratio LF [ms^2]/HF[ms^2]
Analysis of Entire 24 Hours		
Total power (TP)	ms^2	Variance of all normal-to-normal (NN) intervals
ULF	ms^2	Power in the ultra low frequency (ULF) range
VLF	ms^2	Power in the VLF range
LF	ms^2	Power in the LF range
HF	ms^2	Power in the HF range
α		Slope of the linear interpolation of the spectrum in a log-log scale

HRV reflects the changes in HR caused by fluctuations in sympathetic and parasympathetic function under steady conditions. Clinically, HRV was interpreted through time-domain analysis of NN intervals and frequency-domain analysis or the HRV signals. Time-domain analysis of HRV is determined by the HR at any point in time or the intervals between successive normal complexes. SDNN is the simplest variable to calculate the standard deviation of NN intervals, the square root of variance; while RMSSD is the square root of the mean squared difference of successive NN intervals (Hillebrand et al. 2013). Frequency-domain analysis of HRV, so-called power spectral density analysis, provides information about how variance distributes as a function of frequency; TP, LF (frequency range: 0.04-0.15 Hz), and HF (frequency range: 0.15-0.4 Hz) are useful in separating sympathetic from parasympathetic abnormalities (Hillebrand et al. 2013).

During the 1970's, Ewing et al. (1985) devised simple bedside tests of short-term RR differences to detect cardiac autonomic neuropathy (CAN) in normal subjects and patients with diabetes mellitus (DM). Akselrod et al. (1981) introduced frequency-domain measures (power spectral analysis) of the HRV (Task Force of the European Society of Cardiology and the North American Society of Pacing and Electrophysiology 1996). Frequency-domain measures have been reported to provide clinical assessment of parasympathetic and sympathetic activities and balances and quantify the rhythmic oscillations in HRV (Pagani et al. 1988). In early stages, CAN could be detected by decreased HRV with deep breathing, which implies parasympathetic dysfunction in patients with type 2 DM (T2DM) (American Diabetes Association 2017). Takase et al. (2002) reported that HF power, as a direct measure of vagal nerve integrity, was reduced in early CAN in patients with T2DM. Thus, frequency-domain measures of HRV might be a diagnostic method to detect early CAN in patients with T2DM (Takase et al. 2002). Previous studies reported the association between impaired HRV and CAN in patients with and without type 1 diabetes mellitus (T1DM) and T2DM (Bellavere et al. 1992; Ziegler et al. 2001; Serhiyenko and Serhiyenko 2018; Serhiyenko and Serhiyenko 2019).

The significant relationship between altered glucose metabolism and HRV may explain the deleterious general metabolic effects on both parasympathetic and sympathetic activity. Interestingly, blood glucose levels were associated with both an increase in LF (sympathetic) and HF (parasympathetic), as well as SDNN (sympathetic) and RMSSD (parasympathetic), which may appear contradictory. In healthy individuals, parasympathetic activity is triggered by an increase in blood glucose levels through insulin responses (Stockhorst et al. 2011; Benichou et al. 2018).

Cardiovascular autonomic dysfunction results from damage to the nerves and blood vessels innervating the heart and can lead to dysfunctional HR control and abnormal vascular dynamics. Cardiac autonomic dysfunction is a frequent but underdiagnosed complication of T2DM that is associated with arrhythmia, MI, and SCD (Vinik et al. 2003; Maser et al. 2003). One of the earliest subclinical manifestations of cardiac autonomic dysfunction is reduction in HRV with parasympathetic loss preceding sympathetic dysfunction (Vinik and Ziegler 2007).

Compared with their peers without DM, adolescents with T2DM have early signs of cardiac and vascular abnormalities including left ventricular (LV) hypertrophy and diastolic dysfunction (Shah et al. 2013) as well as vascular stiffness and thickness (Dabelea et al. 2017; Shah et al. 2018). However, cardiac autonomic function and HRV have been evaluated less frequently in young adults with youth-onset T2DM (Shah and Urbina 2017). The SEARCH for Diabetes in Youth (SEARCH) study found cardiac autonomic dysfunction in 17% of their cohort of young adults with youth-onset T2DM (compared with a lean control group), a higher prevalence than that seen in young adults with T1DM (17% *vs* 12%) (Jaiswal et al. 2018). However, SEARCH was not able to identify risk factors associated with the presence of autonomic dysfunction, perhaps because of inadequate statistical power (Dabelea et al. 2018; Shahg et al. 2019).

Hyperglycemia is thought to be associated with abnormal signaling of autonomic neurons *via* accumulation of advanced glycation end products (AGE's), activation of polyol pathway, and ischemia-induced atrophy of the autonomic nerve fibers innervating the cardiac and vascular tissue. Both divisions of the ANS are typically affected, with parasympathetic impairment preceding the sympathetic dysfunction (Serhiyenko and Serhiyenko 2018; Serhiyenko et al. 2018; Goldberger et al. 2019). Loss of HRV is one of the earliest manifestations of this process. In the Framingham Heart Study, HRV was found to be inversely associated with the risk of mortality (Tsuji et al. 1996). Similarly, the Atherosclerosis Risk In Communities (ARIC) study found decreased HRV was independently associated with the risk of developing coronary heart disease (CHD) (Liao et al. 1997).

Another convincing point is that the relationships between glycated hemoglobin A1c (HbA1c) and HRV is logical. Chen-Scarabelli and Scarabelli (2006) demonstrated that higher levels of HbA1c were associated with shorter RR intervals, which were associated with an increased risk of ventricular arrhythmias. There was also a tendency for higher LF/HF ratio (i.e., decreased HRV) with higher levels of HbA1c. Furthermore, time from diagnosis of T2DM was linked with a higher level of SDNN. Despite Goit et al. (2012) demonstrated a decreased SDNN in T2DM patients; the meta-regression is not contradictory but may simply highlight the fact that cardiac parasympathetic activity in T2DM is affected before sympathetic activity. Time from diagnosis of T2DM was also linked with a lower level of TP, but not with LF and HF. Thus Benichou et al. (2018) can hypothesize that VLF could be decreased in T2DM. Despite few studies assessing VLF in DM, interesting relationships were reported between VLF and obstructive sleep apnea (OSA) in diabetics (Amra et al. 2017).

Similarly, total cholesterol was associated with both an increase in LF and HF, and high-density *lipoprotein* cholesterol (*HDL-C)* was associated with an increase in SDNN and RMSSD. Interestingly, some studies showed that a decrease in high-density *lipoprotein* cholesterol (LDL-C) by statin therapy could improve HRV parameters (Pehlivanidis et al. 2001; Badea et al. 2014; Benichou et al. 2018). Benichou et al. (2018) demonstrated that an increase in systolic BP (SBP) was linked with shorter RR intervals and a decrease in HF. Despite no study previously assessing this relationship in DM, conflicting results were reported in the general population, with either high BP associated with an increase in all spectral parameters (Askin et al. 2018), or a decrease in HRV (de Andrade et al. 2017). It has also been suggested that the decrease in autonomic nervous function precedes the development of clinical hypertension (HT) (Schroeder et al. 2003). Moreover, Benichou et al. (2018) found a significant relationship between body mass index (BMI) and HRV. Such relationships have been either found (Al-Trad et al. 2015; Mustafa et al. 2017) or not (Yoo et al. 2016; Yadav et al. 2017). However, the severity of obesity and related diseases is not directly linked to the accumulation of total body fat but rather to its distribution, and particularly to visceral localization (Dutheil et al. 2018). HRV parameters have been previously correlated with sagittal abdominal diameter, anterior forearm skinfold thickness (Rastović et al. 2017) and waist hip ratio (Yoo et al. 2016; Benichou et al. 2018).

Measuring HRV is challenged by environmental noise, mental stress, and physical activity during daytime. Night-time HRV during sleep may be a more valid tool to measure CAN and therefore may improve prediction of CV events in low-risk people with T2DM (Hadad et al. 2021). Copenhagen Holter Study included 678 community dwelling subjects aged 55-75 yrs who were free of previous CVD. Day- and night-time HRV were available for 653 participants. The population included 133 people with well-controlled T2DM and newly recognized T2DM. Cardiovascular events were defined as CV death, MI, stroke, or coronary revascularization. Twenty-four-hrs HRV was not associated with CV events, but with all-cause mortality in people with T2DM. Conventional risk factors had a Receiver Operating Characteristic (ROC) value of 0.704 (95% CI 0.602-0.806) to predict CV events in people with T2DM. The prediction of CV events by conventional risk factors was improved in people with T2DM by the addition of night-time SDNN. Thus, reduced night-time HRV predicts increased risk of CV events in people with well-controlled T2DM, thus night-time HRV may add to traditional risk factors in predicting CV events in people with T2DM (Hadad et al. 2021).

In T2DM's cardiac autonomic dysfunction, expressed as reduced vagal activity, leads to HR acceleration and thus to diastole shortening, but it seems that cardiac autonomic dysfunction may shorten diastole duration (DD) *per se*, independently of the effect on HR. Since DD influences strongly subendocardial myocardial viability index (SVI), cardiac autonomic dysfunction plays a primary role in addition to arterial stiffness in the impairment of SVI and may thus worsen CV prognosis (Bianchi et al. 2016). Chorepsima et al. (2017) showed that beyond BP, impaired cardiac autonomic function assessed by determination of HRV was a significant determinant of abnormal pulse wave velocity (PWV) in patients with T2DM. Furthermore, lower values of the frequency-dependent domains of the HRV were independently associated with higher odds of abnormal PWV. It has been shown that low HRV is associated with increased mortality in patients with CHD or DM. HRV is also a sensitive indicator of baroreflex sensitivity (BRS) control, specifically the vagal control. Therefore, arterial stiffness may affect BRS and thereby, HRV.

Increased arterial stiffness evaluated by PWV and/or ambulatory arterial stiffness index (AASI) has been associated with the presence of coronary atherosclerosis and worse CV prognosis both in general population and specific disease groups, including DM. Decreased HRV in the uncomplicated diabetes patients highlights the obscure process of CAN in diabetic patients that begins even before clinical atherosclerotic CVD's becomes apparent (Bagherzadeh et al. 2013). It has also been shown that surrogate atherosclerosis markers were associated with lower HRV, and increased carotid intima media thickness (CIMT) in T2DM participants was significantly associated with decreased HRV, independent from conventional CV risk factors. Therefore, the presence of CAN should be considered much earlier in the course of DM, rather than after the development of clinical CVD's [process of CAN in diabetic patients that begins even before clinical atherosclerotic CVD's becomes apparent (Bagherzadeh et al. 2013)].

Thus, there is strong evidence of a decrease in HRV in T2DM patients. Both sympathetic and parasympathetic activities were decreased, which can be explained by the deleterious effects of altered glucose metabolism on HRV (Benichou et al. 2018). The benefits of an HRV evaluation in assessing and monitoring the severity of T2DM should be further studied, given its potential as a noninvasive, reliable and pain free measurement.

Diabetes Mellitus and Circadian Rhythms of Arterial Stiffness

Autonomic dysfunction is a risk factor for CVD's; however, the exact mechanism linking autonomic dysfunction to CVD's is not known (Ulleryd et al. 2017). Cardiac autonomic dysfunction is a well-established risk factor for CV morbidity. Decreased HRV and low BRS are considered to be early markers of cardiac autonomic dysfunction (Pop-Busui 2010; Spallone et al. 2011; Ziegler et al. 2014; Serhiyenko and Serhiyenko 2015; Pop-Busui et al. 2017). CAN is a major risk factor for CV injury resulting in heart attack, congestive heart failure, stroke as well as sudden arrhythmic death. The latter is associated with myocardial structural remodeling that follows HT, such as hypertrophy and fibrosis. This remodeling accompanies changes in expression, distribution, and function of cell membrane ion channels, intercellular gap junction connexin-43 channels, Ca^{2+}-cycling proteins, and extracellular matrix composition (Serhiyenko and Serhiyenko 2015; Pop-Busui et al. 2017).

Recently, the possibility of using a new CVD's risk factor, which could be more promising in terms of prognosis than metabolic syndrome (MetS), namely, "syndrome of vascular aging," is actively discussed. The processes of increasing the rigidity of the arteries due to the decrease in the content of elastin and the increase in the concentration of collagen in the walls of the vessels, as well as qualitative changes in the arterial wall due to the violation of endothelial-mediated vasodilation, are known in physiological aging, but a number of factors are able to accelerate this process (Agnoletti et al. 2013; Bagherzadeh et al. 2013; Mansour et al. 2013; Xu et al. 2016).

Since aortic stiffness determines its damping function and PWV, and also affects the level of aortic pulse pressure (PPao) and aortic augmentation index (IAxao), these indices may be recommended for its evaluation. Central systolic BP more reflects the blood flow in the coronary and cerebral vessels compared with central diastolic BP and is a more significant predictor of CVD's. Therefore, measurements of PPao and IAxao may be more important for the characterization of the state of the CV system than just the evaluation of the rigidity of the vessel wall in terms of PWV (Agnoletti et al. 2013; Bagherzadeh et al. 2013; Xu et al. 2016).

Several epidemiological studies reported that increased arterial stiffness predicts mortality and morbidity, independently of other CV risk factors. Arterial stiffness measured through carotid-femoral PWV, a gold standard method has been associated with subclinical CVD's. Furthermore, clinical studies have shown that the arterial stiffness increases with aging or various pathological processes associated with HT, MetS, chronic renal disease, and DM (De Oliveira Alvimet al. 2013).

Diabetes may enhance arterial stiffness through pathological changes in the vascular bed, such as reduced nitric oxide (NO) bioavailability, increased oxidative stress (OS), chronic low-grade inflammation, increased sympathetic tone and changes in type or structure of elastin and/or collagen in the arterial wall (Gaszner et al. 2012; De Oliveira Alvimet al. 2013; Okamoto et al. 2016).

Many of the pathophysiological mechanisms responsible for vascular dysfunction in DM are determined by hyperglycemia, which is associated with the activation of pro-inflammatory transcription factors and increased arterial stiffness, leading to vasculopathy. Increased AGE's levels may alter the important matrix of molecules in the vessel wall. In addition, some studies show endothelial and vascular smooth muscle cell (VSMC's) dysfunction in diabetic individuals compared to controls indicating that T2DM may both reduce the bioavailability of endothelial NO and attenuate sensitivity of the VSMC's to NO. All these pathways appear to be involved in mediating the hyperglycemia-associated arterial stiffness. Patients without DM but with primary autonomic failure have stiffer aortas when compared with healthy age- and sex-matched control individuals. These findings imply that there is a pathophysiological link between cardiac autonomic dysfunction and arterial stiffness and that the preservation of the elastic properties of the arteries strongly depends on the integrity of the ANS (De Oliveira Alvimet al. 2013; Chawla et al. 2014; Chorepsima et al. 2017; Bakkar et al. 2020; Cseh et al. 2020).

The path physiological link between aortic stiffness and autonomic dysfunction and whether impaired cardiac autonomic function induces arterial stiffness or whether increased arterial stiffness leads to the impairment of the autonomic function remains obscure. Both arterial stiffness and cardiac autonomic dysfunction share common pathogenetic pathways including chronic hyperglycemia and hyperinsulinemia (HI), formation of AGE's and protein kinase C (PKC) activation, low grade-inflammation and endothelial dysfunction (ED) (Madamanchi et al. 2005; Chorepsima et al. 2017; Jia et al. 2016; Jia et al. 2018).

The hyperglycemia in diabetes leads to vascular and neuronal damages by causing nonenzymatic glycosylation, increased OS and activation of polyol and PKC pathways. The mechanisms triggered by hyperglycemia result in ED and arterial stiffness, which are responsible for the microvascular and macrovascular complications of diabetes (Ciftel et al. 2014; Seko 2020). Hyperglycemia has been shown to trigger ED through decrease in NO synthesis, increase in free radicals levels, and deterioration of antioxidant defense mechanisms. Chronic glycemic exposure also induces VSMC's proliferation and chronic inflammation, increases generation of AGE's and enhances collagen cross-linking within the arterial wall, up-regulates matrix metalloproteinase-2 and -9 expression (enzymes degrading elastin), augments the generation of angiotensin-2 in vascular tissue and increases endothelial permeability (Zhao et al. 2015). ED is considered a key event in the initiation of the atherosclerotic process, yet it also leads to "functional" stiffening of arteries, as a continuous NO release by endothelium contributes to the functional regulation of arterial elasticity, aimed at adapting peripheral conduit artery mechanics to changes in blood flow (Kozakova and Palombo 2016).

Several prospective studies have demonstrated that insulin resistance (IR), as assessed by various techniques, is related to CVD's in both non-diabetic and diabetic subjects, independently of established risk factors. The risk of CVDs events increased across quintiles of the homeostasis model assessment of insulin resistance (HOMA-IR); the association between HOMA-IR index and CVD's was demonstrated. IR has been shown to be associated with decreased synthesis/release of NO and enhanced generation of reactive oxygen species, as well as with an excessive free fatty acids (FFA's) release from adipose tissue. Increased circulating levels of FFA's may impair endothelial function and induce a low-grade inflammation (through activation of nuclear factor kappa-light-chain-enhancer of activated B cells (NF-κB) (Barazzoni et al. 2012; Kozakova and Palombo 2016). Hyperinsulinemia augments hepatic very-*LDL's* synthesis, increases cholesterol transport/synthesis in *cultured arterial* smooth muscle cells (SMC's), stimulates the proliferation of arterial SMC's, augments collagen synthesis and turns on multiple genes involved in inflammation (Kozakova and Palombo 2016). In subjects with T2DM, IR and the use of alcohol were associated with increased arterial stiffness which supports the hypothesis that increased arterial stiffness can act as a mediating factor in the association between T2DM and increased risk of atherothrombotic disease (van Dijk et al. 2003).

It has been shown that fasting insulin levels were associated to mean CIMT; insulin sensitivity was negatively associated with CIMT, and this effect was partly explained by traditional CV risk factors, glucose tolerance and adiposity; the association between HOMA-IR index and CIMT in non-diabetic subjects was fully explained by established CV risk factors, above all by HT; the relationship between HOMA-IR index and carotid atherosclerosis was mostly dependent on the clustered expression of the components of the MetS; the association between lower insulin sensitivity was mediated by circulating FFA's and adipocytokines. These results imply that IR *per se* has no strong influence on carotid atherosclerosis, and that its effect on carotid wall is mediated by other metabolic, cellular and hemodynamic abnormalities related to the IR syndrome and DM, like dyslipidemia (DLP), FFA's, adipocytokines, chronic inflammation and HT (Kozakova and Palombo 2016).

Moreover, a novel relationship between arterial stiffness, HI and CAN in a T2DM population has been shown in a study which signifies their pathogenic roles in the development of CVD's in diabetic patients (Bagherzadeh et al. 2013). The mechanisms linking the autonomic imbalance to atherosclerosis are still elusive. Recent studies suggest an association between autonomic function and inflammation in patients with CVD's, showing an inverse relationship between autonomic activity, measured by HRV, and plasma levels of inflammatory markers (Ulleryd et al. 2017). Inflammation is a key mediator in the pathophysiology of atherosclerosis involving both the innate and the adaptive immune system. Diabetes mellitus course represents a spectrum, so drawing clear lines between these stages may be inaccurate, and probably the endothelial inflammation and microcirculatory dysfunction starts with the first rises in blood glucose levels (Song et al. 2014; Çakar et al. 2015; Çakar et al. 2017; Ulleryd et al. 2017). Tumor necrosis factor-alpha (TNF-alpha), interleukin-1, and some bacterial components like lipopolysaccharides trigger the transcription of a molecule of the selectin family (CD62E) in an

NF-kB-dependent signal cascade. CD62E is associated with endothelium of the blood vessel in different inflammatory situations. Previously published data have demonstrated that E-selectin (CD62) is a proatherogenic and proinflammatory cytokine associated with IR, obesity, and CVD's (Adamska et al. 2014). It has been suggested that circulating CD62 concentration may be a biomarker for indicating the subsequent development of metabolic diseases and in particular CVD's from a healthy state (Mochizuki et al. 2013). Prediabetes patients have higher E-selectin levels as their serum glucose rise, and it looks as they get highest with the DM diagnosis. This is compatible with a hypothesis of rising endothelial inflammation with progression to DM (Çakar et al. 2015; Çakar et al. 2017).

Previous studies suggest a relationship between autonomic dysfunction and inflammation, as well as between inflammation and atherosclerosis. These findings made us hypothesize that inflammation mediates the atherogenic effects of autonomic dysfunction. Few studies have investigated the whole pathway (autonomic function-inflammation-atherosclerosis) in the same population or carotid atherosclerosis as the primary end-point. Intima-media thickness (IMT), a marker of preclinical atherosclerosis, was recently associated with increased C-reactive protein and reduced HRV in patients suffering from depression (Ulleryd et al. 2017). Ulleryd et al. (2017) hypothesize that inflammation can be a mediator in the link between autonomic dysfunction and atherosclerosis, leading to CVD's like stroke and MI; reduced autonomic function will increase low-grade inflammation, subsequently worsening the progression of atherosclerosis.

Increased CIMT reflects very early atherosclerotic changes, whereas plaque presence indicates a more advanced atherosclerotic process. T2DM patients have higher CIMT (on average by 130 μm), higher prevalence of carotid plaques and higher plaque volume as compared to controls, and CIMT and carotid plaque prevalence have been shown to be associated with fasting plasma glucose levels, glucose fluctuation or HbA1c, both in non-diabetic and in diabetic populations (Kozakova and Palombo 2016).

Endothelial dysfunction and atherosclerosis play an important role in the pathogenesis of end-organ damage. The endothelial layer has many functions, such as homeostasis, vascular permeability, vasoregulation, angiogenesis and inflammatory response. It is widely accepted that ED is the atherosclerosis precursor (Mancia et al. 2013). DM candidates have an increasing ED and inflammation with rising blood glucose levels, through the diagnosis of T2DM. ED is a mostly recognized part of macrovascular and microvascular disease complications in diabetes. ED may also precede diabetes onset and may promote IR and glucose dysregulation leading to DM, but there is still controversy about the exact role of ED in patients with prediabetes and DM (Çakar et al. 2015).

In diabetic patients, impaired endothelial functions and increased arterial wall thickness potentially lead to diabetic micro- and macrovascular complications. DM leads to arterial stiffness by causing functional and structural changes in large vessels. Therefore, it is crucial to maintain ED and arterial elasticity by means of intensive diabetes treatment and improved glycemic control. Additionally, management of ED and arterial stiffness with certain experimental drugs and/or some medication used in clinical practice is promising regarding prevention of diabetic complications. Furthermore, management of ED and arterial stiffness is likely to decrease the incidence of CVD's, which is an important cause of mortality in DM (Ciftel et al. 2014).

Increased arterial stiffness is a common indicator for the atherosclerotic involvement of the vascular system and is known to occur as a result of atherosclerotic risk factors, such as DM, smoking, HT, DLP, and aging. Increased arterial stiffness is also associated with CHD, cerebrovascular disease and peripheral arterial disease (Mancia et al. 2013; Çakar et al. 2015). These data provide evidence that diabetes candidates have an increasing ED and inflammation with rising blood glucose levels, through the diagnosis of T2DM. ED is a mostly recognized part of macrovascular and microvascular disease complications in DM. ED may also precede DM onset and may promote IR and glucose disregulation leading to DM, but there is still controversy about the exact role of ED in patients with prediabetes and DM (Çakar et al. 2015). PWV is the gold-standard method for the assessment of arterial stiffness and is an independent predictor of CV morbidity and mortality (Loehr et al. 2016). A number of studies have shown that arterial stiffness is increased in people with DM and that PWV independently predicts mortality in this group of patients (Chorepsima et al. 2017). That means IAxao was similar among T2DM patients and nondiabetes controls; however, IAxao was related to diabetes status after multiple adjustments of several risk factors in the regression model. Other studies have reported that IAxao, which was similar between DM and nondiabetes subjects, increased in diabetes patients after adjustment of HR (Yeboah et al. 2016). Yeboah et al. (2016) hypothesize that; compared to age- and gender-match controls, nonhypertensive, nonobese T2DM patients would have higher levels of arterial stiffness.

Blood pressure and impaired cardiac autonomic function are the main determinants of abnormal PWV in people with T2DM, while the association between impaired cardiac autonomic function and arterial stiffness is not mediated by low BRS or increased HR. Chorepsima et al. (2017) shown that CAN activity influences arterial stiffness and it should be monitored and reported in studies examining factors affecting PWV. DM was associated with a lower peripheral arterial stiffness, opposite from the associations seen for carotid-femoral (cfPWV) and brachial-ankle PWV (baPWV) (Loehr et al. 2016).

Despite such stability, in the sixth ten human life, there is a gradual fragmentation of elastin fibers and plates, which is accompanied by changes in the extracellular matrix in the form of overproduction of collagen and deposition of Ca^{2+}. This diffuse degenerative process leads to increased stiffness of the vessels of the elastic type and the growth of pulse pressure (PP) (Christen et al. 2010). The relationship between the effects of age and hypertonic changes on the vascular wall is extremely complicated. This is due to the fact that a number of pathophysiological changes are unidirectional (increase in the content of collagen), while others have multidirectional effects, which complicates the interpretation of the results of the examination of patients. In particular, vascular ductility may increase with HT and decreases with age. In addition to HT, many traditional risk factors may contribute to the increased rigidity of the vessels, including DLP, smoking, hyperglycemia, etc (Christen et al. 2010). Loehr et al. (2016) hypothesize that the arterial wall tissue remodeling that occurs with DM may differ pathophysiologically in muscular versus elastic large arteries and that loss of elasticity and recoil does not necessarily translate into stiffening of the muscular arteries. Tsuchikura et al. (2010) found opposing directions of effect for the association of peripheral and central stiffness with coronary artery disease (CAD) including in a subset of participants with T2DM. Similarly, a recent publication from the ARIC-study reported that the cfPWV and baPWV were positively associated with age and with HbA1c, whereas femoral-ankle PWV was not. These studies support the observation that peripheral stiffness does not have the same associations with CAD, or its risk factors, that are observed for central arterial stiffness (Meyer et al. 2016; Loehr et al. 2016).

One hypothesis is that impaired cardiac autonomic function results in increased arterial stiffness. An explanation could be that patients with CAN present more often with calcification of the tunica media of the arterial wall. It is noteworthy that the main determinant of the extent of arterial calcification is the severity of diabetic CAN. On the other hand, arterial calcification has been suggested as an important determinant of arterial stiffness according to findings in humans and experimental models (Chorepsima et al. 2017). These data reveal that calcification of the arterial wall may be an additional common pathophysiological pathway that could explain the relationship between impaired cardiac autonomic function and arterial stiffness. High sympathetic activity has been associated with arterial stiffness in hypertensive patients with and without T2DM, as well as in healthy individuals. Increases in HR *per se* may lead to arterial stiffness independently of changes in activity of the ANS (Chorepsima et al. 2017).

The other hypothesis is that arterial stiffness may lead to CAN *via* impairment of baroreceptor function induced by stiffening of the arterial wall. Several studies have found a significant association between low BRS and increased arterial stiffness in patients with congestive HR, in older subjects. However, no difference in central BRS was observed among participants with abnormal and normal PWV. This finding may imply that DM *per se* is a strong factor affecting BRS and outweighs the potential effect of other factors on BRS (Chorepsima et al. 2017).

Some data indicate that increased large arterial stiffness appears already in prediabetic conditions. In a treatment-*naïve* and mostly "healthy" population from the ADDITION-Leicester cohort, cfPWV was increased in individuals with impaired fasting glucose or impaired glucose tolerance as compared to those with normal glucose metabolism, and the increase was identical to that of individuals with newly diagnosed T2DM.Indices of carotid stiffness increased with fasting plasma glucose, insulin and HbA1c; cfPWV increased with HbA1c; fasting glucose, HOMA-IR index and HbA1c, together with waist circumference, triglycerides and HDL-C were all predictors of cfPWV after a follow-up of 17 years. Finally, a recent study in middle-aged subjects free of CVD demonstrated the relationships of cfPWV with HOMA-IR index (direct) and telomere length (inverse) and suggested that IR linked with chronic inflammation can enhance telomere shortening (a marker of cellular senescence), and thus induce accelerated vascular aging (Kozakova and Palombo 2016).

Diabetes Mellitus and Circadian Rhythms of Blood Pressure

Ideal 24-h BP control not only consists of control of average daily BP, but also its diurnal and nocturnal variability. Most hypertensive patients are monitored by clinic-based BP measurement and their daily BP variability is not available. However, studies have shown that office BP (OBP) measurement is not an ideal method for representing 24-h BP compared with ambulatory BP monitoring (ABPM). For example, OBP monitoring has less sensitivity to predict organ damage associated with hypertension (Hodgkinson et al. 2011; Mancia et al. 2013; Salva et al. 2014; Najafi et al. 2018).

Table 3. Diagnostic criteria for ambulatory blood pressure parameters (O'Brien et al. 2013a; O'Brien et al. 2013b)

Blood pressure thresholds for diagnosing hypertension	
Office BP	> 140/90 mmHg
Ambulatory BP monitoring (mean)	
24-h	>130/80 mmHg
Night-time	>120/70 mmHg
Daytime	>135/85 mmHg
Circadian BP patterns	
Normal dipping	A night-time BP fall > 10% but < 20% relative to daytime mean (night to day BP ratio 0.8 < ratio < 0.9)
Extreme dipping	A night-time BP fall > 20% in relative to daytime BP (night to day BP ratio < 0.8)
Non-dipping	No night-time BP fall or a night-time BP fall < 10% (night to day BP ratio 0.9 < ratio < 1)
Reverse dipping	BP rise at nighttime (night to day BP ratio > 1.0)
Hypertensive syndromes	
White coat hypertension/effect†	Office BP > 140/90 and awake ABPM < 135/85 and 24-h ABPM < 130/80 and sleep ABPM < 120/70
Masked hypertension/masked uncontrolled hypertension‡	Normal office BP < 140/90 and 24-h mean ABPM > 130/80 and/or awake mean ABPM > 135/85 and/or sleep mean ABPM > 120/70

†White coat effect: in treated patient. ‡Masked uncontrolled hypertension: in treated patient.

In clinical practice, ABPM in the general population is useful for diagnostic purposes and provides unique and additional information for risk stratification with regard to hypertension-related organ damage and CV events, and for the extent of BP response to treatment (Fagard et al. 2008; Spallone et al. 2011; Mancia et al. 2013). The European Society of Hypertension acknowledges that ABPM may improve predictions of CV risk in hypertensive patients and recommends that 24-h ABPM should be considered in the presence of either noticeable variability of OBP values or a marked discrepancy between office and home BP values, and in case of resistance to drug treatment or suspected hypotensive episodes (Mancia et al. 2013). Table 3 shows the diagnostic criteria for ABPM parameters (O'Brien et al. 2013a; O'Brien et al. 2013b).

ABPM is a standard tool in HT research and management with regard to diagnostic, prognostic, and therapeutic issues. It allows the assessment of the diurnal BP pattern, which is mainly regulated by sleep-awake changes in the autonomic cardiovascular function. ABPM may be used for research purposes too (Mancia et al. 2013):

- evaluate the circadian BP pattern and its abnormalities (e.g., non-dipping, nocturnal hypertension, extreme dipping, morning surge);
- study its relationship with autonomic dysfunction, sleep disturbances, and kidney function;
- assess the 24-h BP response to treatment;
- evaluate the longer-term prognostic implications of circadian BP abnormalities.

Arterial BP displayed distinct diurnal variation, characterized by lowest value during sleep (around 3 a.m.), a slow increase during late sleep, an abrupt rise upon morning awakening, and two daytime peaks (first one around 9 a.m. and second one around 7 p.m.); the latter is separated by a small mid-afternoon drop (around 3 p.m.), and a gradual decline thereafter till the nadir during sleep (Smolensky et al. 2017). The BP during the night-time mean is normally 10-20% lower compared to the daytime mean, which is also called a dipping pattern. Abnormal BP rhythm patterns include: non-dipping (nocturnal BP fall is less than 10% of daytime BP mean), reverse dipping (nocturnal BP mean is higher than daytime BP mean), and extreme dipping (nocturnal BP fall is greater than 20% of daytime BP mean) (Hou 2018). Among the different abnormal circadian BP patterns, reverse dipping, in which the nighttime BP is higher than the daytime BP, seems to have the worst CV prognosis. Arterial BP undergoes a 24-h oscillation that peaks in the active day and reaches a nadir at night during sleep in humans. Reduced nocturnal BP fall (also known as non-dipper) is the most common disruption of BP circadian rhythm and is associated with increased risk of untoward CV events and target organ injury (Palatini et al. 1992; Grossman E 2013; O'Brien et al. 2013).

Research in BP variability has led to the discovery of the normal nocturnal dipping pattern of BP. In healthy subjects, BP at night (sleep) is 10% to 20% less than diurnal (waking) BP. However, abnormal BP dipping patterns either by the having less nocturnal dipping (0%-10% decrease) or even rise in BP in sleep or by having exaggerated dipping of BP at nights (> 20% decrease) are associated with higher CV morbidity and mortality and damages to organs such as brain and kidney (Wang et al. 2013). Morning blood pressure surge (MBPS) is another component of BP variability, of which high values have been shown to independently increase risk of CVDs (Verdecchia et al. 2012). There are 3 definitions for MBPS including sleep-trough surge, prewaking surge, and rising surge. Sleep through surge is one of the dynamic diurnal surges during the specific period from sleep to early morning. Prewaking surge is the BP change occurring 4-h before and after arising. The rising surge may detect the morning risk just after arising (Kario 2010; Bhardwaj et al. 2014; Ates et al. 2015; Irigoyen et al. 2016; Najafi et al. 2018).

It is well-known that the ANS is an important regulator of BP homeostasis. The ANS is also critical to the BP circadian rhythm. The activity of the sympathetic and parasympathetic nerve system (SNS) and (PSNS), two efferent arms of ANS, exhibit day and night difference. The sympathetic nerve activity, in accordance with BP, is typically higher during the active or awake period and lower during sleep. The circadian variation of ANS activity is altered in non-dippers as compared to normal individuals. The baroreflex reflects the reciprocal responses between SNS and PSNS: when the BP increases, baroreceptors activate and trigger the inhibition of SNS and the activation of PSNS, resulting in decreased HR and total peripheral resistance (TPR), and thus lower BP. An inversed reflex will be triggered when BP decreases (Hou 2018). The BRS is used to quantify baroreflex function. The BRS has a daily rhythm that is opposite to BP rhythm in both humans (Takakuwa et al. 2001) and mice (Xie et al. 2015).

Blood pressure is regulated by multiple hormones, among which the renin-angiotensin-aldosterone system (RAAS) is the most important hormonal system for BP homeostasis. The RAAS regulates BP through modulating vasoconstriction and extracellular fluid volume. Interestingly, the components of the RAAS all exhibit circadian rhythm. Besides RAAS, BP is also regulated by other hormones, such as cortisol, thyroid and atrial natriuretic peptide etc. (Hou 2018).

Thereby, BP circadian rhythm is regulated by multiple factors and pathways. And these factors and pathways are interconnected and cannot be separated. For instance, sleep deprivation leads to derangement of central nervous system and endothelial dysfunction (Tobaldini et al. 2017). There is a close interaction between SNS and RAAS: SNS stimulate renin and aldosterone release while angiotensin II activates SNS and modulates baroreflex control of HR. Both the vascular tone and kidney function are influenced by SNS, RAAS, glucocorticoid and thyroid hormones (Nguyen Dinh Cat and Touyz 2011; Amiya et al. 2014; Hou 2018).

Alternations of these factors and pathways have been observed in diabetes. For example, people with DM have increased TPR compared to nondiabetic subjects (Colosia al. 2013). Findings from the Sleep Heart Health Study (SHHS) showed that diabetes is associated with periodic breathing. Over 86% of obese patients with T2DM have OSA (Foster et al. 2009). T2DM is a disease involving multiple organs and is known to expose patients to HT. Approximately 35% to 75% of DM complications are estimated to be attributed to HT. In addition to blood glucose control, BP control remains the cornerstone of the management of micro and macrovascular complications of diabetes (Ryden et al. 2013). It has been demonstrated that diabetic patients tend to have higher nocturnal BP and higher MBPS. Thus, high-risk hypertensive diabetic patients are indicated for ABPM (Sun et al. 2016).

As with non- or reverse dipping, an increased prevalence of nocturnal systolic HT (SHT) or higher nocturnal SBP in relation to daytime BP has been described in T2DM patients (Ayala et al. 2013, Draman et al. 2015). Ayala et al. (2013) reported a prevalence of 89.2% nocturnal HT in type 2 diabetics with poor BP control. Draman et al. (2015) demonstrated a higher nocturnal SBP in relation to daytime BP in 859 T2DM patients in a case control study. The significant association of nocturnal HT and non-dipping with the use of three or more anti-hypertensive medications was also probably the reflection of poor BP control in these patients (Gunawan et al. 2019).

The results of a cross-sectional study in patients with T2DM showed that non-dipping, reverse dipping, nocturnal SHT and masked phenomenon are highly prevalent in patients with T2DM with or without a known history of hypertension. Compared with non-dipping, nocturnal SHT may be a stronger predictor of the end of the organ damage. The reliability of OBP for assessing BP in T2DM is only modest. Patients with T2DM are likely to benefit from routine ABPM (Gunawan et al. 2019). The study confirms a high prevalence of non- or reverse dipping, nocturnal SHT and masked phenomenon in patients with T2DM with or without a known history of HT. The findings also suggest that office BP is only moderately sensitive and specific for assessing BP in patients with T2DM. The findings further demonstrate a significant association between nocturnal SHT and microvascular complications independent of daytime SBP control (Gunawan et al. 2019).

De la Sierra et al. (2009) demonstrated that T2DM was independently associated with non-dipping in a cohort of 34,563 treated and 8384 untreated hypertensive patients in a registry-based study. Ukkola et al. (2009) demonstrated this association in patients with impaired glucose tolerance and Lyhne et al. (2015) in newly diagnosed T2DM. The prevalence of non-dipping in the latter study was 40.2%. Fogari et al. (1993) in a case control study that included 47 normotensive and 49 hypertensive type 2 diabetics demonstrated that the association between T2DM and non- or reverse dipping, particularly for SBP, was independent of the history of HT. Ayala et al. (2013) confirmed that this association was independent of chronic kidney disease, OSA and treatment of HT in a sample of 2954 type 2 diabetic and 9811 non-diabetic hypertensive patients. The prevalence of non-dipping and reverse dipping for SBP in this study was 62.1% and 19.1%, respectively. Kalaycioglu et al. (2014) in a cohort of 86 hypertensive diabetics described non-dipping in 59.3% and

Draman et al. (2015) found 47.4% to have non-dipping for SBP in a cohort of 859 type 2 diabetics (Gunawan et al. 2019). In patients with DM, reduced nocturnal BP fall is also associated with increased risk of CVD's and target organ damage. In a four-year prospective study, the incidence of fatal and nonfatal vascular (cerebrovascular, CV, peripheral vascular arteries, and retinal artery) events is higher in T2DM patients with reverse-dipping BP than in those with dipping BP. The IMT and LV mass index, which are found to be increased in non-dippers, are also greater in diabetic non-dippers than dippers (Di Flaviani et al. 2011). The nocturnal fall in SBP, independent of 24-h BP, is negatively associated with PWV, lower albumin: creatinine ratio and positively associated with glomerular filtration rate in T2DM patients (Jennersjo et al. 2011; Hou 2018).

At night, health subjects exhibit a predominance of vagal tone and decreased sympathetic tone, associated with reduction in nocturnal BP. In diabetic CAN this pattern is altered, resulting in sympathetic predominance during sleep and subsequent nocturnal hypertension. This is associated with a higher frequency of LV hypertrophy and both fatal and severe nonfatal CV events in diabetic CAN subjects (Pop-Busui 2010; Spallone et al. 2011). Non-dipping and reverse dipping patterns were associated with CAN, which was the major determinant of the circadian variation in BP. Several observations in both diabetic and non-diabetic patients linked non-dipping to a disruption of the circadian variation in sympathovagal activity, i.e., a diminished increase in vagal activity and a sympathetic predominance during the night. The day-night difference in SBP was a moderately accurate diagnostic tool for CAN, and reverse dipping as a specific (95%) - albeit insensitive (25%) - marker of CAN (Spallone et al. 2011). Thus, in patients with CAN, ABPM may be particularly useful in detecting non-dipping or reverse dipping conditions, daytime postural BP changes, and postprandial hypotension, and in achieving BP control for the whole 24-h period. Conversely, in clinical practice, the presence of reverse dipping in ABPM may suggest the presence of CAN and thus requires CAN testing (Spallone et al. 2011).

Diabetic Cardiac Autonomic Neuropathy and Circadian Rhythm's Disorders

Cardiac autonomic neuropathy is a significant risk factor for cardiovascular injury resulting in heart attack, congestive heart failure, stroke and sudden arrhythmic death. This issue is associated with myocardial structural remodeling that follows HT, such as hypertrophy and fibrosis. This remodeling accompanies changes in expression, distribution, and function of cell membrane ion channels, Ca^{2+}-cycling proteins, intercellular gap junction connexin-43 channels and extracellular matrix composition (Serhiyenko and Serhiyenko 2015; Serhiyenko and Serhiyenko 2019; Pop-Busui et al. 2017).

There is a lot of evidence that CAN plays a crucial role in the pathogenesis of progressive vessel atherosclerosis (Shah and Brownlee 2016; Mala et al. 2017; Nattero-Chavez et al. 2019). Moreover, diabetic autonomic neuropathy is closely associated with subclinical myocardial dysfunction, interstitial myocardial fibrosis, and metabolic changes (Didangelos et al. 2014; Shang et al. 2019). Additionally, CAN also significantly increases the incidence of CVD's and their related mortality (Liu et al. 2020). However, the independent relationship of CAN with the severity of coronary lesions in T2DM patients has not been established (Flotats and Carrió 2010).

A new CVD's risk factor, which could be more helpful in terms of prognosis in particular, "syndrome of vascular aging," is needed and the possibility of its using is actively discussed. The results of several epidemiological studies reported that increased arterial stiffness predicts mortality and morbidity, independently of other CV risk factors. Additionally, results from clinical studies have shown that arterial stiffness increases with aging or several pathological conditions associated with HT, MetS, chronic renal disease, and DM (Chorepsima et al. 2017). Diabetes may increase arterial stiffness through pathological changes in the vascular bed, such as changes in structure or type of elastin and/or collagen in the arterial wall, chronic low-grade inflammation, increased

OS, reduced NO bioavailability, increased activity of SNS (Gaszner et al. 2012; De Oliveira Alvim et al. 2013).

Cardiac autonomic dysfunction has been considered as a significant risk factor in the acceleration of atherosclerosis and serves as a trigger for cerebrovascular or CV events. The pathophysiology of CAN is complex and involves the endocrine and vegetative nervous system with associated metabolic, inflammatory and hemostatic abnormalities (Chinnaiyan 2019). The pathophysiological link between autonomic dysfunction and aortic stiffness and whether impaired cardiac autonomic function induces the arterial stiffness or increased arterial stiffness leads to autonomic function's impairment remains obscure. Both the arterial stiffness and cardiac autonomic dysfunction share common pathogenetic pathways including chronic hyperglycemia and HI, formation of AGE's and PKC activation, low grade-inflammation and ED (Chorepsima et al. 2017). Insulin resistance and HI increase systemic metabolic disorders and activate the SNS; then activate RAAS; prompt OS, mitochondrial dysfunction, and endoplasmic reticulum stress; and impair Ca^{2+} homeostasis. These effects result in cardiac fibrosis, hypertrophy, cardiomyocyte death, dysfunction of the coronary microcirculation, and eventually heart failure (Jia et al. 2018).

Our study was aimed to examine the insulin resistance, HRV, arterial stiffness and ABPM parameters in patients with type 2 diabetes mellitus and cardiac autonomic neuropathy.

The study involved 107 patients with T2DM, among them 19 patients without CAN; 44 patients with subclinical CAN; 44 patients with definite stage of CAN and 15 healthy persons. All patients observed were residents of Lviv, they were examined and treated by endocrinologist and cardiologist at the endocrinological department of Danylo Halytsky Lviv National Medical University, which is located in Lviv Regional State Clinical Treatment and Diagnostical Endocrinological Center.

Data acquisition and analyses have been approved by the Danylo Halytsky Lviv Medical University Ethics Committee (protocol №2 from 18 February 2013), are in accordance with the Declaration of Helsinki's principles of 1975 as revised in 2014 (World Medical Association

Declaration of Helsinki 2014) and with good clinical practice guidelines. All persons signed informed consent before their enrollment. The study corresponded to the generally accepted norms of morality and observance of the rights, interests and personal dignity. The standard hypoglycemic treatment of T2DM included a dietary regime, appropriate physical activity and oral antihyperglycemic drugs. The therapy remains unchanged during the treatment period.

Study inclusion criteria. Age: individuals recruited were adults 50-60 years old and had been diagnosed with T2DM based on the American Diabetes Association criteria (American Diabetes Association 2017); T2DM with optimal or suboptimal glycemic control; T2DM patients with confirmed CAN; BMI within 20-30 kg/m^2; diabetic polyneuropathy; consent to adhere to the dietary regime and to maintain appropriate physical activity during the study.

Study exclusion criteria: persons with uncontrolled T2DM, ketosis, peripheral vascular diseases, ischemic diabetic foot, distal neuropathy caused by other factors than DM (neurological disease, chronic alcoholism, pharmacological agents), neoplasms, hypothyroidism, women during pregnancy and lactation, patients with T1DM; atrial fibrillation or flutter, pacemaker, history of severe kidney [estimated glomerular filtration rate < 30 ml/min/1.73 m^2] or liver disease, heart failure NYHA III or IV (The New York Heart Association 1994) and acute illness or hypoglycemia in the previous 24 hrs. To minimize the confounding effect of drugs on HRV, arterial stiffness and ABPM parameters, we excluded persons treated with antidepressants and antihistamines, antiarrhythmic drugs other than beta-blockers.

The examination was done between 07.30 and 09.30 hrs in the morning in a room with temperature (22-24°C). All participants abstained from food or drink except water for 12 hrs before and received their medications after the end of examination. We performed a complete physical examination and used established questionnaires to evaluate previous disease history, current disease, and medication use. Height, weight and waist circumference were measured, and BMI was calculated. BP was measured in the sitting position at the brachial artery using an appropriate cuff size three times at 5-min intervals. The mean value of the last two measurements was used in the analysis. Arterial hypertension was defined according to current guidelines (de Boer et al. 2017; Williams et al. 2018) if level of SBP was $\geq$ 140 mmHg and/or level of diastolic BP (DBP) $\geq$ 90 mmHg, or if patient takes antihypertensive medications.

The examination plan included determination of glucose, HbA1c, ECG, five cardiovascular autonomic reflex tests (CART's), Holter-ECG, artery stiffness, ABPM and echocardiographic parameters. ECG was performed using a 12-channel electrocardiograph "UCARD-200" (UTAS, Ukraine) and Holter-ECG - using [ECG "The EC-3H" (Labtech, Hungary)]. The severity of CAN was determined according to the results of five CART's. Physiological values were evaluated as "0" score, borderline values - as "0.5" score, pathological values - as "1" score. The severity of autonomic dysfunction was evaluated according to the sum of five CART's (Spallone et al. 2011; Ziegler et al. 2014; Pop-Busui et al. 2017; Bamira and Picard 2018).

Coronary artery disease was defined as a documented history of MI, tonsillitis, percutaneous transluminal coronary angioplasty or coronary artery bypass grafting. The peripheral arterial disease was defined as a history of intermittent claudication, revascularization procedures, or an ankle-brachial pressure index < 0.9. Cerebrovascular disease was defined as a documented carotid arterial disease with 50% stenosis, history of stroke or revascularization at the carotid stenting or surgical revascularization (Newman et al. 1993; Mancia et al. 2013; Pop-Busui et al. 2017).

Diabetic peripheral neuropathy (DPN) was diagnosed using neuropathy disability score (NDS) and neuropathy symptom score (NSS). The criteria for diagnosing DPN were NDS $\geq$ 6 irrespective of the NSS values, or NDS = 3-5 with NSS $\geq$ 5 (Pop-Busui et al. 2017). Blood was drawn in the morning after 12-hrs of fasting. The blood samples were centrifuged, and then serum was collected for measuring the biochemical parameters. In all participants, fasting blood glucose and 2-h postprandial glucose levels were measured. The glucose concentration in the blood was determined by the glucose oxidase method, while HbA1c was assessed using a highly sensitive ion-exchange liquid chromatography with D-10 analyzer and BIO-RAD reagents (USA). Serum immunoreactive insulin (IRI) level was measured using commercial kits from immunogen insulin immunoradiometric assay reagents (the Czech Republic). HOMA-IR was calculated by following formula: fasting IRI (mcIU/mL) x fasting glucose (mmol/L)/22.5 (Matthews et al. 1985).

Evaluation of HRV was performed in a quiet and temperature-controlled room according to the guidelines of the Task Force for Pacing and Electrophysiology (Camm et al. 1996). Participants were advised to abstain from caffeinated food and beverages on the day of assessment. Repeated assessments were performed at precisely the same time of day after 48-hrs. After 15-min of supine rest with a regular and calm breathing pattern, a continuous 10-min ECG recording was collected using an applanation tonometer interface with HRV software. The HF, LF and VLF components of HRV (measured in absolute units; i.e., ms^2) were obtained (Malliani et al. 1991; Camm et al. 1996).

From the ECG recording, the following statistical and geometric time domain indices were calculated from RR intervals: SDNN, RMSSD. Frequency domain variables including total, HF, and LF powers and LF/HF ratio were derived from spectral analysis of successive RR intervals (Task Force of the European Society of Cardiology and the North American Society of Pacing and Electrophysiology 1996).

Evaluation of the arterial stiffness parameters was performed using the device TensioMedTM™Arteriograph [monitor BP "ABPM-04" ("Meditech"Hungary)] (Horváth et al. 2010; TensioClinicTensioMed™). The Arteriograph cuff was applied on the "non-working" arm over the brachial artery to detect arterial wall oscillations in the upper arm using the "stop-flow" principle (Horváth et al. 2010). In the absence of asymmetry of BP, monitoring was performed on the "non-working" arm of the patient. With asymmetry, BP is greater than or equal to 10 mmHg studies were performed on the arm with higher rates. The cuff of the appropriate type was placed on the patient's shoulder so that the lower edge was 2 cm above the elbow. Before starting the measurement, a series of control BP measurements were performed, and the obtained data were compared with hardware data. The standard measurement frequency was set (interval during the active period of the day - 15 min, during the passive period - 30 min). To determine PWV, the Arteriograph uses the wave reflection's physiological behavior; the ejected direct (first systolic) pulse wave is reflected mostly from the aortic bifurcation. Arteriograph measures the time interval between the peaks of the direct (first) and reflected (late) systolic wave (return time). For both the invasive and non-invasive aortic PWV computations, the aortic length is measured as the distance from the sternal notch (jugulum) to the upper edge of the pubic bone (symphysis). Stiffer arteries are less elastic and transmit pulse waves faster, reflected earlier at the abdominal aortic bifurcation, resulting in decreased pulse return time and higher PWV (Sugawara et al. 2010). Upon completion of the examination, the data were transferred to a personal computer for further analysis. The daily parameters of arteriography were determined: aorta augmentation index (AIxao), brachialaugmentation index (AIxbr), PWV and ambulatory arterial stiffness index. The program orders the values of the AIxbr and PWV into ranges as follows: optimal values: AIxbr > -30%, PWV < 7 m/s; normal values: -30% < AIxbr < -10%, 7 m/s < PWV < 10 m/s; elevated values: -10% < AIxbr < 9.8%,9.8 m/s < PWV < 12 m/s; pathological values: AIxbr > 10%, PWV > 12 m/s (Townsend et al. 2015).

Office blood pressure. Prior to commencing 24-h ABPM, OBP was obtained under standardized conditions by one of the investigators using patients' non-dominant arm. Blood pressure was measured using a calibrated oscillometric automated digital BP apparatus and an appropriate size cuff after a minimum rest of five minutes in a seated position. The mean of three BP measurements measured at least two minutes apart was taken for analysis as the final OBP reading (O'Brien et al. 2013).

Ambulatory blood pressure monitoring. All participants underwent 24-hrs ABPM (using the device TensioMedTM™Arteriograph [monitor BP "ABPM-04" ("Meditech" Hungary)] (Horváth et al. 2010; TensioClinicTensioMed™). Participants were commonly instructed to begin bed rest or try to sleep before midnight and to rise after morning. During the monitoring period, BP was measured at 30-min intervals during the daytime, and at 60-min intervals during the night-time. Night-time (for the 60-min BP monitoring interval) was defined as from the time patients went to bed until the time each got out of bed in the morning, and the daytime interval (for the 30-min BP measurements) as the rest of the day. The 24-h BP changes could be divided into four patterns on the basis of the daytime and night-time BP: dipper, non-dipper, extreme dipper and night-peaker (reverse or invert dipper). The normal decline in nocturnal BP by 10-20% is defined as the dipper pattern. Non-dippers and extreme dippers are defined as those with a night-time BP reduction of < 10 or > 20% of daytime BP, respectively. Night-peakers have night-time BP levels that are higher than daytime levels (Chang et al. 2018; Gunawan et al. 2019). Standard ABPM diagnostic criteria were used to define BP control and abnormal ABPM patterns (Table 3) (O'Brien et al. 2013).

Statistical analysis. Data were expressed as mean ± SEM. One-way ANOVA or repeated-measures ANOVA followed by Tukey post hoc correction for multiple comparisons to localize significant effects were used to assess between-group differences. *P* values less than 0.05 were considered statistically significant. All analyses were performed using ANOVA (MicroCal Origin v 8.0) software (Lang and Secic 2006). The SPSS statistical software package (version 18.0 for Windows; SPSS Inc. Chicago, IL) was used for data analysis. Clinical and anthropometric characteristics of study participants are presented in Table 4.

There were no significant differences between patients of all studied groups in age, gender and BMI parameters. As shown in Table 4 among patients with T2DM and subclinical CAN were found higher HR levels [+5.2% ($P < 0.01$)] compared to patients without CAN. In patients with definite CAN statistically significant levels of SBP, DBP and HR compared to patients without and with subclinical CAN were found. HbA1c, IRI and HOMA-IR parameters in the blood among the studied groups are presented in Table 5.

Table 4. Clinical characteristic of patients with T2DM and CAN (M ± SEM)

Parameters	Patients with T2DM (n = 107)		
	Without CAN (n = 19)	Subclinical (n = 44)	Definite (n = 44)
	1st group	2nd group	3rd group
Age (yrs)	52.84 ± 0.81	53.82 ± 0.73	54.03 ± 0.48
Gender			
Male (%)	9 (47.4%)	25 (56.8%)	24 (54.5%)
Female (%)	10 (52.6%)	19 (43.2%)	20 (45.5%)
Diabetes duration (yrs)	3.53 ± 0.45	4.43 ± 0.35	4.37 ± 0.39
BMI (kg/m^2)	27.7 ± 0.58	28.1 ± 0.21	27.9 ± 0.35
HR (bpm)	82.9 ± 1.52	87.2 ± 1.03**	90.2 ± 0.54$^{***\#}$
SBP (mmHg)	140.7 ± 1.88	140.6 ± 1.76	149.8 ± 2.12$^{**\#\#}$
DBP (mmHg)	88.8 ± 1.89	84.1 ± 1.98	90.3 ± 2.04$^{*\#\#}$
History of HT (%)	14 (77.8%)	33 (75.0%)	34 (77.3%)
History of CAD (%)	8 (42.1%)	16 (36.4%)	17 (39.5%)
Medications			
ACE inhibitors/ARBs (%)	15 (79.0%)	32 (72.7%)	37 (84.1%)
β-blockers (%)	5 (26.3%)	12 (27.3%)	13 (29.3%)
Metformin (%)	9 (47.4%)	21 (47.7%)	18 (40.8%)
Combined hypoglycemic therapy (%)	10 (52.6%)	23 (52.3%)	26 (59.1%)

ACE: angiotensin-converting enzyme; ARBs: angiotensin II receptor blockers. $^{*}P< 0.05$ compared to patients of 1st group. $^{**}P< 0.01$ compared to patients of 1st group. $^{***}P< 0.001$ compared to patients of 1st group. $^{\#}P< 0.05$ compared to patients of 2nd group. $^{\#\#}P< 0.01$ compared to patients of 2nd group.

Table 5. HbA1c, IRI and HOMA-IR in the blood among the studied groups (M ± SEM)

Parameters	Control (n=15)	Patients with T2DM (n = 107)		
		Without CAN (n = 19)	Subclinical CAN (n = 44)	Definite CAN (n = 44)
	1st group	2nd group	3rd group	4th group
HbA1c (%)	5.45 ± 0.16	6.53 ± 0.14***	6.69 ± 0.13***	7.04 ± 0.08$^{***\#\dagger}$
IRI (mU/L)	11.23 ± 0.67	16.12 ± 1.15***	20.1 ± 1.3$^{***\#}$	26.4 ± 0.88$^{***\#\#\#\dagger\dagger}$
HOMA-IR	2.38 ± 0.15	4.36 ± 0.47***	5.73 ± 0.6**	8.23 ± 0.4$^{***\#\#\#\dagger}$

Note to Tables 5-13: $^{*}P < 0.05$ compared to patients of 1st group. $^{**}P < 0.01$ compared to patients of 1st group. $^{***}P < 0.001$ compared to patients of 1st group. $^{\#}P < 0.05$ compared to patients of 2nd group. $^{\#\#}P < 0.01$ compared to patients of 2nd group. $^{\#\#\#}P < 0.001$ compared to patients of 2nd group. $^{\dagger}P < 0.05$ compared to patients of 3rd group. $^{\dagger\dagger}P < 0.01$ compared to patients of 3rd group. $^{\dagger\dagger\dagger}P< 0.001$ compared to patients of 3rd group.

In patients with T2DM without verified CAN, was observed an increase in HbA1c [+19.82% ($P < 0.001$)], IRI [+43.54% ($P < 0.001$), and HOMA-IR [+83.19% ($P < 0.001$)] compared with the control group. Among patients with subclinical CAN was found statistically significant increase in the content of IRI [+24.69% ($P < 0.05$)] compared to patients without verified CAN. The definite stage was characterized by a significant increase in IRI [+31.3% ($P < 0.01$)], HbA1c [+5.23% ($P < 0.05$)], and HOMA-IR [+43.6% ($P < 0.05$)] compared to the subclinical stage of CAN (Table 5). Time domain parameters of HRV parameters during 24-h ECG monitoring in patients of studied groups are presented in Table 6.

Table 6. Time-domain parameters of heart rate variability in patients with T2DM and CAN (M ± SEM)

Parameters	**Control (n = 15)**	**Patients with T2DM (n = 107)**		
		Without CAN (n = 19)	Subclinical CAN (n = 44)	Definite CAN (n = 44)
	1st group	2nd group	3rd group	4th group
SDNN (ms)	149.7 ± 3.83	141.1 ± 4.94	136.1 ± 5.87	111.6 ± 3.11$^{***\#\#\#\dagger\dagger\dagger}$
SDANNi (ms)	136.7 ± 4.8	128.2 ± 4.89	124.1 ± 6.06	92.8 ± 2.81$^{***\#\#\#\dagger\dagger\dagger}$
RMSSD (ms)	41.1 ± 2.55	40.5 ± 2.55	33.4 ± 2.04$^{*\#}$	22.8 ± 0.79$^{***\#\#\#\dagger\dagger\dagger}$
pNN50 (%)	12.6 ± 1.59	11.2 ± 1.48	9.23 ± 0.99	6.39 ± 0.48$^{***\#\#\#\dagger\dagger\dagger}$

Patients without verified CAN showed no significant changes in HRV time-domain parameters compared to control group ($P > 0.05$). A specific characteristic of the subclinical stage was a statistically significant decrease in RMSSD [-17.5% ($P < 0.05$)] compared to patients without CAN, while no dynamics for SDNN, SDANNi, and pNN50 were observed. The definite stage was characterized by a pronounced decrease in SDNN [-18.0% ($P < 0.001$)], SDANNi [-25.2% ($P < 0.001$)]; RMSSD [-31.7% ($P < 0.001$)] and pNN50 [-30.8% ($P < 0.001$)] compared to data, obtained among patients with subclinical CAN (Table 6). Spectral parameters of HRV during the active period in patients with T2DM and CAN are presented in Table 7.

Table 7. Spectral parameters of heart rate variability during the active period in patients with T2DM and CAN (M ± SEM)

Parameters	Control (n = 15)	Patients with T2DM (n = 107)		
		Without CAN (n = 19)	Subclinical CAN (n = 44)	Definite CAN (n = 44)
	1st group	2nd group	3rd group	4th group
VLF (мс^2)	1521.3 ± 71.8	1503.7 ± 69.52	1350.3 ± 55.3	1232.7 ± 31.9$^{**\#\#}$
LF(мс^2)	712.5 ± 11.65	694.3 ± 39.58	623.3 ± 37.1	509.1 ± 19.8$^{***\#\#\dagger\dagger}$
HF (мс^2)	392.3 ± 10.98	419.7 ± 32.36	334.2 ± 20.1$^{\#\#}$	296.9 ± 7.8$^{***\#\#\#\dagger}$
LF/HF ratio	1.83 ± 0.03	1.7 ± 0.06	1.98 ± 0.11$^{\#}$	1.72 ± 0.03*††

Analysis of the spectral parameters of HRV during the active period of the 24-h interval revealed that in patients without verified CAN, no statistically significant changes in the spectral parameters of HRV were detected. The subclinical stage was characterized by a significant inhibition of the HF [-20.4% ($P < 0.01$)] and an increase in LF/HF ratio [+16.5% ($P < 0.05$)] compared to patients without CAN; no dynamics for VLF and LF were observed. A specific characteristic of the definite stage was a statistically significant inhibition of the LF [-18.3% ($P < 0.01$)] and HF [-11.2% ($P < 0.05$)], and LF/HF ratio [-13.1% ($P < 0.01$)] compared to patients with subclinical CAN (Table 7). Spectral parameters of HRV during the passive period in patients with T2DM and CAN are presented in Table 8.

Table 8. Spectral parameters of heart rate variability during the passive period in patients with T2DM and CAN (M ± SEM)

Parameters	Control (n – 15)	Patients with T2DM (n = 107)		
		Without CAN (n = 19)	Subclinical CAN (n = 44)	Definite CAN (n = 44)
	1st group	2nd group	3rd group	4th group
VLF (мс^2)	1982.2 ± 62.0	1955.5 ± 100.73	1773.5 ± 71.9	1589.1 ± 48.0$^{*\#\#\dagger}$
LF(мс^2)	911.4 ± 11.88	918.0 ± 56.22	858.8 ± 58.4	672.6 ± 24.7$^{***\#\#\#\dagger\dagger\dagger}$
HF (мс^2)	526.1 ± 12.87	570.4 ± 40.9	426.1 ± 28.0$^{*\#\#}$	374.2 ± 10.8$^{***\#\#\#}$
LF/HF ratio	1.74 ± 0.03	1.63 ± 0.04*	2.11 ± 0.14$^{*\#}$	1.75 ± 0.03$^{\#\dagger}$

During the passive period of the day a decrease in LF/HF ratio [-6.3% ($P < 0.05$)] was detected; no dynamics for VLF, LF, and HF were observed in patients without verified CAN. The subclinical CAN was characterized by a pronounced inhibition of HF [-25.3% ($P < 0.01$)]; an increase in LF/HF [+29.5% ($P < 0.05$)] compared to patients without CAN; no changes in VLF and LF were observed. A specific characteristic of the definite stage is a pronounced inhibition of VLF [-10.4% ($P < 0.05$)]; LF [-21.7% ($P < 0.001$)], and a decrease in LF/HF [-17.1% ($P < 0.05$)]; no dynamics for HF compared to patients with subclinical CAN were observed. Therefore, in T2DM patients with CAN, a decrease in the LF and HF components of HR is observed (Table 8).

24-h ambulatory parameters of the artery stiffness during the active period in patients with T2DM and CAN are presented in Table 9.

Table 9. 24-h ambulatory parameters of the artery stiffness during the active period in patients with T2DM and CAN (M ± SEM)

Parameters	**Control (n = 15)**	**Patients with T2DM (n = 107)**		
		Without CAN (n = 19)	Subclinical CAN (n = 44)	Definite CAN (n = 44)
	1st group	2nd group	3rd group	4th group
AIxao (%)	17.97 ± 1.5	26.3 ± 1.75**	30.0 ± 0.96$^{***\#}$	31.7 ± 0.72$^{***\#}$
AIxbr (%)	-37.4 ± 2.72	-26.0 ± 1.52***	-15.2 ± 2.06$^{***\#\#\#}$	-10.9 ± 1.32$^{***\#\#\#}$
PWV (м/s)	7.9 ± 0.31	9.08 ± 0.24**	10.2 ± 0.14$^{***\#\#\#}$	10.95 ± 0.3$^{***\#\#\dagger\dagger}$

It was established that among persons of control group optimal values for AIxao were found in 60% of patients, normal -in 40%, and for PWV – in 33.33% and 66.67%, respectively. In patients without verified CAN, AIxao value was optimal in 15.79% of patients, normal in 78.95%, and elevated in 5.26%; PWV value was optimal in 5.26% of patients, normal in 84.21%, and elevated in 10.53%.

Among patients with subclinical CAN values of AIxao was optimal in 4.55% of patients, normal in 56.82%, affected in 38.63%; among them: elevated in 36.36%, pathological in 2.27%. We have found out that in patients with subclinical CAN average value of PWV exceeds the physiological one and it is considered as high. Among the patients of this group in 43.2% we observed the normal values of PWV and in 56.8% elevated. Among the patients with definite CAN the value of AIxao was optimal in 5.09%, normal in 47.14%, affected in 47.77%; among them: elevated in 32.48%, pathological in 15.29%. The value of PWV was normal in 19.6%, elevated in 47.8% and pathological in 32.6% of cases.

Therefore, in patients without verified CAN, a statistically significant increase in AIxao in the active period of the 24-h interval [+46.36% ($P < 0.01$)], AIxbr [+30.48% ($P < 0.001$)], and PWV [+14.94% ($P < 0.01$)] was observed. The subclinical CAN was characterized by a pronounced increase in AIxao [+14.07% ($P < 0.05$)], AIxbr [+41.54% ($P < 0.001$)], and PWV [+12.34% ($P < 0.001$)] compared to patients without CAN. The definite stage was associated with a further statistically significant increase in PWV [+7.35% ($P < 0.01$)] compared to subclinical CAN; no changes in AIxao and AIxbr were observed (Table 9). 24-hrs ambulatory parameters of the artery stiffness during the passive period in patients with T2DM and CAN are presented in Table 10.

Table 10. 24-h ambulatory parameters of the artery stiffness during the passive period in patients with T2DM and CAN (M ± SEM)

Parameters	**Control (n = 15)**	**Patients with T2DM (n = 107)**		
		Without CAN (n = 19)	Subclinical CAN (n = 44)	Definite CAN (n = 44)
	1st group	2nd group	3rd group	4th group
AIxao (%)	20.3 ± 1.75	26.7 ± 1.44**	33.8 ± 1.37***##	35.5 ± 0.73***###
AIxbr (%)	-32.1 ± 3.4	-21.6 ± 1.52**	-9.2 ± 2.77***###	-5.1 ± 1.26***###
PWV (м/s)	6.9 ± 0.28	8.8 ± 0.23**	10.1 ± 0.2***###	11.07 ± 0.19***###††

It was established that in patients without verified CAN, a significant increase in AIxao in the passive period [+31.53% ($P < 0.01$)], AIxbr [+32.71% ($P < 0.01$)], and PWV [+27.54% ($P < 0.001$)] was observed. The subclinical stage was characterized by an increase in AIxao [+26.59% ($P < 0.01$)], AIxbr [+57.41% ($P < 0.001$)], and PWV [+14.77% ($P < 0.001$)]. The definite CAN was associated with a statistically significant increase in PWV [+9.01% ($P < 0.01$)] compared to patients with subclinical CAN and [+25.11% ($P < 0.001$)] compared to patients without CAN; no changes in

AIxao and AIxbr were observed. Results of 24-h SBP and DBP ABPM parameters in patients with T2DM and CAN are presented in Table 11.

Table 11. Ambulatory blood pressure parameters in patients with T2DM and CAN (M ± SEM)

Parameters	**Control (n = 15)**	**Patients with T2DM (n = 107)**		
		Without CAN (n = 19)	Subclinical CAN (n = 44)	Definite CAN (n = 44)
	1st group	2nd group	3rd group	4th group
24-h SBP (mmHg)	118.9 ± 2.4	120.6 ± 0.9	132.7 ± 2.8$^{***###}$	141.9 ± 3.0$^{***###†††}$
Daytime SBP (mmHg)	129.1 ± 2.9	128.8 ± 1.3	140.8 ± 1.7$^{***###}$	149.4 ± 2.1$^{***###†††}$
Night-time SBP (mmHg)	108.2 ± 2.1	115.1 ± 2.3*	127.7 ± 1.8$^{***###}$	137.6 ± 2.9$^{***###†††}$
24-h DBP (mmHg)	71.9 ± 1.8	73.1 ± 1.6	76.3 ± 1.1*	90.4 ± 2.11$^{***###†††}$
Daytime DBP (mmHg)	74.4 ± 1.8	78.3 ± 1.8	80.9 ± 2.1*	94.6 ± 2.1$^{***###†††}$
Night-time DBP (mmHg)	64.6 ± 0.73	66.9 ± 0.8*	71.7 ± 1.74$^{***##}$	85.3 ± 2.16$^{***###†††}$

It was established that in patients without verified CAN, a significant increase in night-time SBP [+6.38% ($P < 0.05$)] and night-time DBP [+3.56% ($P < 0.05$)] was observed. The subclinical stage was characterized by an increase in 24-h SBP [+10.03% ($P < 0.001$)], daytime SBP [+9.32% ($P < 0.001$)], and night-time SBP [+10.94% ($P < 0.001$)] and night-time DBP [+7.18% ($P < 0.01$)] compared to patients without CAN. Among patients with subclinical CAN violated blood pressure patterns were found in 43.2% of cases, namely in 9.1% - "night-peaker" and in 34.1% - "non-dipper." The definite CAN was associated with a statistically significant increase in 24-h SBP and DBP, as well as in day and night values of SBP and DBP compared to patients with subclinical CAN ($P < 0.001$). Among patients with definite CAN violated blood pressure patterns were found in 61.4% of cases, namely in 15.9% - "night-peaker," in 4.6% - "extreme dipper" and in 40.9% - "non-dipper." Results of variability of SBP parameters in patients with T2DM and CAN are presented in Table 12.

Table 12. Variability of the systolic blood pressure parameters in patients with T2DM and CAN (M ± SEM)

Parameters	**Control (n = 15)**	**Patients with T2DM (n = 107)**		
		Without CAN (n = 19)	Subclinical CAN (n = 44)	Definite CAN (n = 44)
	1st group	2nd group	3rd group	4th group
24-h SBP	12.0 ± 0.29	12.9 ± 0.47	$13.5 \pm 0.23^{**}$	$15.7 \pm 0.43^{***\#\#\#\dagger\dagger\dagger}$
Daytime SBP	12.8 ± 0.22	13.0 ± 0.96	$13.7 \pm 0.21^{*}$	$15.9 \pm 0.43^{***\#\#\#\dagger\dagger\dagger}$
Night-time SBP	11.5 ± 0.24	11.8 ± 0.66	$12.8 \pm 0.37^{**\#\#}$	$14.5 \pm 0.36^{***\#\#\#\dagger\dagger\dagger}$

It was found, that 24-h SBP, as well as daytime and nighttime SBP variability in patients without CAN were comparable to data, obtained in control group ($P > 0.05$). Development of subclinical CAN was accompanied by increase of 24-h SBP variability [+12.5% ($P < 0.01$)], daytime SBP variability [+7.0% ($P < 0.05$)] and night-time SBP variability [+11.3% ($P < 0.01$)] compared to control group and increase in night-time SBP variability [+8.48% ($P < 0.01$)] compared to patients without CAN. Development of definite CAN was associated with increase in 24-hour SBP variability [+16.3% ($P < 0.001$)], daytime SBP variability [+16.1% ($P < 0.001$)] and night-time SBP variability [+13.3% ($P < 0.001$)] compared to data of patients with subclinical CAN (Table 12). Results of variability of the DBP parameters in patients with T2DM and CAN are presented in Table 13.

Table 13. Variability of the diastolic blood pressure parameters in patients with T2DM and CAN (M ± SEM)

Parameters	**Control (n = 15)**	**Patients with T2DM (n = 107)**		
		Without CAN (n = 19)	Subclinical CAN (n = 44)	Definite CAN (n = 44)
	1st group	2nd group	3rd group	4th group
24-h DBP	9.3 ± 0.19	9.6 ± 0.63	$10.5 \pm 0.36^{**}$	$12.7 \pm 0.38^{***\#\#\#\dagger\dagger\dagger}$
Daytime DBP	9.6 ± 0.27	10.3 ± 0.44	$10.8 \pm 0.29^{***}$	$13.4 \pm 0.32^{***\#\#\#\dagger\dagger\dagger}$
Night-time DBP	8.4 ± 0.29	8.9 ± 0.62	$10.1 \pm 0.43^{***}$	$11.8 \pm 0.34^{***\#\#\#\dagger\dagger\dagger}$

It was found, that 24-hrs DBP, as well as daytime and night-time DBP variability in patients without CAN were comparable to data, obtained in control group ($P > 0.05$). Development of subclinical CAN was accompanied by increase in 24-h DBP variability [+12.9% ($P < 0.01$)], daytime DBP variability [+12.5% ($P < 0.001$)] and night-time DBP variability [+20.2% ($P < 0.001$)] compared to control group. Development of definite CAN was associated with increase in 24-h DBP variability [+20.96% ($P < 0.001$)], daytime DBP variability [+24.1% ($P < 0.001$)] and night-time SBP variability [+16.8% ($P < 0.001$)] compared to data of patients with subclinical CAN (Table 13).

Conclusion

It is reported that reduction of elasticity of large arteries and dysfunction of ANS appear already at the time of verification of DM. Questions arise: 1) What is the primary: vegetative dysfunction or increased rigidity of large arteries? 2) What kind of cause-effect relationship exists between the dysfunction of the ANS and the rigidity of large arteries? Two hypotheses are proposed: arterial stiffness can lead to dysfunction of the myocardium, or, conversely, diabetic CAN promotes attachment and/or progression of the rigidity of the wall of the large arteries. In addition, the processes of reducing the elasticity of large arteries and CAN develop in parallel, taking into account the effects of aging and “toxic” effects of hyperglycemia (Christen et al. 2010).

One of the possible mechanisms that can explain the first hypothesis is the violation of the BRS, which are induced by increased arterial stiffness. On the other hand, dysfunction of the cardiac ANS can change the elasticity of the artery wall, affecting the vascular tone of large arteries. In particular, animal experiments have shown that the integrity of the ANS plays an important role in preserving the elastic properties of the aorta (Stoner et al. 2012). However, it is known that experimental studies are accompanied by acute modulation of changes in the state of the ANS, while in patients with DM, a chronic, gradual, cumulative effect is observed. Another mechanism that may be involved in the processes of induction of arterial stiffness and the development of autonomic dysfunction is the increase in heart rate. Indeed, an increase in HR *per se,* regardless of the changes in the activity of the ANS, contributes to the arterial stiffness development. However, it has been found that the normalization of HR in patients with T1DM does not affect the correlation between of LV diastolic function and PWV. Consequently, the association between parasympathetic dysfunction and arterial stiffness is not mediated by an increase in heart rate. Finally, changes in vegetative tone can increase the arterial stiffness due to trophic effects, which contribute to changes in the structure of the vessels (Christen et al. 2010; Stoner et al. 2012).

A great amount of data demonstrates that hyperglycemia and IR activate number of mechanisms triggering the structural and functional changes in the arterial wall, which are likely to contribute to accelerated vascular aging and increased CV risk in T2DM (Prenner et al. 2015). Moreover, the stiffness gradient between the proximal elastic arteries and more distal muscular arteries decreases, and therefore, increases the transmission of pressure to the microcirculation that may be already damaged by diabetic microvascular disease. Moreover, T2DM is associated with other metabolic and systemic abnormalities, like atherogenic DLP, HT and obesity that may cause atherosclerosis, arterial stiffness or both (Smulyan et al. 2016; Ecobici and Stoicescu 2017; Elias et al. 2017). Furthermore, these pathophysiological changes in cardiomyocytes underlie the risk factors for IR and HI, resulting in a potentially vicious cycle, which prompts the following question: does heart failure cause DM? (Negishi 2018).

However, future work is needed to further clarify whether large arterial stiffness and the pulsatile hemodynamic changes that accompany it are involved in the pathogenesis of DM, and whether interventions targeting arterial stiffness are associated with improved clinical outcomes in diabetes mellitus.

ACKNOWLEDGMENTS

The research was conducted according to the Declaration of Helsinki (2004) and the Ethics Committee of Danylo Halytsky Lviv National Medical University approved the study before the initiation (protocol №2 from 18 February 2013). The work was performed within the framework of the State task of Danylo Halytsky National Medical University, Lviv, Ukraine. The authors would like to express gratitude to all physicians who approached the patients with type 2 diabetes mellitus and CAN and for following the protocol.

REFERENCES

Adamska A, Karczewska-Kupczewska M, Nikołajuk A, Otziomek E, Górska M, Kowalska I, Strączkowski M (2014) Relationships of serum soluble E-selectin concentration with insulin sensitivity and metabolic flexibility in lean and obese women. *Endocrine* 45(3): 422-429.

Agnoletti D, Zhang Y, Borghi C, Blacher J, Safar ME (2013) Effects of antihypertensive drugs on central blood pressure in humans: a preliminary observation. *Am J Hypertens* 26(8): 1045-1052.

Akselrod S, Gordon D, Ubel FA, Shannon DC, Berger AC, Cohen RJ (1981) Power spectrum analysis of heart rate fluctuation: a quantitative probe of beat-to-beat cardiovascular control. *Science* 213(4504): 220-222.

Al-Trad BA, Faris M, Al-Smadi M, Bashir A, Mansi M, Alaraj M, Al-Hazimi A (2015) Cardiac autonomic dysfunction in young obese males is not associated with disturbances in pituitary-thyroid axis hormones. *Eur Rev Med Pharmacol Sci* 19(9): 1689-1695.

American Diabetes Association (2017) Standards of Medical Care in Diabetes-2017. *Diabetes Care* 40: S11-S24.

Amiya E, Watanabe M, Komuro I (2014) The relationship between vascular function and the autonomic nervous system. *Ann Vasc Dis* 7(2): 109-119.

Amra B, Behjati M, Penzel T, Fietze I, Schoebel C, Sarrafzadegan N (2017) Nocturnal heart rate variation in diabetic and non-diabetic patients with sleep apnea syndrome. *Sleep Med* 29: 57-60.

Askin L, Cetin M, Turkmen S (2018) Ambulatory blood pressure results and heart rate variability in patients with premature ventricular contractions. *Clin Exp Hypertens* 40(3): 251-256.

Ateş İ, Altay M, Kaplan M, Ozkayar N, Toprak G, Alagüney ME, Özkara A (2015) Relationship between socioeconomic level, and the prevalence of masked hypertension and asymptomatic organ damage. *Med Sci Monit* 21: 1022-1030.

Ayala DE, Moyá A, Crespo JJ, Castiñeira C, Domínguez-Sardiña M, Gomara S, Sineiro E, Mojón A, Fontao MJ, Hermida RC (2013) Hygia Project Investigators. Circadian pattern of ambulatory blood pressure in hypertensive patients with and without type 2 diabetes. *Chronobiol Int* 30(1-2): 99-115.

Badea AR, Nedelcu L, Valeanu M, Zdrenghea D (2014) The relationship between serum lipid fractions and heart rate variability in diabetic patients with statin therapy. *Clujul Med* 87(3): 152-158.

Bagherzadeh A, Nejati-Afkham A, Tajallizade-Khoob Y, Shafiee A, Sharifi F, Esfahani MA, Badamchizade Z, Alatab S, Fakhrzadeh H (2013) Association of cardiac autonomic neuropathy with arterial stiffness in type 2 diabetes mellitus patients. *J Diabetes Metab Disord* 12(1): 55.

Bakkar NZ, Dwaib HS, Fares S, Eid AH, Al-Dhaheri Y, El-Yazbi AF (2020) Cardiac autonomic neuropathy: A progressive consequence of chronic low-grade inflammation in type 2 diabetes and related metabolic disorders. *Int J Mol Sci* 21(23): 9005.

Bamira D, Picard MH (2018) Imaging: echocardiology-assessment of cardiac structure and function. In: Vasan RS, Sawyer DB, editors. *Encyclopedia of Cardiovascular Structure and Function:* Elsevier, Amsterdam, Oxford, Cambridge. pp. 35-54.

Barazzoni R, Zanetti M, Cappellari G, Semolic A, Boschelle M, Codarin E, Pirulli A, Cattin L, Guarnieri G (2012) Fatty acids acutely enhance insulin-induced oxidative stress and cause insulin resistance by increasing mitochondrial reactive oxygen species (ROS) generation and nuclear factor-kappa B inhibitor (Ikappa B)-nuclear factor-kappa B (NF-kappa B) activation in rat muscle, in the absence of mitochondrial dysfunction. *Diabetologia* 55(12): 773-782.

Bellavere F, Balzani I, De Masi G, Carraro M, Carenza P, Cobelli C, Thomaseth K (1992) Power spectral analysis of heart-rate variations improves assessment of diabetic cardiac autonomic neuropathy. *Diabetes* 41(5): 633-640.

Benichou T, Pereira B, Mermillod M, Tauveron I, Pfabigan D, Maqdasy S, Dutheil F (2018) Heart rate variability in type 2 diabetes mellitus: A systematic review and meta-analysis. *PLoS One* 13(4): e0195166.

Bhardwaj S, Verma N, Anjum B, Bhardwaj K (2014) Variations in 7-day/24-h circadian pattern of ambulatory blood pressure and heart rate of type 2 diabetes patients. *J Diabetes Investig* 5(6): 728-733.

Bianchi L, Chiheb S, Banu I, Rezki A, Cosson E, Valensi P (2016) Influence of cardiac autonomic dysfunction and arterial stiffness on subendocardial myocardial viability in patients with type 2 diabetes. *Diabetes Metab* 42(4): 297-298.

Black N, D'Souza A, Wang Y, Piggins H, Dobrzynski H, Morris G, Boyett MR (2019) Circadian rhythm of cardiac electrophysiology, arrhythmogenesis, and the underlying mechanisms. *Heart Rhythm* 16(2): 298-307.

Çakar M, Akhan M, Doğan T, Taşkın G, Öztürk K, Çınar M, Arslan E, Yılmaz S (2017) Investigation of the arterial stiffness and associated factors in patients with familial Mediterranean fever. *Anatol J Cardiol* 17(2): 132-138.

Çakar M, Balta Ş, Şarlak H, Akhan M, Demirkol S, Karaman M, Ay SA, Kurt Ö, Çayci T, İnal S, Demirbaş Ş (2015) Arterial stiffness and endothelial inflammation in prediabetes and newly diagnosed diabetes patients. *Arch Endocrinol Metab* 59(5): 407-413.

Camm A, Malik M, Bigger J, Günter B, Cerutti S, Choen R (1996) Heart rate variability: standards of measurement, physiological interpretation and clinical use. Task force of the European Society of Cardiology and the North American Society of Pacing and Electrophysiology. *Circulation* 93(5): 1043-1065.

Chang J, Hou YP, Wu JL, Fang XY, Li SL, Liu MB, Sun QM (2018) Blood pressure circadian rhythms and adverse outcomes in type 2 diabetes patients diagnosed with orthostatic hypotension. *J Diabetes Investig* 9(2): 383-388.

Chawla D, Bansal S, Banerjee BD, Madhu SV, Kalpa PO, Tripathi AK (2014) Role of advanced glycation end product (AGE)-induced receptor (RAGE) expression in diabetic vascular complications. *Microvasc Res* 95: 1-6.

Chen-Scarabelli C, Scarabelli TM (2006) Suboptimal glycemic control, independently of QT interval duration, is associated with increased risk of ventricular arrhythmias in a high-risk population. *Pacing Clin Electrophysiol* 29(1): 9-14.

Chinnaiyan KM (2019) Role of stress management for cardiovascular disease prevention. *Curr Opin Cardiol* 34(5): 531-535.

Chorepsima S, Eleftheriadou I, Tentolouris T, Moyssakis I, Protogerou A, Kokkinos A, Sfikakis PP, Tentolouris N (2017) Pulse wave velocity and cardiac autonomic function in type 2 diabetes mellitus. *BMC Endocrine Disord* 17(1): 27.

Christen AI, Armentano RL, MirandaA, Graf S, Santana DB, Zócalo Y, Baglivo HP, Sánchez RA (2010) Arterial wall structure and dynamics in type 2 diabetes mellitus: methodological aspects and pathophysiological findings. *Curr Diabetes Rev* 6(6): 367-377.

Ciftel M, Ertug H, Parlak M, Akcurin G, Kardelen F (2014) Investigation of endothelial dysfunction and arterial stiffness in children with type 1 diabetes mellitus and the association with diastolic dysfunction. *Diab Vasc Dis Res* 11(1): 19-25.

Colosia AD, Palencia R, Khan S (2013) Prevalence of hypertension and obesity in patients with type 2 diabetes mellitus in observational studies: a systematic literature review. *Diabetes Metab Syndr Obes* 6: 327-338.

Cseh D, Climie RE, Offredo L, Guibout C, Thomas F, Zanoli L, Danchin N, Sharman JE, Laurent S, Jouven X, Boutouyrie P, Empana JP (2020) Type 2 diabetes mellitus is independently associated with decreased neural baroreflex sensitivity: The Paris Prospective Study III. *Arterioscler Thromb Vasc Biol* 40(5): 1420-1428.

Dabelea D, Stafford JM, Mayer-Davis EJ, D'Agostino R Jr, Dolan L, Imperatore G, Linder B, Lawrence JM, Marcovina SM, Mottl AK, Black MH, Pop-Busui R, Saydah S, Hamman RF, Pihoker C (2017) SEARCH for Diabetes in Youth Research Group. Association of Type 1 Diabetes vs Type 2 Diabetes Diagnosed during Childhood and Adolescence with Complications during Teenage Years and Young Adulthood. *JAMA* 317(8): 825-835.

Davidson AJ, London B, Block GD, Menaker M (2005) Cardiovascular tissues contain independent circadian clocks. *Clin Exp Hypertens* 27(2-3): 307-311.

de Andrade PE, do Amaral JAT, Paiva LDS, Adami F, Raimudo JZ, Valenti VE, Abreu LC, Raimundo RD (2017) Reduction of heart rate variability in hypertensive elderly. *Blood Press* 26(6): 350-358.

de Boer IH, Bangalore S, Benetos A, Davis AM, Michos ED, Muntner P, Rossing P, Zoungas S, Bakris G (2017) Diabetes and Hypertension: A Position Statement by the American Diabetes Association. *Diabetes Care* 40(9): 1273-1284.

de la Sierra A, Redon J, Banegas JR, Segura J, Parati G, Gorostidi M, de la Cruz JJ, Sobrino J, Llisterri JL, Alonso J, Vinyoles E, Pallarés V, Sarría A, Aranda P, Ruilope LM (2009) Spanish Society of Hypertension Ambulatory Blood Pressure Monitoring Registry Investigators. Prevalence and factors associated with circadian blood pressure patterns in hypertensive patients. *Hypertension* 53(3): 466-472.

de Oliveira Alvim R, Santos PCJL, Musso MM, de Sá Cunha R, Krieger JE, Mill JG, Pereira AC (2013) Impact of diabetes mellitus on arterial stiffness in a representative sample of an urban Brazilian population. *Diabetol Metab Syndr* 5(1): 45.

Di Flaviani A, Picconi F, Di Stefano P, Giordani I, Malandrucco I, Maggio P, Palazzo P, Sgreccia F, Peraldo C, Farina F, Frajese G, Frontoni S (2011) Impact of glycemic and blood pressure variability on surrogate measures of cardiovascular outcomes in type 2 diabetic patients. *Diabetes Care* 34(7): 1605-1609.

Didangelos TP, Arsos G, Karamitsos T, Iliadis F, Papageorgiou A, Moralidis E, Athyros V (2014) Left ventricular systolic and diastolic function in normotensive type 2 diabetic patients with or without autonomic neuropathy: a radionuclide ventriculography study. *Angiology* 65(10): 877-882.

Draman MS, Dolan E, van der Poel L, Tun TK, McDermott JH, Sreenan S, O'Brien E (2015) The importance of night-time systolic blood pressure in diabetic patients: Dublin Outcome Study. *J Hypertens* 33(7): 1373-1377.

Durgan DJ, Hotze MA, Tomlin TM, Egbejimi O, Graveleau C, Abel ED, Shaw CA, Bray MS, Hardin PE, Young ME (2005) The intrinsic circadian clock within the cardiomyocyte. *Am J Physiol Heart Circ Physiol* 289(4): H1530-H1541.

Durgan DJ, Young ME (2010) The cardiomyocyte circadian clock: emerging roles in health and disease. *Circ Res* 106(4): 647-658.

Dutheil F, Gordon BA, Naughton G, Crendal E, Courteix D, Chaplais E, Thivel D, Lac G, Benson AC (2018) Cardiovascular risk of adipokines: a review. *J Int Med Res* 46(6): 2082-2095.

Ecobici M, Stoicescu C (2017) Arterial stiffness and hypertension-which comes first? *Maedica (Buchar)* 12(3): 184-190.

Elias MF, Crichton GE, Dearborn PJ, Robbins MA, Abhayaratna WP (2017) Associations between type 2 diabetes mellitus and arterial stiffness: a prospective analysis based on the Maine-Syracuse Study. *Pulse* 5: 88-98.

Ewing DJ, Martyn CN, Young RJ, Clarke BF (1985) The value of cardiovascular autonomic function tests: 10 years experience in diabetes. *Diabetes Care* 8(5): 491-498.

Fagard RH, Celis H, Thijs L, Staessen JA, Clement DL, De Buyzere ML, De Bacquer DA (2008) Daytime and nighttime blood pressure as predictors of death and cause-specific cardiovascular events in hypertension. *Hypertension* 51(1): 55-61.

Flotats A, Carrió I (2010) Is cardiac autonomic neuropathy the basis of nonischemic diabetic cardiomyopathy? *JACC Cardiovasc Imaging* 3(12): 1216-1218.

Fogari R, Zoppi A, Malamani GD, Lazzari P, Destro M, Corradi L (1993) Ambulatory blood pressure monitoring in normotensive and hypertensive type 2 diabetes. Prevalence of impaired diurnal blood pressure patterns. *Am J Hypertens* 6(1): 1-7.

Foster GD, Sanders MH, Millman R, Zammit G, Borradaile KE, Newman AB, Wadden TA, Kelley D, Wing RR, Sunyer FX, Darcey V, Kuna ST (2009) Sleep ahead Research Group. Obstructive sleep apnea among obese patients with type 2 diabetes. *Diabetes Care* 32(6): 1017-1019.

Gaszner B, Lenkey Z, Illyés M, Sárszegi Z, Horváth IG, Magyari B, Molnár F, Kónyi A, Cziráki A (2012) Comparison of aortic and carotid arterial stiffness parameters in patients with verified coronary artery disease. *Clin Cardiol* 35(1): 26-31.

General Assembly of the World Medical Association (2014) World Medical Association Declaration of Helsinki: ethical principles for medical research involving human subjects. *J Am Coll Dent* 81(3): 14-148.

Goit RK, Khadka R, Sharma SK, Limbu N, Paudel BH (2012) Cardiovascular autonomic function and vibration perception threshold in type 2 diabetes mellitus. *J Diabetes Complicat* 26(4): 339-342.

Goldberger JJ, Arora R, Buckley U, Shivkumar K (2019) Autonomic nervous system dysfunction: JACC focus seminar. *J Am Coll Cardiol* 73(10): 1189-1206.

Grossman E (2013) Ambulatory blood pressure monitoring in the diagnosis and management of hypertension. *Diabetes Care* 2: S307-S311.

Gunawan F, Ng HY, Gilfillan C, Anpalahan M (2019) Ambulatory Blood Pressure Monitoring in Type 2 Diabetes Mellitus: A Cross-sectional Study. *Curr Hypertens Rev* 15(2): 135-143.

Hadad R, Larsen BS, Weber P, Stavnem D, Kristiansen OP, Nielsen OW, Haugaard SB, Sajadieh A (2021) Night-time heart rate variability identifies high-risk people among people with uncomplicated type 2 diabetes mellitus. *Diabet Med* 13: e14559.

Halberg F, Cornélissen G, Katinas G, Tvildiani L, Marina Gigolashvili M, Janashia K, Toba T, Revilla M, Regal P, Sothern RB, Wendt HW, Wang Z, Zeman M, Jozsa R, Singh RB, Mitsutake G, Chibisov SM, Lee J, Holley D, Holte JE, Sonkowsky RP, Schwartzkopff O, Delmore P, Otsuka K, Bakken EE, Czaplicki J, the International BIOCOS Group (2006) Chronobiology's progress. Part I, season's appreciations 2004-2005: time-, frequency-, phase-, variable-, individual-, age- and site-specific chronomics. *J Appl Biomed* 4: 1-38.

Halberg F, Cornélissen G, Wilson D, Singh RB, De Meester F, Watanabe Y, Otsuka K, Khalilov E (2009) Chronobiology and chronomics: detecting and applying the cycles of nature. *Biologist (London)* 56(4): 209-214.

Hansen TW, Li Y, Boggia J, Thijs L, Richart T, Staessen JA (2011) Predictive role of the nighttime blood pressure. *Hypertension* 57(1): 3-10.

Electrophysiology TF (1996) Heart rate variability: standards of measurement, physiological interpretation and clinical use. *Circulation* 93(5): 1043-1065.

Hillebrand S, Gast KB, de Mutsert R, Swenne CA, Jukema JW, Middeldorp S, Rosendaal FR, Dekkers OM (2013) Heart rate variability and first cardiovascular event in populations without known cardiovascular disease: meta-analysis and dose-response meta-regression. *Europace* 15(5): 742-749.

Hodgkinson J, Mant J, Martin U, Guo B, Hobbs FD, Deeks JJ, Heneghan C, Roberts N, McManus RJ (2011) Relative effectiveness of clinic and home blood pressure monitoring compared with ambulatory blood pressure monitoring in diagnosis of hypertension: systematic review. *BMJ* 342.

Horváth IG, Németh A, Lenkey Z, Alessandri N, Tufano F, Kis P, Gaszner B, Cziráki A (2010) Invasive validation of a new oscillometric device (Arteriograph) for measuring augmentation index, central blood pressure and aortic pulse wave velocity. *J Hypertens* 28(10): 2068-2075.

Hou T (2018) *Mechanisms and potential therapy on disrupted blood pressure circadian rhythm in diabetes.* Theses and Dissertations-Pharmacology and Nutritional Sciences. 26.

Hufnagel C, Chambres P, Bertrand PR, Dutheil F (2017) The Need for objective measures of stress in Autism. *Front Psychol* 8: 64.

Irigoyen MC, De Angelis K, Dos Santos F, Dartora DR, Rodrigues B, Consolim-Colombo FM (2016) Hypertension, blood pressure variability, and target organ lesion. *Curr Hypertens Rep* 18(4): 31.

Jaiswal M, Divers J, Urbina EM, Dabelea D, Bell RA, Pettitt DJ, Imperatore G, Pihoker C, Dolan LM, Liese AD, Marcovina S, Linder B, Feldman EL, Pop-Busui R (2018) SEARCH for Diabetes in Youth Study Group. Cardiovascular autonomic neuropathy in adolescents and young adults with type 1 and type 2 diabetes: The search for Diabetes in Youth Cohort Study. *Pediatr Diabetes* 19(4): 680-689.

Jennersjö PE, Wijkman M, Wiréhn AB, Länne T, Engvall J, Nystrom FH, Östgren CJ (2011) Circadian blood pressure variation in patients with type 2 diabetes-relationship to macro- and microvascular subclinical organ damage. *Prim Care Diabetes* 5(3): 167-173.

Jia G, DeMarco VG, Sowers JR (2016) Insulin resistance and hyperinsulinaemia in diabetic cardiomyopathy. *Nat Rev Endocrinol* 12(3): 144-153.

Jia G, Whaley-Connell A, Sowers JR (2018) Diabetic cardiomyopathy: a hyperglycaemia- and insulin-resistance-induced heart disease. *Diabetologia* 61(1): 21-28.

Kalaycıoğlu E, Gökdeniz T, Aykan AÇ, Gül İ, Uğur M, Gürsoy OM, Boyacı F, Çelik Ş (2014) The influence of dipper/nondipper blood pressure patterns on global left ventricular systolic function in hypertensive diabetic patients: a speckle tracking study. *Blood Press Monit* 19(5): 263-270.

Kario K (2010) Morning surge in blood pressure and cardiovascular risk: evidence and perspectives. *Hypertension* 56(5): 765-773.

Kozakova M, Palombo C (2016) Diabetes mellitus, arterial wall, and cardiovascular risk assessment. *Int J Environl Res Public Health* 13(2): 201.

Lang TA, Secic M (2006) *How to report statistics in medicine: annotated guidelines for authors, editors, and reviewers.* American College of Physicians, Philadelphia, USA, pp. 45-74.

Liao D, Cai J, Rosamond WD, Barnes RW, Hutchinson RG, Whitsel EA, Rautaharju P, Heiss G (1997) Cardiac autonomic function and incident coronary heart disease: a population-based case-cohort study. The ARIC Study. Atherosclerosis Risk in Communities Study. *Am J Epidemiol* 145(8): 696-706.

Liu L, Wu Q, Yan H, Chen B, Zheng X, Zhou Q (2020) Association between cardiac autonomic neuropathy and coronary artery lesions in patients with type 2 diabetes. *Dis Markers* 2020: 6659166.

Loehr LR, Meyer ML, Poon AK, Selvin E, Palta P, Tanaka H, Pankow JS, Wright JD, Griswold ME, Wagenknecht LE, Heiss G (2016) Prediabetes and diabetes are associated with arterial stiffness in older adults: The ARIC Study. *Am J Hypertens* 29(9): 1038-1045.

Lyhne JM, Laugesen E, Høyem P, Cichosz S, Christiansen JS, Knudsen ST, Hansen KW, Hansen TK, Poulsen PL (2015) Morning blood pressure surge and target organ damage in newly diagnosed type 2 diabetic patients: a cross sectional study. *BMC Endocr Disord* 15: 77.

Madamanchi NR, Vendrov A, Runge MS (2005) Oxidative stress and vascular disease. *Arterioscler Thromb Vasc Biol* 25(1): 29-38.

Mala S, Potockova V, Hoskovcova L, Pithova P, Brabec M, Kulhankova J, Keil R, Riedlbauchova L, Broz J (2017) Cardiac autonomic neuropathy may play a role in pathogenesis of atherosclerosis in type 1 diabetes mellitus. *Diabetes Res Clin Pract* 134: 139-144.

Malliani A, Pagani M, Lombardi F, Cerutti S (1991) Cardiovascular neural regulation explored in the frequency domain. *Circulation* 84(2): 482-492.

Mancia G, Fagard R, Narkiewicz K, Redón J, Zanchetti A, Böhm M, Christiaens T, Cifkova R, De Backer G, Dominiczak A, Galderisi M, Grobbee DE, Jaarsma T, Kirchhof P, Kjeldsen SE, Laurent S, Manolis AJ, Nilsson PM, Ruilope LM, Schmieder RE, Sirnes PA, Sleight P, Viigimaa M, Waeber B, Zannad F, Task Force Members (2013) 2013 ESH/ESC Guidelines for the management of arterial hypertension: the Task Force for the management of arterial hypertension of the European Society of Hypertension (ESH) and of the European Society of Cardiology (ESC). *J Hypertens* 31(7): 1281-1357.

Mansour AS, Yannoutsos A, Majahalme N, Agnoletti D, Safar ME, Ouerdane S, Blacher J (2013) Aortic stiffness and cardiovascular risk in type 2 diabetes. *J Hypertens* 31(8): 1584-1592.

Martino T, Arab S, Straume M, Belsham DD, Tata N, Cai F, Liu P, Trivieri M, Ralph M, Sole MJ (2004) Day/night rhythms in gene expression of the normal murine heart. *J Mol Med (Berl)* 82(4): 256-264.

Maser RE, Mitchell BD, Vinik AI, Freeman R (2003) The association between cardiovascular autonomic neuropathy and mortality in individuals with diabetes: a meta-analysis. *Diabetes Care* 26(6): 1895-1901.

Matthews DR, Hosker JP, Rudenski AS, Naylor BA, Treacher DF, Turner RC (1985) Homeostasis model assessment: insulin resistance and beta-cell function from fasting plasma glucose and insulin concentrations in man. *Diabetologia* 28(7): 412-419.

Meyer ML, Tanaka H, Palta P, Cheng S, Gouskova N, Aguilar D, Heiss G (2016) Correlates of segmental pulse wave velocity in older adults: the Atherosclerosis Risk in Communities (ARIC) Study. *Am J Hypertens* 29(1): 114-122.

Mochizuki K, Inoue S, Miyauchi R, Misaki Y, Shimada M, Kasezawa N, Tohyama K, Goda T (2013) Plasma sE-selectin level is positively correlated with neutrophil count and diastolic blood pressure in Japanese men. *J Nutr Sci Vitaminol (Tokyo)* 59(5): 447-453.

Mohawk JA, Green CB, Takahashi JS (2012) Central and peripheral circadian clocks in mammals. *Annu Rev Neurosci* 35: 445-462.

Monfredi O, Lyashkov AE, Johnsen AB, Inada S, Schneider H, Wang R, Nirmalan M, Wisloff U, Maltsev VA, Lakatta EG, Zhang H, Boyett MR (2014) Biophysical characterization of the underappreciated and important relationship between heart rate variability and heart rate. *Hypertension* 64(6): 1334-1343.

Mustafa G, Kursat FM, Ahmet T, Alparslan GF, Omer G, Sertoglu E, Ekran S, Ediz Y, Turker T, Ayhan K (2017) The relationship between erythrocyte membrane fatty acid levels and cardiac autonomic function in obese children. *Rev Port Cardiol* 36(7-8): 499-508.

Najafi MT, Khaloo P, Alemi H, Jaafarinia A, Blaha MJ, Mirbolouk M, Mansournia MA, Afarideh M, Esteghamati S, Nakhjavani M, Esteghamati A (2018) Ambulatory blood pressure monitoring and diabetes complications: Targeting morning blood pressure surge and nocturnal dipping. *Medicine (Baltimore)* 97(38): e12185.

Nattero-Chávez L, Redondo López S, Alonso Díaz S, GarnicaUreña M, Fernández-Durán E, Escobar-Morreale HF, Luque-Ramírez M (2019) Association of cardiovascular autonomic dysfunction with peripheral arterial stiffness in patients with type 1 diabetes. *J Clin Endocrinol Metab* 104(7): 2675-2684.

Negishi K (2018) Echocardiographic feature of diabetic cardiomyopathy: where are we now? *Cardiovasc Diagn Ther* 8(1): 47-56.

Newman AB, Siscovick DS, Manolio TA, Polak J, Fried LP, Borhani NO, Wolfson SK (1993) Ankle-arm index as a marker of atherosclerosis in the Cardiovascular Health Study. Cardiovascular Heart Study (CHS) Collaborative Research Group. *Circulation* 88(3): 837-845.

Nguyen Dinh Cat A, Touyz RM (2011) Cell signaling of angiotensin II on vascular tone: novel mechanisms. *Curr Hypertens Rep* 13(2): 122-128.

O'Brien E, Parati G, Stergiou G (2013a) Ambulatory blood pressure measurement: what is the international consensus? *Hypertension* 62(6): 988-994.

O'Brien E, Parati G, Stergiou G, Asmar R, Beilin L, Bilo G, Clement D, de la Sierra A, de Leeuw P, Dolan E, Fagard R, Graves J, Head GA,

Imai Y, Kario K, Lurbe E, Mallion JM, Mancia G, Mengden T, Myers M, Ogedegbe G, Ohkubo T, Omboni S, Palatini P, Redon J, Ruilope LM, Shennan A, Staessen JA, vanMontfrans G, Verdecchia P, Waeber B, Wang J, Zanchetti A, Zhang Y (2013b) European Society of Hypertension Working Group on Blood Pressure Monitoring. European Society of Hypertension position paper on ambulatory blood pressure monitoring. *J Hypertens* 31(9): 1731-1768.

Ogawa M, Zhou S, Tan AY, Song J, Gholmieh G, Fishbein MC, Luo H, Siegel RJ, Karagueuzian HS, Chen LS, Lin SF, Chen PS (2007) Left stellate ganglion and vagal nerve activity and cardiac arrhythmias in ambulatory dogs with pacing-induced congestive heart failure. *J Am Coll Cardiol* 50(4): 335-343.

Okamoto M, Nakamura F, Musha T, Kobayashi Y (2016) Association between novel arterial stiffness indices and risk factors of cardiovascular disease.*BMC Cardiovasc Disord* 16: 211.

Pagani M, Malfatto G, Pierini S, Casati R, Masu AM, Poli M, Guzzetti S, Lombardi F, Cerutti S, Malliani A (1988) Spectral analysis of heart rate variability in the assessment of autonomic diabetic neuropathy. *J Auton Nerv Syst* 23(2): 143-153.

Palatini P, Penzo M, Racioppa A, Zugno E, Guzzardi G, Anaclerio M, Pessina AC (1992) Clinical relevance of nighttime blood pressure and of daytime blood pressure variability. *Arch Intern Med* 152(9): 1855-1860.

Pehlivanidis AN, Athyros VG, Demitriadis DS, Papageorgiou AA, Bouloukos VJ, Kontopoulos AG (2001) Heart rate variability after long-term treatment with atorvastatin in hypercholesterolaemic patients with or without coronary artery disease. *Atherosclerosis* 157(2): 463-469.

Pop-Busui R (2010) Cardiac autonomic neuropathy in diabetes: a clinical perspective. *Diabetes Care* 33(2): 434-441.

Pop-Busui R, Boulton AJ, Feldman EL, Bril V, Freeman R, Malik RA, Sosenko JM, Ziegler D (2017) Diabetic Neuropathy: A Position Statement by the American Diabetes Association. *Diabetes Care* 40(1): 136-154.

Portaluppi F, Hermida RC (2007) Circadian rhythms in cardiac arrhythmias and opportunities for their chronotherapy. *Adv Drug Deliv Rev* 59(9-10): 940-951.

Prenner SB, Chirinos JA (2015) Arterial stiffness in diabetes mellitus. *Atherosclerosis* 238(2): 370-379.

Rastović M, Srdić-Galić B, Barak O, Stokić E (2017) Association between anthropometric measures of regional fat mass and heart rate variability in obese women: Anthropometry and heart rate variability in obese women. *Nutr Diet* 74(1): 51-60.

Rydén L, Grant PJ, Anker SD, Berne C, Cosentino F, Danchin N, Deaton C, Escaned J, Hammes HP, Huikuri H, Marre M, Marx N, Mellbin L, Ostergren J, Patrono C, Seferovic P, Uva MS, Taskinen MR, Tendera M, Tuomilehto J, Valensi P, Zamorano JL; ESC Committee for Practice Guidelines (CPG), Zamorano JL, Achenbach S, Baumgartner H, Bax JJ, Bueno H, Dean V, Deaton C, Erol C, Fagard R, Ferrari R, Hasdai D, Hoes AW, Kirchhof P, Knuuti J, Kolh P, Lancellotti P, Linhart A, Nihoyannopoulos P, Piepoli MF, Ponikowski P, Sirnes PA, Tamargo JL, Tendera M, Torbicki A, Wijns W, Windecker S, Document Reviewers, De Backer G, Sirnes PA, Ezquerra EA, Avogaro A, Badimon L, Baranova E, Baumgartner H, Betteridge J, Ceriello A, Fagard R, Funck-Brentano C, Gulba DC, Hasdai D, Hoes AW, Kjekshus JK, Knuuti J, Kolh P, Lev E, Mueller C, Neyses L, Nilsson PM, Perk J, Ponikowski P, Reiner Z, Sattar N, Schächinger V, Scheen A, Schirmer H, Strömberg A, Sudzhaeva S, Tamargo JL, Viigimaa M, Vlachopoulos C, Xuereb RG (2013) ESC Guidelines on diabetes, pre-diabetes, and cardiovascular diseases developed in collaboration with the EASD: the Task Force on diabetes, pre-diabetes, and cardiovascular diseases of the European Society of Cardiology (ESC) and developed in collaboration with the European Association for the Study of Diabetes (EASD). *Eur Heart J* 34(39): 3035-3087.

Salwa P, Gorczyca-Michta I, Kluk M, Dziubek K, Wożakowska-Kapłon B (2014) Variability of circadian blood pressure profile during 24-hour ambulatory blood pressure monitoring in hypertensive patients. *Kardiol Pol* 72(5): 432-437.

Scheer FA, Hu K, Evoniuk H, Kelly EE, Malhotra A, Hilton MF, Shea SA (2010) Impact of the human circadian system, exercise, and their interaction on cardiovascular function. *Proc Natl Acad Sci USA* 107(47): 20541-20546.

Schroeder EB, Liao D, Chambless LE, Prineas RJ, Evans GW, Heiss G (2003) Hypertension, blood pressure, and heart rate variability: the Atherosclerosis Risk in Communities (ARIC) study. *Hypertension* 42(6): 1106-1111.

Seko Y (2020) A novel and dominant factor that mediates oxidative stress-induced apoptotic signaling-autocrine/paracrine mechanism of the secreted form of eukaryotic translation initiation factor 5A. *Vessel Plus* 4: 22.

Serhiyenko VA, Serhiyenko AA (2015) Diabetic cardiac autonomic neuropathy: Do we have any treatment perspectives? *World J Diabetes* 6(2): 245-258.

Serhiyenko VA, Serhiyenko AA (2018) Cardiac autonomic neuropathy: Risk factors, diagnosis and treatment. *World J Diabetes* 9(1): 1-24.

Serhiyenko VA, Serhiyenko LM, Serhiyenko AA (2018) Diabetic cardiac autonomic neuropathy and arterial stiffness-which comes first? *Diabetes Updates* 1(1): 1-5.

Serhiyenko VA, Serhiyenko AA (2019) Diabetic cardiac autonomic neuropathy. In: Saldaña JR, editor. *Diabetes Textbook: Clinical Principles, Patient Management and Public Health Issues:* Springer, Cham. Springer Nature Switzerland AG, Basel. pp. 825-850.

Shah AS, El Ghormli L, Gidding SS, Bacha F, Nadeau KJ, Levitt Katz LE, Tryggestad JB, Leibel N, Hale DE, Urbina EM (2018) Prevalence of arterial stiffness in adolescents with type 2 diabetes in the TODAY cohort: Relationships to glycemic control and other risk factors. *J Diabetes Complicat* 32(8): 740-745.

Shah AS, Urbina EM (2017) Vascular and endothelial function in youth with type 2 diabetes mellitus. *Curr Diab Rep* 17(6): 36.

Shah MS, Brownlee M (2016) Molecular and cellular mechanisms of cardiovascular disorders in diabetes. *Circ Res* 118(11): 1808-1829.

Shah RV, Abbasi SA, Neilan TG, Hulten E, Coelho-Filho O, Hoppin A, Levitsky L, de Ferranti S, Rhodes ET, Traum A, Goodman E, Feng H, Heydari B, Harris WS, Hoefner DM, McConnell JP, Seethamraju R, Rickers C, Kwong RY, Jerosch-Herold M (2013) Myocardial tissue remodeling in adolescent obesity. *J Am Heart Assoc* 2(4): e000279.

Shang Y, Zhang X, Leng W, Lei X, Chen L, Zhou X, Chow K, Shi Y, Dong J, Liang Z, Wang J (2019) Increased fractal dimension of left ventricular trabeculations is associated with subclinical diastolic dysfunction in patients with type-2 diabetes mellitus. *Int J Card Imaging* 35(4): 665-673.

Sheward WJ, Naylor E, Knowles-Barley S, Armstrong JD, Brooker GA, Seckl JR, Turek FW, Holmes MC, Zee PC, Harmar AJ (2010) Circadian control of mouse heart rate and blood pressure by the suprachiasmatic nuclei: behavioral effects are more significant than direct outputs. *PLoS One* 5(3): e9783.

Smolensky MH, Hermida RC, Portaluppi F (2017) Circadian mechanisms of 24-hour blood pressure regulation and patterning. *Sleep Med Rev* 33: 4-16.

Smulyan H, Lieber A, Safar ME (2016) Hypertension, diabetes type II, and their association: role of arterial stiffness. *Am J Hypertens* 29(1): 5-13.

Song MK, Davies NM, Roufogalis BD, Huang TH-W (2014) Management of cardiorenal metabolic syndrome in diabetes mellitus: a phytotherapeutic perspective. *J Diabetes Res* 2014: 313718.

Spallone V, Ziegler D, Freeman R, Bernardi L, Frontoni S, Pop-Busui R, Stevens M, Kempler P, Hilsted J, Tesfaye S, Low P, Valensi P (2011) Cardiovascular autonomic neuropathy in diabetes: clinical impact, assessment, diagnosis, and management. *Diabetes Metab Res Rev* 27(7): 639-653.

Stockhorst U, Huenig A, Ziegler D, Scherbaum WA (2011) Unconditioned and conditioned effects of intravenous insulin and glucose on heart rate variability in healthy men. *Physiol Behav* 103(1): 31-38.

Stoner L, Young JM, Fryer S (2012) Assessments of arterial stiffness and endothelial function using pulse wave analysis. *Int J Vasc Med* 2012: 903107.

Sugawara J, Hayashi K, Yokoi T, Tanaka H (2010) Carotid-femoral pulse wave velocity: impact of different arterial path length measurements. *Artery Res* 4(1): 27-31.

Sun L, Yan B, Gao Y, Su D, Peng L, Jiao Y, Wang Y, Han D, Wang G (2016) Relationship between blood pressure reverse dipping and type 2 diabetes in hypertensive patients. *Sci Rep* 6: 25053.

Sztajzel J (2004) Heart rate variability: a noninvasive electrocardiographic method to measure the autonomic nervous system. *Swiss Med Wkly* 134(35-36): 514-522.

Takahashi JS (2017) Transcriptional architecture of the mammalian circadian clock. *Nat Rev Genet* 18(3): 164-179.

Takakuwa H, Ise T, Kato T, Izumiya Y, Shimizu K, Yokoyama H, Kobayashi KI (2001) Diurnal variation of hemodynamic indices in non-dipper hypertensive patients. *Hypertens Res* 24(3): 195-201.

Takase B, Kitamura H, Noritake M, Nagase T, Kurita A, Ohsuzu F, Matsuoka T (2002) Assessment of diabetic autonomic neuropathy using twenty-four-hour spectral analysis of heart rate variability: a comparison with the findings of the Ewing battery. *Jpn Heart J* 43: 127-135.

Task Force of the European Society of Cardiology and the North American Society of Pacing and Electrophysiology (1996) Heart rate variability: standards of measurement, physiological interpretation and clinical use. *Circulation* 93(5): 1043-1065.

TensioClinicTensioMed™. Arteriograph and TensioMed™ program. *User's manual*. Type: TL1 TensioMed™ Arteriograph-5-01, v1. 9. Available from: http://www.arteriograph.nl/?act=download&fid=173. [last accessed on 26 Feb 2021].

The Criteria Committee of the New York Heart Association (1994) *Nomenclature and Criteria for Diagnosis of Diseases of the heart and great vessels* (9th ed.). Little, Brown & Co, Boston, USA. pp. 253-256.

Thorp AA, Schlaich MP (2015) Relevance of sympathetic nervous system activation in obesity and metabolic syndrome. *J Diabetes Res* 2015: 341583.

Tobaldini E, Costantino G, Solbiati M, Cogliati C, Kara T, Nobili L, Montano N (2017) Sleep, sleep deprivation, autonomic nervous system and cardiovascular diseases. *Neurosci Biobehav Rev* 74: 321-329.

Tong M, Watanabe E, Yamamoto N, Nagahata-Ishiguro M, Maemura K, Takeda N, Nagai R, Ozaki Y (2013) Circadian expressions of cardiac ion channel genes in mouse might be associated with the central clock in the SCN but not the peripheral clock in the heart. *Biol Rhythm Res* 44(4): 519-530.

Townsend RR, Wilkinson IB, Schiffrin EL, Avolio AP, Chirinos JA, Cockcroft JR, Heffernan KS, Lakatta EG, McEniery CM, Mitchell GF, Najjar SS, Nichols WW, Urbina EM, Weber T (2015) American Heart Association Council on Hypertension. Recommendations for Improving and Standardizing Vascular Research on Arterial Stiffness: A Scientific Statement from the American Heart Association. *Hypertension* 66(3): 698-722.

Tsuchikura S, Shoji T, Kimoto E, Shinohara K, Hatsuda S, Koyama H, Emoto M, Nishizawa Y (2010) Central versus peripheral arterial stiffness in association with coronary, cerebral and peripheral arterial disease. *Atherosclerosis* 211(2): 480-485.

Tsuji H, Larson MG, Venditti FJ Jr, Manders ES, Evans JC, Feldman CL, Levy D (1996) Impact of reduced heart rate variability on risk for cardiac events. The Framingham Heart Study. *Circulation* 94(11): 2850-2855.

Ukkola O, Vasunta RL, Kesaniemi YA (2009) Non-dipping pattern in ambulatory blood pressure monitoring is associated with metabolic abnormalities in a random sample of middle-aged subjects. *Hypertens Res* 32(11): 1022-1027.

Ulleryd MA, Prahl U, BoErsbo J, Schmidt C, Nilsson S, Bergstrom G, Johansson ME (2017) The association between autonomic dysfunction, inflammation and atherosclerosis in men under investigation for carotid plaques. *PLoS One* 12(4): e0174974.

Van Dijk RA, Bakker SJ, Scheffer PG, Heine RJ, Stehouwer CD (2003) Associations of metabolic variables with arterial stiffness in type 2 diabetes mellitus: focus on insulin sensitivity and postprandial triglyceridaemia. *Eur J Clin Invest* 33(4): 307-315.

Verdecchia P, Angeli F, Mazzotta G, Garofoli M, Ramundo E, Gentile G, Ambrosio G, Reboldi G (2012) Day-night dip and early-morning surge in blood pressure in hypertension: prognostic implications. *Hypertension* 60(1): 34-42.

Vinik AI, Maser RE, Mitchell BD, Freeman R (2003) Diabetic autonomic neuropathy. *Diabetes Care* 26(5): 1553-1579.

Vinik AI, Ziegler D (2007) Diabetic cardiovascular autonomic neuropathy. *Circulation* 115(3): 387-397.

Wang C, Zhang J, Liu X, Li C, Ye Z, Peng H, Chen Z, Lou T (2013) Reversed dipper blood-pressure pattern is closely related to severe renal and cardiovascular damage in patients with chronic kidney disease. *PloS One* 8(2): e55419.

Williams B, Mancia G, Spiering W, AgabitiRosei E, Azizi M, Burnier M, Clement DL, Coca A, de Simone G, Dominiczak A, Kahan T, Mahfoud F, Redon J, Ruilope L, Zanchetti A, Kerins M, Kjeldsen SE, Kreutz R, Laurent S, Lip GYH, McManus R, Narkiewicz K, Ruschitzka F, Schmieder RE, Shlyakhto E, Tsioufis C, Aboyans V, Desormais I (2018) ESC Scientific Document Group. ESC/ESH Guidelines for the management of arterial hypertension. *Eur Heart J* 39(33): 3021-3104.

Xie Z, Su W, Liu S, Zhao G, Esser K, Schroder EA, Lefta M, Stauss HM, Guo Z, Gong MC (2015) Smooth-muscle BMAL1 participates in blood pressure circadian rhythm regulation. *J Clin Invest* 125(1): 324-336.

Xu M, Huang Y, Xie L, Peng K, Ding L, Lin L, Wang P, Hao M, Chen Y, Sun Y, Qi L, Wang W, Ning G, Bi Y (2016) Diabetes and risk of arterial stiffness: a Mendelian randomization analysis. *Diabetes* 65(6): 1731-1740.

Yadav RL, Yadav PK, Yadav LK, Agrawal K, Sah SK, Islam MN (2017) Association between obesity and heart rate variability indices: an intuition toward cardiac autonomic alteration - a risk of CVD. *Diabetes Metab Syndr Obes* 10: 57-64.

Yeboah K, Antwi DA, Gyan B (2016) Arterial stiffness in nonhypertensive type 2 diabetes patients in Ghana. *Int J Endocrinol* 2016: 6107572.

Yoo HJ, Hwang SY, Choi KM, Baik SH, Lee EM, Kim EJ, Rha SW, Park CG, Oh DJ, Seo HS (2016) Clinical implication of body size phenotype on heart rate variability. *Metabolism* 65(11): 1589-1596.

Zhao XY, Wang XF, Li L, Zhang L, Shen DL, Li DH, Jin QS, Zhang JY (2015) Effects of high glucose on human umbilical vein endothelial cell permeability and myosin light chain phosphorylation. *Diabetol Metab Syndr* 7: 98.

Ziegler D, Keller J, Maier C, Pannek J (2014) German Diabetes Association Diabetic neuropathy. *Exp Clin Endocrinol Diabetes* 122(7): 406-415.

Ziegler D, Laude D, Akila F, Elghozi JL (2001) Time- and frequency-domain estimation of early diabetic cardiovascular autonomic neuropathy. *Clin Auton Res* 11(6): 369-376.

ABOUT THE EDITOR

Rajeshwar P. Sinha, DAAD Fellowship Awardee and Fellow, Society for Applied Biotechnology, India, is a Professor of Molecular Biology, Centre of Advanced Study in Botany, Banaras Hindu University (BHU), Varanasi, India. He received his PhD (Biotechnology) from BHU, Varanasi and visited countries such as Argentina, Austria, Belgium, Canada, China, Germany, Greece, France, Italy, Japan, Luxembourg, Norway, Poland, Republic of Korea, Spain, Switzerland, The Netherlands,

UK and USA, in the field of academics/research. He is working on the physiological, biochemical, molecular and computational biology aspects of ultraviolet radiation's effects on the organisms of aquatic and terrestrial ecosystems e.g., circadian rhythms, DNA damage and repair, phycobiliproteins, mycosporine-like amino acids and scytonemin. He has published over 400 research papers/reviews/book chapters/conference proceedings and edited/authored 8 books. He is a lifetime member of several national and international scientific societies and an editorial board member of various national and international journals. He has over 12,000 citations with an h-index of 52 and i10-index of 126. He is among the top 2% most cited scientists in the world.

INDEX

#

A

B

C

D

E

I

J

K

R

S

T

U

V

W

Y